Bionanomaterials for Biosensors, Drug Delivery, and Medical Applications

This book covers advances in nanostructured materials across a variety of biomedical applications as the field evolves from development of prototype devices to real-world implementation. It provides an in-depth look at the current state of the art in oxide nanostructures, carbon nanostructures, and 2D material fabrication and highlights the most important biomedical applications and devices of nanomaterials, including drug delivery, medical imaging, gene therapy, biosensors, and diagnostics.

FEATURES

- Presents the findings of cutting-edge research activities in the field of nanomaterials, with a particular emphasis on biological and pharmaceutical applications
- Details finished and ongoing toxicity evaluations of emerging nanomaterials
- Offers a multidisciplinary perspective

This book is recommended for senior undergraduate and graduate students, professionals, and researchers working in the fields of bioengineering, materials science and engineering, and biotechnology.

Emerging Materials and Technologies

Series Editor: Boris I. Kharissov

The Emerging Materials and Technologies series is devoted to highlighting publications centered on emerging advanced materials and novel technologies. Attention is paid to those newly discovered or applied materials with potential to solve pressing societal problems and improve quality of life, corresponding to environmental protection, medicine, communications, energy, transportation, advanced manufacturing, and related areas.

The series takes into account that, under present strong demands for energy, material, and cost savings, as well as heavy contamination problems and worldwide pandemic conditions, the area of emerging materials and related scalable technologies is a highly interdisciplinary field, with the need for researchers, professionals, and academics across the spectrum of engineering and technological disciplines. The main objective of this book series is to attract more attention to these materials and technologies and invite conversation among the international R&D community.

Wastewater Treatment with the Fenton Process: Principles and Applications
Dominika Bury, Piotr Marcinowski, Jan Bogacki, Michal Jakubczak, and Agnieszka Jastrzebska

Mechanical Behavior of Advanced Materials: Modeling and Simulation
Edited by Jia Li and Qihong Fang

Shape Memory Polymer Composites: Characterization and Modeling
Nilesh Tiwari and Kanif M. Markad

Impedance Spectroscopy and its Application in Biological Detection
Edited by Geeta Bhatt, Manoj Bhatt, and Shantanu Bhattacharya

Nanofillers for Sustainable Applications
Edited by N.M. Nurazzi, E. Bayraktar, M.N.F. Norrahim, H.A. Aisyah, N. Abdullah, and M.R.M. Asyraf

Bionanomaterials for Biosensors, Drug Delivery, and Medical Applications
Edited by Won-Chun Oh and Suresh Sagadevan

For more information about this series, please visit: www.routledge.com/Emerging-Materials-and-Technologies/book-series/CRCEMT

Bionanomaterials for Biosensors, Drug Delivery, and Medical Applications

Edited by Won-Chun Oh and Suresh Sagadevan

CRC Press is an imprint of the
Taylor & Francis Group, an informa business

First edition published 2024
by CRC Press
2385 NW Executive Center Drive, Suite 320, Boca Raton FL 33431

and by CRC Press
4 Park Square, Milton Park, Abingdon, Oxon, OX14 4RN

CRC Press is an imprint of Taylor & Francis Group, LLC

© 2024 selection and editorial matter, Won-Chun Oh and Suresh Sagadevan; individual chapters, the contributors

Reasonable efforts have been made to publish reliable data and information, but the author and publisher cannot assume responsibility for the validity of all materials or the consequences of their use. The authors and publishers have attempted to trace the copyright holders of all material reproduced in this publication and apologize to copyright holders if permission to publish in this form has not been obtained. If any copyright material has not been acknowledged please write and let us know so we may rectify in any future reprint.

Except as permitted under U.S. Copyright Law, no part of this book may be reprinted, reproduced, transmitted, or utilized in any form by any electronic, mechanical, or other means, now known or hereafter invented, including photocopying, microfilming, and recording, or in any information storage or retrieval system, without written permission from the publishers.

For permission to photocopy or use material electronically from this work, access www.copyright.com or contact the Copyright Clearance Center, Inc. (CCC), 222 Rosewood Drive, Danvers, MA 01923, 978–750–8400. For works that are not available on CCC please contact mpkbookspermissions@tandf.co.uk

Trademark notice: Product or corporate names may be trademarks or registered trademarks and are used only for identification and explanation without intent to infringe.

ISBN: 978-1-032-54554-7 (hbk)
ISBN: 978-1-032-54557-8 (pbk)
ISBN: 978-1-003-42542-7 (ebk)

DOI: 10.1201/9781003425427

Typeset in Times
by Apex CoVantage, LLC

Contents

SECTION C Medical Applications

Preface

Recently, with the improvement of human quality of life, many efforts are being made to live a healthy, pleasant, and comfortable life. Sensors are one of the essential factors for improving the quality of life, and high-performance intelligent sensors are required, so demand for them is expected to increase further in the future. Recently, functional materials that give functions to sensors are becoming more important. In particular, nanofunctional materials existing in various components and forms are attracting more attention because they have unique physical, chemical, mechanical, and optical properties differentiated from existing bulk materials. Recently, research on nanotubes, including Carbon nanotubes (CNTs) and nanocomposites, has been very active to further develop existing nanoparticle research and applications as sensors. This book examines the latest nanosensor R&D trends such as gas sensors, water quality sensors, biosensors, light sensors, and physical sensors, and it is predicted that high functionality and miniaturization of sensors by nanomaterial will become very important.

Research on nanosensors using nanomaterials such as metals, inorganic, organic, bio, and composite materials with a size less than or equal to 100 nm in various compositions and forms such as nanoparticles, nanowires, and nanotubes has been actively reported. This is due to the unique physical, chemical, mechanical, and optical properties of nanomaterials. For example, when a nano-sensing material with a large specific surface area is used in a gas sensor, a low-concentration air pollution source may be detected even at a low operating temperature, thereby lowering power consumption and ultimately miniaturizing a sensor, and thus, a material with a very large specific surface area such as nanoparticles, nanowires, and graphene is used. In this book, there are two main points. The first point is about how to manufacture nanostructured materials such as nanoparticles and nanowires, and the second point is about the latest nanosensor technology trends using nanomaterials. In fact, many nanosensor research results using nanomaterials are being published, so the contents covered in this book will be part of them, and it can be used as a reference for researchers and developers who want to conduct related research and development.

Trends in sensor R&D technology using various nanomaterials are described. Although the vast amount of research worldwide deals with only a few research cases, nanomaterials are developed in various ways in the form of nanoparticles, nanowires, nanotubes, and nanocomposites, and it is confirmed that they are expanding to various applications such as gas sensors, water sensors, biosensors, light sensors, and physical sensors. From this, furthermore, it can be predicted that future state-of-the-art sensor technologies will act as more essential factors for high functionality and miniaturization of nanomaterials. However, sensors using nanowires have the advantage of being able to manufacture nanowires of a wide variety of compositions and structures compared to the top-down method, but more research is needed on the technology for assembling and aligning nanowires in two dimensions as in the top-down method. In addition, although research has already been conducted at advanced research institutes in various countries, research on the stability of ultrafine nanomaterials, such as toxicity to the human body and the environment, is expected to develop into a technology that can bring about sustainable R&D and industrialization.

Won-Chun Oh
Suresh Sagadevan

Editors

Won-Chun Oh is a Professor in the Department of Advanced Materials and Engineering at Hanseo University in Korea and the School of Materials Science and Engineering at Anhui University of Science and Technology in China. He also is a guest professor at universities in China, Thailand, and Indonesia. He has received the Research Front Award from the Korean Carbon Society, the Yangsong Award from the Korea Ceramic Society, the Excellent Paper Award from the *Korea Journal of Material Research,* the Best Paper Award from the *Journal of Industrial and Engineering Chemistry* for his pioneering work, and Awards of Appreciation from ICMMA2011, ICMMA2014, and ICMMA2019. He is an ICMMA committee board member and was appointed as a conference chairman and vice chairman from 2007 to the present. Dr. Oh was appointed one of the Top 100 Scientists in the World at IBC, UK, and one of the Top 2% Scientists at Stanford University, USA. He is the author or a coauthor of 930 papers in domestic and international journals and has delivered speeches at conferences as a special lecturer, plenary lecturer, and keynote speaker. Dr. Oh is the Editor-in-Chief of the *Journal of Multifunctional Materials and Photoscience* and *Asian Journal of Materials Chemistry* and an advisory board member of the *Asian Journal of Chemistry* and *Nanomaterials.*

Suresh Sagadevan is an Associate Professor in the Nanotechnology and Catalysis Research Centre, University of Malaya. His outstanding research findings and novel publications include more than 350 research papers in ISI top-tier journals and Scopus, refereed internationally. Dr. Sagadevan has authored 12 international book series and 40 book chapters. He was selected as one of the Top 2% of Scientists by Stanford University in 2020 and 2021. In 2021, he was recognized for his outstanding research contributions as Fellow of the Royal Society of Chemistry (FRSC). He is a guest editor and editorial board member of many ISI journals. Dr. Sagadevan is a member of several professional bodies at the national and international level. He is a recognized reviewer for many reputed journals. Dr. Sagadevan works in various fields, including nanofabrication, functional materials, crystal growth, graphene, polymeric nanocomposite, glass materials, thin films, bio-inspired materials, drug delivery, tissue engineering, supercapacitors, optoelectronics, photocatalytics, green chemistry, and biosensor applications.

Editors

Contributors

Ashfaq Ahmad
School of Materials Science and Engineering
Shanghai Jiaotong University
Shanghai, China

Hassan Akbar
Department of Physics
Abbottabad University of Science and Technology (AUST)
Havelian, Khyber Pakhtunkhwa, Pakistan
and
College of Environmental Science and Engineering
North China Electric Power
Beijing China

Asghar Ali
Department of Physics
University of Lahore
Lahore, Pakistan

Amalia Kurnia Amin
Research Center for Chemistry
National Research and Innovation Agency (BRIN)
B. J. Habibie Science and Technology Area
South Tangerang, Banten, Indonesia

Saravanakumar Arthanari
Department of Pharmaceutics
Vellalar College of Pharmacy
Erode, Tamil Nadu, India

Budhijanto Budhijanto
Department of Chemical Engineering
Faculty of Engineering
Universitas Gadjah Mada
Yogyakarta, Indonesia

Xiao Chen
School of Materials Science and Engineering
Anhui University of Technology
Huainan, Anhui, China

Ruey-an Doong
Institute of Analytical and Environmental Sciences
National Tsing Hua University
Hsinchu, Taiwan

Ganjar Fadillah
Chemistry Department
Universitas Islam Indonesia
Yogyakarta, Indonesia

Is Fatimah
Chemistry Department
Universitas Islam Indonesia
Yogyakarta, Indonesia

Latifah Hauli
Research Center for Chemistry
National Research and Innovation Agency (BRIN)
B. J. Habibie Science and Technology Area
South Tangerang, Banten, Indonesia

Saba Iqbal
Department of Physics
University of Lahore
Lahore, Pakistan

Hilda Ismail
Department of Pharmaceutical Chemistry
Faculty of Pharmacy
Universitas Gadjah Mada
Yogyakarta, Indonesia

Zuhong Ji
Department of Chemical Engineering
Hanseo University
Chungnam, Korea

Venkateshwaran Krishnaswami
Department of Pharmaceutics
S. A. Raja Pharmacy College
Tirunelveli, Tamil Nadu, India

Xinyu Liu
School of Materials Science and Engineering
Anhui University of Technology
Huainan, Anhui, China

Ayesha Malik
Department of Physics
University of Lahore
Lahore, Pakistan

Won-Chun Oh
Department of Advanced Materials and Engineering
Hanseo University
Chungnam-do, Korea

Zambaga Otgonbayar
Department of Advanced Materials Science and Engineering
Hanseo University
Chungnam-do, Korea

Amin Ur Rashid
Department of Applied Physical and Material Sciences
University of Swat
Swat, Khyber-Pakhtunkhwa, Pakistan

Suresh Sagadevan
Chemistry Department
Universitas Islam Indonesia
Yogyakarta, Indonesia and Nanotechnology and Catalysis Research Center (NANOCAT)
Universiti Malaya
Kuala Lumpur, Malaysia

Wahyu Dita Saputri
Research Center for Quantum Physics
National Research and Innovation Agency (BRIN)
B. J. Habibie Science and Technology Area
South Tangerang, Banten, Indonesia

Aldino Javier Saviola
Department of Chemistry
Faculty of Mathematics and Natural Sciences
Universitas Gadjah Mada
Yogyakarta, Indonesia

Chenwei Shang
School of Materials Science and Engineering
Anhui University of Technology
Huainan, Anhui, China

Siwaluk Srikrajang
Department of Physical Therapy
Faculty of Medicine
Prince of Songkla University
Songkla, Thailand

Yu Tian
School of Materials Science and Engineering
Anhui University of Technology
Huainan, Anhui, China

Salah Uddin
Department of Applied Physical and Material Sciences
University of Swat
Swat, Khyber-Pakhtunkhwa, Pakistan

Kefayat Ulla
Department of Applied Physical and Material Sciences
University of Swat
Swat, Khyber-Pakhtunkhwa, Pakistan

Jing Wang
School of Materials Science and Engineering
Anhui University of Technology
Huainan, Anhui, China

Karna Wijaya
Department of Chemistry
Faculty of Mathematics and Natural Sciences
Universitas Gadjah Mada
Yogyakarta, Indonesia

Ika Yanti
Chemistry Department
Universitas Islam Indonesia
Yogyakarta, Indonesia

Chang-Min Yoon
Department of Chemical and Biological Engineering
Hanbat National University
Daejeon, Korea

Lei Zhang
School of Materials Science and Engineering
Anhui University of Technology
Huainan, Anhui, China

Section A

Biosensors

1 Electric Biosensors Based on Carbon Nanotubes

Jing Wang, Lei Zhang, Xiao Chen, Chenwei Shang, Yu Tian, and Xinyu Liu

1.1 INTRODUCTION

Diseases are closely related to biomolecules. Early prevention and timely detection of anomalies are of great significance in the field of life sciences. In medicine, early diagnosis can be done before the disease has worsened, thus saving many lives. The majority of changes caused by diseases in the body can be seen in the amount of metabolites. Certain bodily chemicals have been associated with lung cancer, toluene, uric acid (UA), and gout.[1] Moreover, unhealthy amounts of metabolites can have an impact on health. Conversely, too little leads to scurvy.[2] During the patient's medication process, it is also necessary to monitor changes in drug content in the body.[3,4] Moreover, molecules such as H_2O_2 present in the body have also been implicated in breast cancer and lung disease.[5]

In the beginning, researchers examined the differences between healthy and diseased organisms to look for pathogenic elements or disease indicators. At the moment, laboratory testing is frequently utilized in[6] medical testing. For instance, researchers must draw blood from patients and analyze the proteins, antibodies, and genes present. Typical processes in the laboratory testing procedure include sample preparation, separation, detection, and data processing. Also, this approach takes time. The testing apparatus is also highly expensive. Consequently, it is important for medical testing procedures to balance cost savings with increased effectiveness.

New opportunities for disease analysis[7,8] in biosensors are presented by the quick development of nanotechnology and semiconductor materials manufacturing technologies. Carbon nanotubes (CNTs), one of the isomers of carbon, such as graphene, are a key component of carbon-based materials used in the production of biosensors. In 1991,[9] Dr. Mishima discovered carbon nanotubes, and related research has been deepening and expanding over the years. Crimped graphene layers form the cylindrical structures known as carbon nanotubes. Depending on the number of layers, CNTs can be divided into single-walled carbon nanotubes (SWCNTs) and multiwall carbon nanotubes (MWCNTs). While MWCNTs possess a strong corrosion resistance, these two CNTs are studied for their tensile strength and elastic modulus as biosensors. Carbon nanotubes are essential materials because of their special qualities. First, CNTs excel in mechanical applications. A carbon nanotube has a

DOI: 10.1201/9781003425427-2

diamond-like elastic modulus. Moreover, CNTs exhibit excellent electrical conductivity as a result of sharing graphene's sheet-like structure. The Thermoconductive and light-modulating properties of carbon nanotubes are also very promising. Moreover, CNTs have strong chemical stability, with good electrochemical capability and large specific surface area, which is very promising in the field of detection of medicinal biomolecules.[10,11] Carbon nanotubes provide many reaction sites that are able to interact with a lot of biological molecules.

Moreover, CNT conductivity is sensitive to analyte absorption, allowing for sensitive biosensing. There has been research on CNTs in a variety of areas, primarily in the areas of the environment,[12,13] farming,[14,15] food,[16] energy,[17,18] and medicine.[19] Because of their excellent selectivity and mobility, carbon nanotubes have been extensively exploited, particularly in the medical sector. Furthermore, sodium chloride and lactic acid have been detected in sweat using carbon nanotube sensors in wearable technology.[20] Even new sensors that can be used to feel the pulse in traditional Chinese medicine have them added.[21]

Carbon nanotube-based biosensors can be categorized as electrochemical, optical, semiconductor, calometric, and other types of biosensors depending on their technical design.[22] Electrochemical sensors have great advantages in many aspects and are a very hot[23] type.

This review summarizes the modern developments in nanobiosensors. Glucose, DNA, proteins, and neurotransmitters are just a few examples of the various biomolecules that CNT-based biosensors are used to detect (Figure 1.1). We conclude by summarizing the present research on carbon nanotube-based biosensors' drawbacks and prospective outcomes.

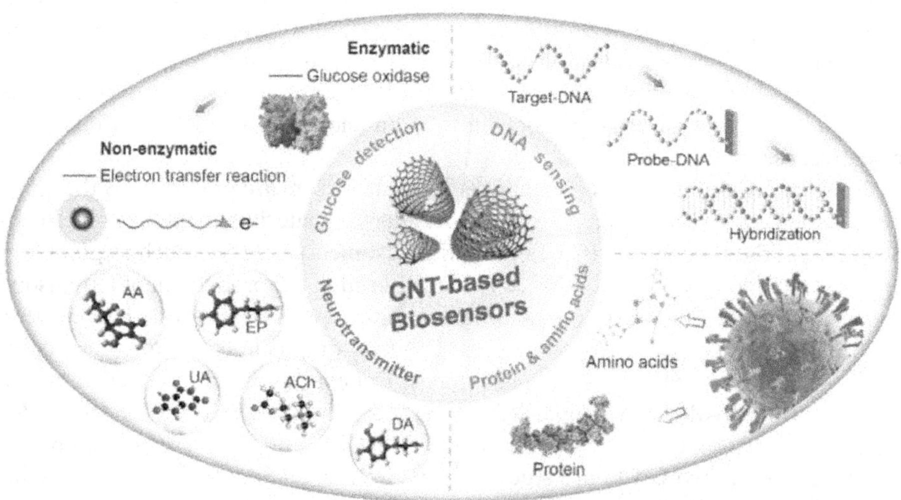

FIGURE 1.1 CNT-based biosensors for the detection of biomolecules like amino acids, glucose, DNA, and neurotransmitters.

1.2 STRUCTURE AND PROPERTIES OF CARBON NANOTUBES

Since their discovery by researchers, carbon nanotubes, also known as bartubes, have garnered a lot of attention due to evidence that they are a novel nanostructure with distinct structural, electrical, and mechanical properties.[24,25] According to studies conducted in the last ten years, carbon nanotubes are a novel type of carbon material that has unexpected uses in a wide range of industries, including energy storage and conversion, electromechanical actuators, chemical sensors, and others.[26,27] A structure diagram is shown in Figure 1.2. Generally speaking, carbon nanotubes are divided into two types: single-wall (SWCNT) and multi-wall (MWCNT). The former is a graphene-wound cylinder, formed by the smooth rolling of the (m, n) lattice vector. The primary factor governing tube metallicity and chirality is the (m, n) index. Coaxial multilayer graphene tubes make up the latter. The diameter of SWCNT varies from 0.4 to 3 nm, and that of MWCNT varies from 1.4 to 100 nm.

With unique structural characteristics and electronic properties, carbon nanotubes are a novel kind of carbon-based material that outperforms existing carbon-based materials employed in electrochemistry, that is, glass carbon (GC), diamond, and graphite. Therefore, in addition to their amazing applications in other fields, the unique properties of carbon nanotube electrochemistry studies shown so far have basically led them to apply to most electrochemical studies, such as electrocatalysis, transparent electrodes, direct protein electrochemistry and capacitors, and electrochemical sensors and transistors (Figure 1.3).

The size of a CNT is in the transition area at the junction of microscopic objects and macroscopic objects represented by atoms and molecules, making it neither a typical microscopic system nor a typical macro system, Therefore, it mainly has volume, macroscopic quantum tunnel, quantum, surface, and other effects. Carbon

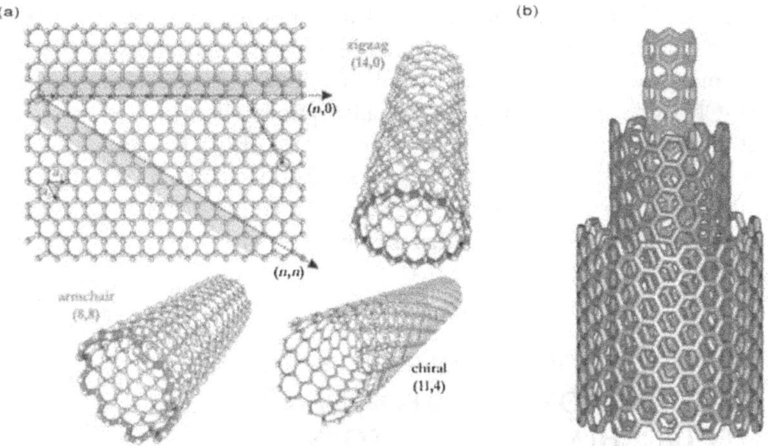

FIGURE 1.2 (a) A graphene sheet is rolled up to produce three different forms of SWCNTs; (b) a multi-walled carbon nanotube is composed of three shells with various chiralities.[28]

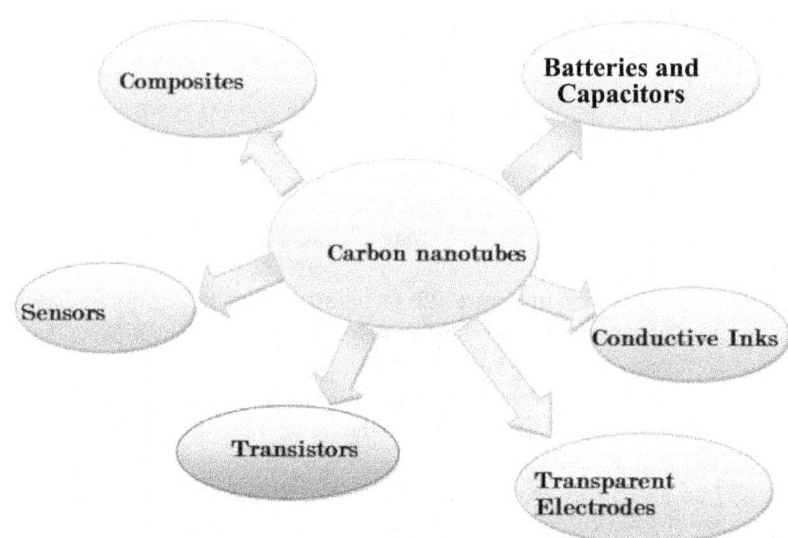

FIGURE 1.3 Applications of carbon nanotubes.

nanotubes are a one-dimensional nanomaterial that are light and have a fully connected regular hexagonal structure. They have unique mechanical properties, excellent electrical properties, and stable chemical and physical characteristics. Using them to modify an electrode can reduce the overpotential of chemical redox reaction and improve the redox reversibility of biological molecules. CNTs have a special tubular structure and extremely small size, which can easily cross the cell wall as an efficient mass transfer unit, and thus they have great advantages for application to biosensors.[29] The specific surface area of the CNT is not only conducive to the immobilization of the enzyme, promoting electron transport between the active center and the surface of the electrode, but also makes it easy to adsorb organic molecules. Using it to modify an electrode can improve the selectivity of H + and make electrochemical sensors. Gas sensors can be made by using the electrical conductivity of the CNT and its selectivity for gas adsorption. Microadsorption of oxygen at different temperatures can change the conductivity of CNTs and can be converted between metals and semiconductors. The p-n junction can be formed by locally filling the alkali metal within the CNT. The CNT can be filled with light-sensitive, wet-sensitive, pressure-sensitive, and other materials, which can be made into various nano-scale functional sensors.[30]

1.3 INTRODUCTION TO THE PRINCIPLES AND CLASSIFICATION OF CARBON NANOTUBE ELECTROCHEMICAL BIOSENSORS

Electrochemical biosensors are dual-or three-electrode electrochemical cells that can transmit biological events into electrochemical signals (Figure 1.3). On the electrode, they typically have a biometric component that interacts with the analyte to produce the electrochemical signal.[31] Due to its many advantages, such as high sensitivity, fast response speed, special mobility, and simple operation, the CNT plays a significant role in biosensors. According to the method of identification process, it is divided into catalytic sensors and affinity sensors. While biocatalytic sensors use biometric elements (such as enzymes) that produce electrically active species, bioaffinity sensors monitor binding events between the biometric elements and the analyte.[32,33] bioaffinity electrochemical sensors based on CNT-based enzymatic electrochemical biosensors will be covered in detail in this chapter.

The combination of electrochemical technology and enzyme specificity in enzyme biosensors provides a huge opportunity for early diagnostic strategies.[34] Redox active centers and dielectric-free electrodes, with electron transfer between them, enable the development of enzymatic biosensors. However, The protein coats the enzyme and exists in the hydrophobic cavity, so the direct electrochemistry of the enzyme is difficult to achieve.[35,36] Carbon nanotubes have been employed extensively to facilitate electron transfer between the electrodes and redox centers of the enzyme because of their tiny size, superior electrochemical characteristics, and high specific surface area.[37] The electrochemical detection of therapeutically significant analytes using enzyme biosensors has been reported frequently in recent years. Examples include lactate oxide, horseradish peroxidase (HRP), glucose oxidase (GOx), malate dehydrogenase (MDH), and others. How to bind the enzyme in a stable manner while keeping its biological activity is one of the key design issues for an enzymatic biosensor based on CNTs. Depending on the constructs, there are four cnt-derived enzyme electrodes, as described in the following.[38–42]

Biosensors that recognize and specifically bind two biomolecules, such as DNA biosensors and immunosensors, are examples of biological affinity sensors. The target analyte will be captured by one of the two biomolecules that was initially linked to the sensor. Measurable electrochemical signals produced by molecule recognition are collected by bioaffinity electrochemical sensors. The following paragraphs[43] will explain CNT-based immunosensors and DNA electrochemical sensors.

1.4 CARBON NANOTUBE–BASED ENZYMATIC BIOSENSORS

Enzymatic biosensors rely on the interaction between the enzyme and its substrates. According to the target analyte, substrate detection and enzyme inhibition are the two primary working processes of this type of biosensor, as shown in Figure 1.4. When the enzymatic reaction occurs on the enzyme membrane, the electrode active substance is produced, and the electrode responds to it. Because of a linear relationship between the response signal and the concentration of the substrate, the concentration of the detected substance can be measured as shown in Figure 1.5. Taking a glucose oxidase (Glucoseoxidase, GOD) sensor as an example, its electrocatalytic working principle is as follows:

$$\text{Enzyme layer: glucose} + GOD_2FAD \rightarrow \text{gluconolactone} + GOD_2FADH_2$$
$$GOD_2FADH_2 + O_2 \rightarrow GOD_2FAD + H_2O_2$$
$$\text{electrode: } H_2O_2 \rightarrow 2H{+}{+}O_2{+}2e$$

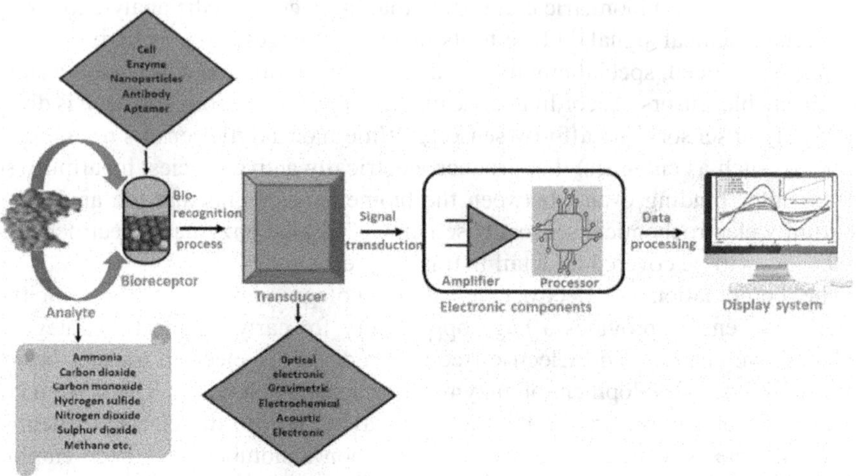

FIGURE 1.4 Sketch of the typical biosensor with its components: analyte, bioreceptor, transducer, electronic components, and display system.

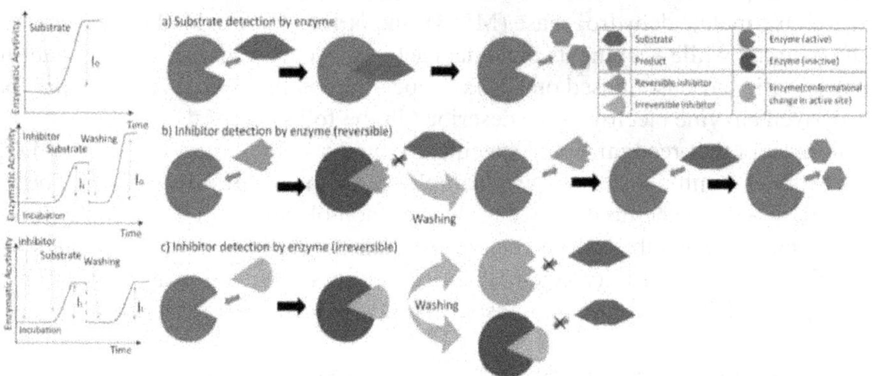

FIGURE 1.5 Scheme of enzymatic biosensor for substrate and inhibitor (reversible and irreversible) detection.

The redox reaction of oxygen on the electrode produces a responsive current, which is related to the concentration of oxygen in the solution and the mass transfer speed. To eliminate the detection error caused by the oxygen concentration change, the hydrogen peroxide[44] produced by the enzyme reaction can be detected.

The results showed that the electrodes modified by CNT had excellent electrocatalytic reduction properties of H_2O_2, which has the potential to use CNT-modified electrodes for the development of enzymatic biosensors. CNT as a fixed material of the enzyme, as well as a modified material of the base electrode, can become a new CNT-modified enzyme sensor. The advantages of such a sensor include the following: (1) the good electrical properties of CNT make it a modified material to effectively promote electronic transmission, improve the detection speed of the enzyme sensor, reduce overpotential, and improve detection sensitivity; (2) CNTs can increase the load and the sensitivity of the sensor; (3) CNT sensors' good biocompatibility is conducive to maintaining the activity of the enzyme, thus improving the stability and service life of the enzyme sensor.

Azamian et al.[45] fixed SWCNTs on a glass carbon electrode by adsorption to make a GOD-SWCNT–modified glass carbon electrode. The article describes the anodic response becomes catalytic on the addition of equilibrated β-D-glucose and the magnitude of this catalytic response is more than 10-fold greater than that observed on the same GC electrode surface in the absence of SWCNTs (for the same substrate concentration) and is aided not only by the high levels of enzyme loading achievable on nanotubes but also by their transducing ability.

Nanoparticles and GOD were deposited in Nafion CNT membranes by Lim et al. A fast responsive GOD-type biosensor was prepared by detecting hydrogen peroxide produced by glucose under the enzyme. The electrode is a glass carbon electrode, and the additional Nafion membrane prevents the interference with glucose detection caused by ascorbic acid and uric acid. This sensor has a linear range of up to 12 mmol/L, with a lower limit of detection of 0.15 mmol/L.

Li et al. also used GOD covalently to modify MWCNT to produce glucose biosensors. Mixed acid –treated MWCNT forms MWCNT amino terminals in the presence of SOCl2 and ethylenediamine and then reacts with the hydroxyl group on the GOD product oxidized by periodate, as shown in Figure 1.1. Covalent binding of GOD on MWCNT maintained the strong biological activity of GOD.

Compared with other analytical methods, an enzyme sensor modified by carbon nanotubes has the advantages of convenient carrying, low cost, good sensitivity, and excellent stability. Compared with general chemical catalysis technology, enzyme catalysis technology possesses the characteristics of higher catalytic efficiency, wider application range, and selective specificity, and the reaction conditions are mild. Enzymes are biocatalysts with high activity, high selectivity, and low energy consumption, coupled with the catalytic and sensitization effects of CNT itself, giving CNT-based modified enzyme sensors broad application prospects.[46]

1.5 CARBON NANOTUBE–BASED GAS BIOSENSORS

A CNT has a special one-dimensional hollow structure, a large specific surface area, and a slightly larger layer spacing than graphite (0.343nm), which can adsorb many substances and fill in them. Because the adsorbed gas molecule interacts with CNT, it changes its Fermi level and greatly changes its macroscopic resistance. It is clear that CNTs have four basic gas adsorption sites: (1) internal site, or the pore inside a tube; the external surface of the tube/bundle is depicted in Figure 1.6 together with (2) the interstitial or channel site between three adjacent bundles; (3) grooves produced with neighboring tubes; and (4) the interstitial or channel site. Therefore, CNTs can be applied to a gas-sensitive sensor to detect the gas composition by measuring the change in its apparent resistance.

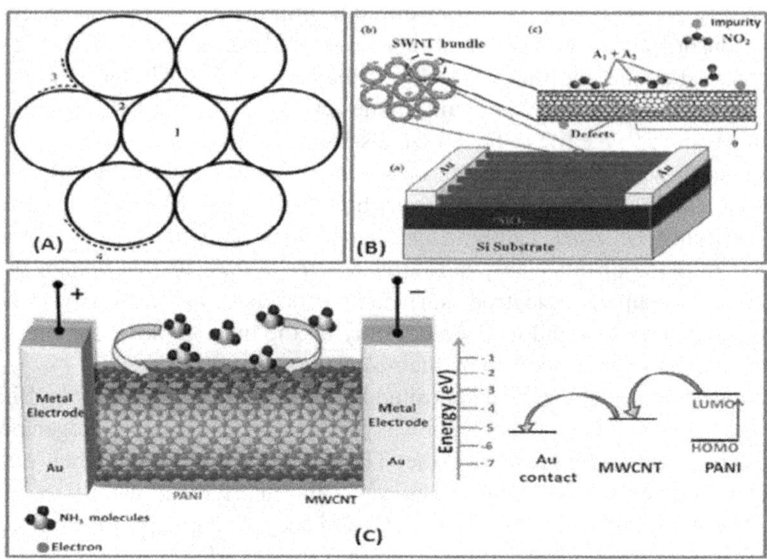

FIGURE 1.6 (a) adsorption site in carbon nanotubes (b), the gas-sensing mechanism based on the SWCNT shows the chemadsorption of the analyte (c), the combined energy map and gas-sensing mechanism shows the adsorption of ammonia and charge transfer after this.

The adsorption of molecular oxygen on the surface of SWCNTs was investigated by Collins et al. The adsorption of molecular oxygen under vacuum can increase CNT conductance. In the same environment, high-temperature–heated CNTs can completely desorb. Kong et al.[47] found that tiny chemical probes made with SWCNTs can produce rapid adsorption and high sensitivity of low concentrations of NH3 and NO_2 molecules at room temperature.

Compared with ordinary gas sensors, CNT gas sensors have many advantages, such as small size, fast reaction, high sensitivity, large surface area; they can operate at room temperature or higher and can be placed in a new environment or reused after heating. CNTs can produce the smallest biomedical molecular-grade gas elements with a response time at least one order of magnitude faster than the currently used similar metal oxide or polymer sensors. The CNT gas sensor is used for patient breathing monitoring, detection, and other aspects and will achieve the effect that is difficult to achieve with a general sensor.

1.6 DNA BIOSENSORS BASED ON CNTS

1.6.1 DNA Biosensors and Research Progress

Waston and Crick put forward the double-coiled spiral structure model of Deoxyri bonucleic Acid in 1953. The DNA molecule is a double-helix structure composed of two deoxyribonucleotide chains coiled around the same axis in an antiparallel direction (As shown in Figure 1.7).

FIGURE 1.7 Double-helix structure of DNA.

With more research on gene structure and function, especially the progress of Human Genome Project (HGP), gene isolation, analysis, and testing play an increasingly important role in the fields of health care, pandemic prevention, and medical diagnosis and development, such as pharmaceutical, environmental science, bioengineering, etc. Because DNA sequencing rules determine the genetic characteristics of genes, the key to decoding the genetic code is DNA sequencing analysis, which is also one of the most important issues in genetic engineering.[48] Nowadays, many genetic diseases are diagnosed early by analyzing whether the DNA base is a mutation sequence. For example: β, The gene of thymus (T) corresponding to globulin point, is the replacement of adenosine (A) which causes amino acid changes, resulting in sickle cell anemia. Hepatitis B is also caused by the replacement of the corresponding gene, which makes a copy of the related protein, leading to disease.[49–50]

In view of the rapid development of genetic engineering, many research groups are currently working to develop highly sensitive, high-precision, portable, and rapid DNA detection to meet the research needs of life science and clinical diagnosis. The emergence of DNA and gene chip sensors can greatly shorten the time to detect the mismatch between targeted DNA and DNA. This is a simple and convenient process, with no pollution, quantity and quality, high sensitivity and excellent selectivity, showing very attractive application probability. It aims to apply DNA to diagnosis and detection in the market today. Experts agree that DNA sensors are the most promising, and they should combine the advantages of being fast, accurate, simple, and cheap. DNA sensors are a common method to study specific sequence DNA by using specific complementary pairing between DNA molecules. Quick analysis is possible. DNA sensing is the combination of molecular biology, microelectronics, optics, and electrochemistry and has become one of the most important technologies in DNA and information analysis. So far, searching for DNA sequences from biosensors has become the frontier in the field of biosensors.

1.6.2 Basic Principles of Electrochemical DNA Biosensors

An Electrochemical DNA sensor is a sensor that uses various electrodes (including metal carbon electrodes, etc.), such as a transformer, to convert DNA when the presence of detectable electrical signals can be detected. In short, electrochemical DNA sensor is a kind of sensor prepared by electrical DNA modification. Wang, Mikklesen, Marrazza, Palecek, and Blum have done detailed reviews of electrochemical DNA biosensors and their research status in the last ten years.[51–54] Xiaohong[55] and others in China have also published many relevant articles. Wang[56–59] also published an overview of the technology from electrochemical nucleic acid sensors to DNA chips, and Chunhai[60] also published an article on DNA chips.

The process of DNA determination by chemical DNA biosensor generally includes the following four steps: (1) The structure of a single-stranded DNA (ssDNA) sensing interface, which connects the ssDNA to the surface of the solid electrode through a certain modification process to form the probe electrode of DNA; (2) The hybridization process to control the appropriate conditions and place the electrode fixed with ssDNA in the cDNA to hybridize and identify the objective DNA; (3) The indicator after hybridization converts the hybridization information into measurable electrochemical signals and feeds them back; and (4) Detection of electrochemical signals, that is current, potential, conductivity, and so on, as detection signals so as to analyze data and draw conclusions.

1.7 ELECTROCHEMICAL BIOSENSORS BASED ON CNTS AND GPES

In recent years, with the rapid growth of carbon nanomaterials, there has been increased interest in developing new biosensors using CNTs and graphene. Because carbon nanotubes and graphene are used as sensor components, their remarkable performance can make sensors more reliable, more accurate, and faster. Several approaches to designing sensor devices might be used depending on the type of target molecule (Figure 1.8). The most sophisticated sensors are still far from having practical applications, despite some advancements in carbon nanotubes and graphene. A multidisciplinary approach incorporating materials science, biology, and electrical engineering must be used for future endeavors.

The special structure of graphene gives it excellent chemical and physical properties. Graphene (Figure 1.9) has excellent properties: high electroconductivity, excellent electrocatalytic performance, gigantic specific surface area, and great biocompatibility, which makes graphene widely used in many areas.

Recently, there is new information about the application of graphene in electrochemical biosensors. Cai[61] used functionalized flexible graphene fiber to synchronously detect ascorbic acid (AA), dopamine (DA), and uric acid (UA), showing good electrochemical performance, and the detection limit reached 50, 0.1, and 0.2 μmol/L, respectively. Gao[62] used pyrrole peroxide/graphene with

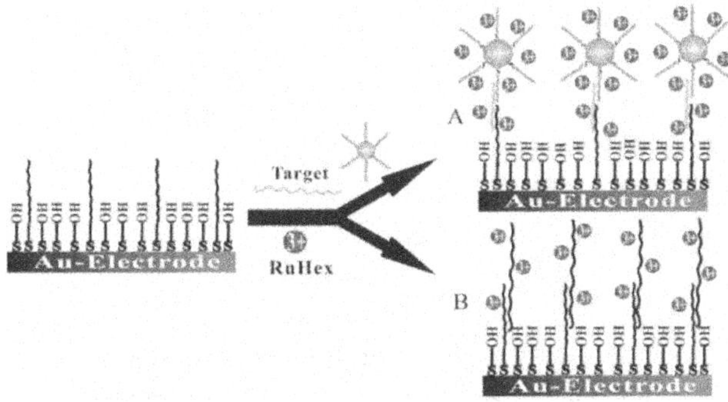

FIGURE 1.8 Novel electrochemical DNA nanoscale biosensors.

excellent electrochemical performance as a new electrode material to simultaneously detect GUA and ADE; the response concentration ranges for GUA and ADE are 0.04–100 and 0.06–100 μmol/L, with detection limits of 0.0 and 0.02 μmol/L, respectively. Zhao[63] used a graphene/tantalum electrode prepared by the hot-wire CVD method modified with magnesium oxide (MgO) nanorods as the sensing electrode to synchronously detect AA, DA, and UA, showing good electrochemical performance. The response range for ascorbic acid, dopamine, and uric acid detection is 5.0–350 μ mol/L, 0.1–7 μ mol/L, and 1–70 μ mol/L, respectively with detection limits of 0.03 μ mol/L, 0.15 μ mol/L, and 0.12 μ mol/L, respectively. It can be seen that composite materials based on nanomaterial–modified graphene have extensive application in research into electrochemical biosensors, and

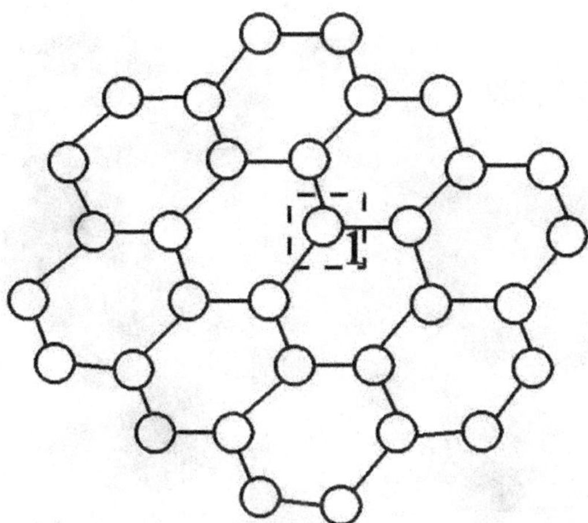

FIGURE 1.9 Graphene structure.

graphene has great research value and application potential in the application of electrochemical biosensors.

1.8 ELECTROCHEMICAL BIOSENSORS BASED ON CNTS AND CQDS

1.8.1 Bioimaging

At present, most materials used for cell imaging (Figure 1.10) and cell labeling are organic dyes and semiconductor quantum dots, but they have a certain toxicity to organisms and are not conducive to cell proliferation. Because of their adjustable photoluminescence property, low toxicity, water solubility, and high chemical/optical stability, carbon quantum dots have become candidate materials for biological imaging.

The solution of fluorescent carbon quantum dots prepared by an excellent calcination method using anthocyanins as carbon precursors exhibits light yellow fluorescence under sunlight and strong blue fluorescence under 365 nm ultraviolet light. It should be noted that the surface of this material contains a large number of hydroxyl groups and has good water solubility. The average particle size is 2.5 nm. Because of their good fluorescence stability and biocompatibility, CQDs can be applied to the biological imaging of zebrafish. In addition, the antioxidant stress property of hydrogen peroxide–induced carbon quantum dots in vivo was studied. The results showed that oxidative damage can be reduced by controlling the production of reactive oxygen species. At the same time, the mRNA expression of related genes can be promoted by carbon quantum dots, thus encoding more antioxidant proteins and preventing oxidative damage to zebrafish. Add water and

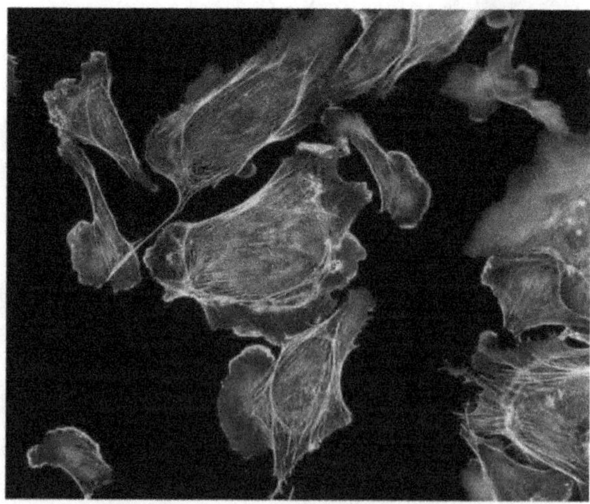

FIGURE 1.10 Analysis of cell imaging by CQD biosensor.

milk in a micro-reactor; stir at 800 rpm for 2 h under 180°C; maintain the system at a constant pressure of 1.2 MPa; ensure that the water in the equipment is in a subcritical state; and obtain carbon quantum dot powder through centrifugation, filtration, and freeze-drying. This material can be easily absorbed by HeLa cells without additional surface functionalization and can be used as a fluorescent nano-probe for biological imaging.

1.8.2 Molecular/Ionic Detection

Fluorescent sensors can be used to detect various inorganic ions or other chemical substances. Methods such as inductively coupled plasma mass spectrometry or atomic absorption spectrometry for the detection and analysis of metal ions and chemical substances cannot achieve real-time detection due to their lack of flexibility and expensive equipment. Due to the low cost, easy operation, and high sensitivity and selectivity, fluorescent sensors are ideal candidates for these applications. Because of their good photostability and chemical stability, carbon quantum dots can be detected by observing the change of their fluorescence intensity.

The Internal filtering effect and fluorescence resonance energy transfer make up the majority of the molecular/ion detection method. When the separation between the donor and receptor molecules is less than 10 nm, resonance energy transfer generally refers to non-radiative energy transfer. A dipole–dipole contact occurs during resonance energy transmission. Chemical, fluorescence, and bioluminescence are the three subtypes of resonance energy transmission. Among them, donor molecules are chemiluminescent compounds, bioluminescent proteins, and fluorophores, and the receptor molecules are fluorophores. The Internal filtering effect refers to the formation of ground state complexes without fluorescence emission by CQDs and quenched substances. When exposed to external light sources, the ground state complexes compete with CQDs for incident light, resulting in the reduction of fluorescence emission intensity of CQDs, Or the absorption spectrum of the quenched substance overlaps with the excitation spectrum or emission spectrum of the CQDs, so the quenched substance will seize the excitation luminous energy or the emission light energy of the CQDs, and the phenomenon in the burning light spectrum is the weakening of the fluorescence emission intensity. Although the internal filtration effect leads to a decrease of intensity, it does not change the fluorescence lifetime. when the distance between the emitter and the absorber is greater than 10 nm, the internal filtering effect can still be produced. This makes the sensor easier to operate.

1.8.3 Light-Emitting Devices

Since a large portion of global energy is used for lighting, it is particularly important to develop efficient lighting technology. The potential applications of CQDs in light-emitting diodes (LEDs) (Figure 1.11) have been extensively studied. Carbon quantum dots are used in two ways in light-emitting diodes. The first method is to cover the LED with carbon quantum dots to change the illumination intensity and emission

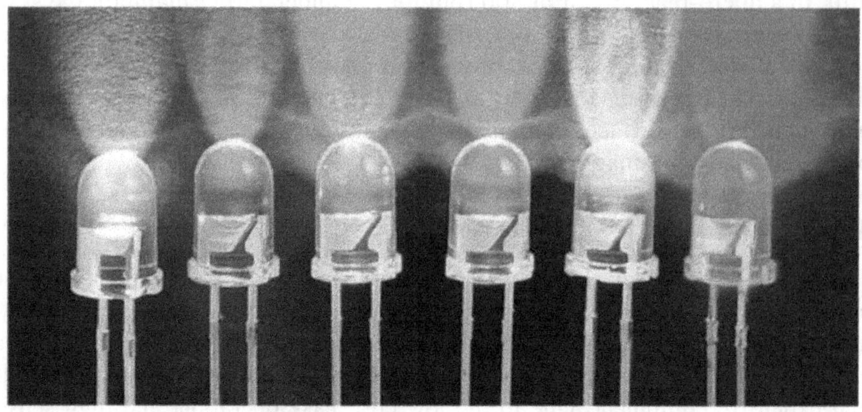

FIGURE 1.11 LED.

wavelength of the LED. The second is to use carbon quantum dots in the internal structure of the LED to improve the LED light-emitting layer.

Due tointeraction or excessive resonance energy transfer, carbon quantum dots can cause fluorescence quenching of the aggregated state, which restricts their utilization in solid fluorescent materials. CQDs with solid state fluorescence were made using citric acid and the precursors 1,4,7,10-tetraazacyclododecane (ring). CQDs efficiently resist the fluorescence quenching effect brought on by aggregation because they generate brilliant yellow solid fluorescence when exposed to 365-nm UV light. CQDs have an emission peak center at 546 nm (0.43, 0.55), which means that they can be paired with blue chips to create white LEDs with an absolute quantum yield of 48% when the excitation is set at 450 nm. This shows that carbon quantum dots have good light conversion ability in the solid state. Because of their excellent optical properties, they may completely replace current LED devices in the future.

1.8.4 Photocatalysis

Carbon quantum dots have become the focus of research in the application of photocatalysis due to their good absorption of near-ultraviolet and visible light. High-temperature pyrolysis was used to create nanoscale zinc oxide (C-ZnO NCs) modified by carbon quantum dots with new visible light catalytic capacity. The gas created by the reagent at a high temperature during the synthesis process can efficiently break down the result into minute particles. The degradation rate of C-ZnO NCs in rhodamine B solution under visible light is 433.3% higher than that of pure zinc oxide as a result of a rise in the absorbance of C-ZnO NCs in the visible light region and the development of additional valence bands. A workable water purification system was also developed by the author using a continuous photodegradation reactor under the condition of maximizing natural light, which significantly decreased the cost and broadened the application field. Because C-ZnO NCs have the advantages of reusability, low cost, visible light feasibility, and environmental friendliness, they can be used as a potential photocatalyst for the effective use of sunlight.

REFERENCES

[1] N.F. Atta, A. Galal, A.R. El-Gohary, Crown ether modified poly(hydroquinone)/carbon nanotubes based electrochemical sensor for simultaneous determination of levodopa, uric acid, tyrosine and ascorbic acid in biological fluids, *J. Electroanal. Chem.* 863 (2020) 114032.

[2] F.S. Lisboa, E.G.C. Neiva, M.F. Bergamini, L.H. Marcolino Junior, A.J.G. Zarbin, Evaluation of carbon nanotubes/polyaniline thin films for development of electrochemical sensors, *J. Braz. Chem. Soc.* 31 (2020) 1093e1100.

[3] P.K. Kalambate, Dhanjai, A. Sinha, Y. Li, Y. Shen, Y. Huang, An electrochemical sensor for ifosfamide, acetaminophen, domperidone, and sumatriptan based on self-assembled MXene/MWCNT/chitosan nanocomposite thin film, *Microchim. Acta* 187 (2020) 402.

[4] E. Mynttinen, N. Wester, T. Lilius, E. Kalso, B. Mikladal, I. Varjos, S. Sainio, H. Jiang, E.I. Kauppinen, J. Koskinen, T. Laurila, Electrochemical detection of oxycodone and its main metabolites with nafion-coated single-walled carbon nanotube electrodes, *Anal. Chem.* 92 (2020) 8218e8227.

[5] L. Long, H. Liu, X. Liu, L. Chen, S. Wang, C. Liu, S. Dong, J. Jia, Co-embedded N-doped hierarchical carbon arrays with boosting electrocatalytic activity for in-situ electrochemical detection of H2O2, *Sensor. Actuator. B Chem.* 318 (2020) 128242.

[6] R. Haldavnekar, K. Venkatakrishnan, D.B. Tan, Boosting the sub-cellular biomolecular cancer signals by self-functionalized tag-free nano sensor, *Biosens. Bioelectron.* 190 (2021) 113407–113407.

[7] L. Qian, S. Durairaj, S. Prins, A. Chen, Nanomaterial-based electrochemical sensors and biosensors for the detection of pharmaceutical compounds, *Biosens. Bioelectron.* 175 (2021) 105609.

[8] A.W. Barnard, M. Zhang, G.S. Wiederhecker, M. Lipson, P.L. McEuen, Realtime vibrations of a carbon nanotube, *Nature* 566 (2019) 89e93.

[9] S. Iijima, Helical microtubules of graphitic carbon, *Nature* 354 (1991) 56e58.

[10] L. Wang, S. Xie, Z. Wang, F. Liu, Y. Yang, C. Tang, X. Wu, P. Liu, Y. Li, H. Saiyin, Functionalized helical fibre bundles of carbon nanotubes as electrochemical sensors for long-term in vivo monitoring of multiple disease biomarkers, *Nat. Biomed. Eng.* 4 (2020) 159e171.

[11] M. Zhao, Y. Chen, K. Wang, Z. Zhang, J.K. Streit, J.A. Fagan, J. Tang, M. Zheng, C. Yang, Z. Zhu, W. Sun, DNA-directed nanofabrication of high-performance carbon nanotube field-effect transistors, *Science* 368 (2020) 878e881.

[12] Y.-D. Dong, H. Zhang, G.-J. Zhong, G. Yao, B. Lai, Cellulose/carbon composites and their applications in water treatment—A review, *Chem. Eng. J.* 405(2021) 126980.

[13] D. Liu, C. Li, J. Wu, Y. Liu, Novel carbon-based sorbents for elemental mercury removal from gas streams: a review, *Chem. Eng. J.* 391 (2020) 123514.

[14] V. Kumar, K. Vaid, S.A. Bansal, K.H. Kim, Nanomaterial-based immunosensors for ultrasensitive detection of pesticides/herbicides: current status and perspectives, *Biosens. Bioelectron.* 165 (2020) 112382.

[15] M. Bilal, H.M.N. Iqbal, Chemical, physical, and biological coordination: an interplay between materials and enzymes as potential platforms for immobilization, *Coord. Chem. Rev.* 388 (2019) 1e23.

[16] A.P.A.D. Carvalho, C.A. Conte Junior, Green strategies for active food packagings: a systematic review on active properties of graphene-based nanomaterials and biodegradable polymers, *Trends Food Sci. Technol.* 103 (2020) 130e143.

[17] A.S.R. Bati, L. Yu, M. Batmunkh, J.G. Shapter, Recent advances in applications of sorted single-walled carbon nanotubes, *Adv. Funct. Mater.* 29 (2019) 1902273.

[18] H. Li, X. Zhang, Z. Zhao, Z. Hu, X. Liu, G. Yu, Flexible sodium-ion based energy storage devices: recent progress and challenges, *Energy Storage Mater.* 26 (2020) 83e104.

[19] K.L. Hess, I.L. Medintz, C.M. Jewell, Designing inorganic nanomaterials for vaccines and immunotherapies, *Nano Today* 27 (2019) 73e98.

[20] S.M. Mugo, J. Alberkant, A biomimetric lactate imprinted smart polymers as capacitive sweat sensors, *IEEE Sensor. J.* 20 (2020) 5741e5749.

[21] Z. Pei, Q. Zhang, Y. Liu, Y. Zhao, X. Dong, Y. Zhang, W. Zhang, S. Sang, A high gauge-factor wearable strain sensor array via 3D printed mold fabrication and size optimization of silver-coated carbon nanotubes, *Nanotechnology* 31(2020) 305501.

[22] C. Sabu, T.K. Henna, V.R. Raphey, K.P. Nivitha, K. Pramod, Advanced biosensors for glucose and insulin, *Biosens. Bioelectron.* 141 (2019) 111201.

[23] J. Wang, X. Huang, J. Xie, Y. Han, Y. Huang, H. Zhang, Exosomal analysis: advances in biosensor technology, *Clin. Chim. Acta* 518 (2021) 142e150.

[24] W. Tong, Z. Fan, G. Luo, et al., A new structure for multi-walled carbon nanotubes reinforced alumina nanocomposite with high strength and toughness [J], *Mater. Lett.* 62(4–5) (2008) 641–644.

[25] M.R. Falvo, G.J. Clary, R.M.I. Taylor, et al., Bending and buckling of carbon nanotubes under large strain [J], *Nature* 389(6651) (1997) 582–584.

[26] A. Abdellah, A. Yaqub, C. Ferrari, B. Fabel, P. Lugli, G. Scarpa, Spray deposition of highly uniform CNT films and their application in gas sensing [A], *2011 11th IEEE International Conference on Nanotechnology* [C], 2011.

[27] A. Todri-Sanial, J. Liang, Power and performance analysis of doped SW/DW CNT for on-chip interconnect application [A], *GRAPHENE* [C], 2017.

[28] M. Ates, A.S. Sarac, Conducting polymer coated carbon surfaces and biosensor applications, *Prog. Org. Coat.* 66 (2009) 337–358.

[29] Z. Chen, Y. Wu, Y. Yang, et al., Multilayered carbon nanotube yarn based optoacoustic transducer with high energy conversion efficiency for ultrasound application [J], *Nano Energy* 46 (2018) 314–321.

[30] J. Li, X. Lan, S. Lei, et al., Effects of carbon nanotube thermal conductivity on optoacoustic transducer performance [J], *Carbon* 145 (2019) 112–118.

[31] N. Gan, H. Jin, T. Li, L. Zheng, Fe(3)O(4)/Au magnetic nanoparticle amplification strategies for ultrasensitive electrochemical immunoassay of alfa-fetoprotein, *Int. J. Nanomedicine* 6 (2011) 3259–3269. https://doi.org/10.2147/IJN.S26212

[32] N.J. Ronkainen, H.B. Halsall, W.R. Heineman, Electrochemical biosensors, *Chem. Soc. Rev.* 39(5) (2010) 1747–1763. https://doi.org/10.1039/b714449k

[33] D. Grieshaber, R. MacKenzie, J. Voros, E. Reimhult, Electrochemical biosensors—sensor principles and architectures, *Sensors* 8(3) (2008) 1400–1458. https://doi.org/10.3390/s8031400

[34] J. Wang, Carbon-nanotube based electrochemical biosensors: a review, *Electroanal* 17(1) (2005) 7–14. https://doi.org/10.1002/elan.200403113

[35] A. Ruhal, J.S. Ruhal, S. Kumar, A. Kumar, Immobilization of malate dehydrogenase on carbon nanotubes for development of malate biosensor, *Cell. Mol. Biol.* 58(1) (2012) 15–20. https://doi.org/10.1170/t915

[36] Y. Shao, J. Wang, H. Wu, J. Liu, I.A. Aksay, Y. Lin, Graphene based electrochemical sensors and biosensors: a review, *Electroanal* 22(10) (2010) 1027–1036. https://doi.org/10.1002/elan.200900571

[37] P. Pandey, M. Datta, B.D. Malhotra, Prospects of nanomaterials in biosensors, *Anal. Lett.* 41(2) (2008) 159–209. https://doi.org/10.1080/00032710701792620
[38] P. Pandey, M. Datta, B.D. Malhotra, Prospects of nanomaterials in biosensors, *Anal. Lett.* 41(2) (2008) 159–209. https://doi.org/10.1080/00032710701792620
[39] C. Cai, J. Chen, Direct electron transfer of glucose oxidase promoted by carbon nanotubes, *Anal. Biochem.* 332(1) (2004) 75–83. https://doi.org/10.1016/j.ab.2004.05.057
[40] Y.M. Lee, O.Y. Kwon, Y.J. Yoon, K. Ryu, Immobilization of horseradish peroxidase on multi-wall carbon nanotubes and its electrochemical properties, *Biotechnol. Lett.* 28(1) (2006) 39–43. https://doi.org/10.1007/s10529-005-9685-8
[41] M.D. Rubianes, G.A. Rivas, Enzymatic biosensors based on carbon nanotubes paste electrodes, *Electroanal* 17(1) (2005) 73–78. https://doi.org/10.1002/elan.200403121
[42] A. Ruhal, J.S. Ruhal, S. Kumar, A. Kumar, Immobilization of malate dehydrogenase on carbon nanotubes for development of malate biosensor, *Cell. Mol. Biol.* 58(1) (2012) 15–20. https://doi.org/10.1170/t915
[43] Z. Zhu, L. Garcia-Gancedo, A.J. Flewitt, H. Xie, F. Moussy, W.I. Milne, A critical review of glucose biosensors based on carbon nanomaterials: carbon nanotubes and graphene, *Sensors* 12(5) (2012) 5996–6022. https://doi.org/10.3390/s120505996
[44] D.T. Tran, V.H. Hoa, L.H. Tuan, N.H. Kim, J.H. Lee, Cu-au nanocrystals functionalized carbon nanotube arrays vertically grown on carbon spheres for highly sensitive detecting cancer biomarker, *Biosens. Bioelectron.* 119 (2018) 134–140.
[45] B.R. Azamian, J.J. Davis, K.S. Coleman, et al., Bioelectrochemical single-walled carbon nanotubes [J], *J. Am. Chem. Soc.* 124(43) (2002) 12664.
[46] P.G. Collins, K. Bradley, M. Ishigami, et al., Extreme oxygen sensitivity of electronic properties of carton nanotubes [J], *Science* 287 (2000) 1801–1804.
[47] J. Kong, N.R. Franklin, C. Zhou, et al. Nanotube molecular wires as chemical sensor [J], *Science* 287(5453) (2000) 622–625.
[48] L.E. Hood, M.W. Hunkapiller, L.M. Smith, Automated DNA sequencing and analysis of the human genome [J], *Genomics* 1(3) (1987) 201–212.
[49] S.S. Zhang, S.Y. Niu, B. Han, W. Cao, Sensitive DNA biosensor improved by Luteolin copper(II) as indicator based on silver nanoparticles and carbon nanotubes modified electrode[J], *Anal. Chim. Atca* 651(1) (2009) 42–47.
[50] B. Tang, F. Li, Y. Feng, et al., Gold nanoparticles modified electrode via a mercaptodiazoaminobenzene monolayer and its development in DNA electrochemical biosensor [J], *Biosens. Bioelectron.* 25(9) (2010) 2084–2088.
[51] S. Palanti, G. Marrazza, Electrochemical DNA probes [J], *Anal. Lett.* 29(13) (1996) 2309–2331.
[52] S.R. Mikklesen, Electrochemical biosensors for DNA sequence detection [J], *Electroanalysis* 8(1) (1996) 15–19.
[53] E. Palecek, M. Fojta, Differential pulsed voltammetric determination of RNA at the picomole level in the presence of DNA and nucleic acid components [J], *Anal. Chem.* 66(9) (1994) 1566–1571.
[54] L.J. Blum, A. Sassolas, B.D. Leca-Bouvier, DNA biosensors and microarrays [J], *Chem. Rev.* 108(1) (2008) 109–139.
[55] X.H. Li, H. Gong, T.Y. Zhong, et al., Unlabeled hairpin DNA probe for electrochemical detection of single-nucleotide mismatches based on Mut S-DNA interactions [J], *Anal. Chem.* 81(20) (2009) 8639–8643.
[56] J. Wang, Electrochemical nucleic acid biosensors [J], *Anal. Chim. Acta.* 469(1) (2002) 63–71.
[57] J. Wang, Towards genoelectronics: electrochemical biosensing of DNA hybridization [J], *Chem. Eur. J.* 5(6) (1999) 1681–1689.

[58] J. Wang, Survey and summary from DNA biosensors to gene chips [J], *Nucleic Acid. Res.* 28(16) (2000) 3011–3016.

[59] J. Wang, Nanoparticle-based electrochemical DNA detection [J], *Anal. Chim. Acta.* 500(1–2) (2003) 247–257.

[60] C.H. Fan, G. Liu, et al., An enzyme-based E-DNA sensor for sequence-specific detection of femtomolar DNA targets [J], *J. Am. Chem. Soc.* 130(21) (2008) 6820–6825.

[61] W. Cai, J. Lai, T. Lai, et al., Controlled functionalization of flexible graphene fibers for the simultaneous determination of ascorbic acid, dopamine and uric acid[J], *Sens. Actuators B Chem.* 224 (2016) 225–232.

[62] Y.S. Gao, J.K. Xu, L.M. Lu, et al., Overoxidized polypyrrole/graphene nanocomposite with good electrochemical performance as novel electrode material for the detection of adenine and guanine[J], *Biosens. Bioelectron.* 62(6) (2014) 261–267.

[63] L. Zhao, H. Li, S. Gao, et al., MgO nanobelt-modified graphene-tantalum wire electrode for the simultaneous determination of ascorbic acid, dopamine and uric acid[J], *Electrochim. Acta* 168 (2015) 191–198.

2 Graphene-Based Semiconducting Nanomaterials for Chemical and Biological Sensing Application Biosensors and Their Applications in Human Life and Agroecosystems

Zuhong Ji and Won-Chun Oh

2.1 BACKGROUND

Conventional sensing methods, such as electrochemical methods, fluorescent microarrays and cross-flow immunometry, enzyme-bonded immunosorption measurement (ELISA), deoxyribonucleic acid (DNA) microarrays, DNA sequencing techniques, and polymerase chain reaction (PCR)–based methods, among others [1–6], require high-precision measurement, quantification, and expensive reagents. Moreover, we cannot quantitatively monitor most responses in real time. Thus, new sensors with inexpensive, simple, and highly unusual sensing properties will enable real-time measurement and evaluation of target biomolecules. This method will enable extensive clinical application. In the field of life sciences and medicine, sensors have been used to improve patient intensive care, diagnose patients, and monitor vital signs [7–10]. Due to the need for minimally invasive detection and diagnosis of diseases and early detection, we need to develop many new sensors. One special focus of sensor development is to develop nanomaterials for manufacturing nanosensors and to realize miniaturization and precision of sensors through their application. The miniaturization and precision properties of nanomaterials can improve patient care by making sensors extremely sensitive and minimally invasive with respect to their unique electrical and chemical properties [10]. Sensitivity of the sensor is important for detecting the target molecule, but this sensitivity is one of the key parameters

DOI: 10.1201/9781003425427-3

because it correlates with the sensor's detection limit and accuracy and affects the negative and positive predictive values. In general, this study reports the range of use and sensitivity of biosensors that can impose detection limits on sensors. However, due to the novelty of recently designed sensors, there are no statistical data or detailed reports on the precision, accuracy, and negative and positive predicted values of these parameters. These important parameters should be considered in future studies. Among the nanomaterials for graphene-based nanomaterials, graphene and nanosensor manufacturing exhibit the greatest potential because they exhibit improved signal response in various sensing applications [11–13]. In addition, graphene-based nanomaterials have excellent biocompatibility with various biomolecules such as proteins, cells, DNA, antibodies, and enzymes and can expand sensitivity and expand application range [13]. By applying a graphene-based detection method (Figure 2.1) to these biomolecules, the scope of development and use of biosensors has been expanded. Various molecules, biomolecules, and even cells can be detected by these biometric sensors [14,15].

High-sensitivity biosensors capable of detecting trace analyte molecules in femtomol or picomol concentrations have many applications in biological and medical fields, such as clinical diagnosis and glucose monitoring as well as agricultural, food industry, and environmental monitoring [17–21]. The healthcare industry urgently needs to develop high-sensitivity devices and new methods to provide efficient point-of-care detection with low cost and high accuracy [16,22,23]. In addition, biocompatibility, selectivity, and sensitivity are important for biosensors in vivo, which are attracting attention because they can monitor target analytes in living cells for a

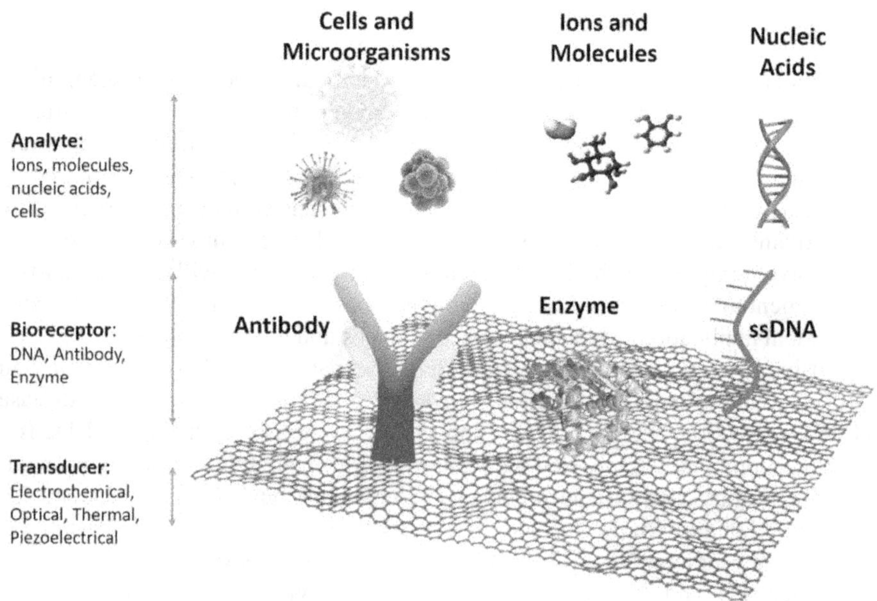

FIGURE 2.1 Examples of biosensors and components on a graphene platform [16].

long time [24,25]. Biosensor research used in medical and biological fields includes extensive sensing capabilities and is considered an important area including protein, nucleic acids, antibodies, cholesterol, glucose, blood oxygen levels, body exercise, blood pressure, heart rate, pulse, cancer cells, heavy metals in drinking water, and food toxins [26–28]. Raman spectroscopy-based platforms, fluorescent biosensors, electrochemical biosensors, potential biosensors, and colorimetric biosensors have been explored in many approaches to increase sensitivity by combining spectroscopic fields [29–33]. Methods based on electrochemical [34,35] and fluorescence [36] provide a simpler, cheaper, and more sensitive detection method for monitoring very high detection sensitivity, fast response-recovery time, and various analytes compared to other detection methods [37]. Due to the unique physical and chemical properties of nanoscale materials since the discovery of Buckminster fullerene (C_{60}) molecules in 1985, the nanotechnology sector focused on the development of new devices and materials on a scale of 1–100 nanometers [38]. We have developed various kinds of nanoscale materials, including zero-dimensional (0D) nanoparticles (such as semiconductors and metal nanoparticles) [39,40], one-dimensional (1D) nanostructures (nanotubes, nanorods, nanowires) [41], transition metal daikalcogenides (TMDs), and graphene nanosheets (GNs). Previous studies have shown significant progress in characterization, potential applications, synthesis, and processing [42], including catalysts, supercapacitors, fuel cells, batteries, solar cells, magnetic disks, communication devices, memory devices, and light-emitting diodes [43,44]. The development of high-sensitivity biosensors to significantly increase sensitivity, coupled with interfaces with various probe biomolecules, suggests a more advanced method [45]. Due to the size and unprecedented physical properties of these nanomaterials, it is possible to develop biosensors that are very small in size and greatly improved in performance, presenting new possibilities for the commercialization and development of next-generation biosensors in the medical and biopharmaceutical sectors [46].

Since the first graphene production, separation, and characterization of graphene from natural graphite by Novoselov and Geim in 2004 [47], the potential application of two-dimensional graphene nanomaterials has shown many results with extensive research and applications [48–51]. The physical properties of graphene (GR) are unique in their thermal conduction, mechanical, optical, and electrical aspects, and graphene with these properties is an atomically thick two-dimensional carbon nanosheet [52]. Graphene carbon nanosheets are easily exfoliated from the abundant graphite present on Earth and are thought to be isotropic with carbon nanotubes and fullerenes. Graphene can be easily processed in various ways into multi-layer, hydrophobic, or single-layer nanosheets; nanoribbons; foams; paper; and stretchable ultrathin films [53–58]. The characteristics of single-layer graphene nanosheets show high mechanical strength with a Young's modulus of 1.1 TPa [55], a large specific surface area (2630 m^2g^{-1}), high optical transparency (2.3% absorption), high charge carrier mobility (200,000 $cm^2V^{-1}s^{-1}s^{-1}$), and 5000 WMK^{-1} [59–62]. Thus, we explored a wide range of applications for graphene-based materials (GBMs), including bulk heterojunction [63], dye-sensitized solar cells [64], energy storage [65], touchscreen panel devices and electronic skin [66], field-effect transistors [67], light-emitting diodes (LEDs) [68], chemical and gas sensors [68], nanomedicine [69], drug delivery

[70], and many other applications [71]. Recently, nano-based materials of graphene and its oxidation derivatives have emerged as potential applications in biosensors, and various oxygen-containing functional groups (carboxyl, hydroxyl epoxy functional groups) such as graphene oxide (GO) contain favorable properties for sensor function [72]. GO nanosheets are highly hydrophilic due to the presence of these functional groups and can aggregate various types of inorganic nanoparticles such as metal oxides, quantum dots (QDs), semiconductor nanoparticles, nanoclusters (NCs), and precious metals, thereby improving the performance of sensors based on them [73,74]. In addition, reducing GO to reduced GO (rGO) compared to CVD-grown graphene increases the density of defects, resulting in high electrochemical activity, which is particularly useful for the development and application of electrochemical biosensors. The unique properties and morphological structures useful for detection, which are the basis of sensors, are correlated with the application of graphene-based nanocomposites [75]. The three-dimensional interconnected layered structure of graphene nanocomposites aids in the diffusion of various types of biomolecules and optimizes biological detection by maintaining biocatalytic functions [76–78]. Graphene-based polymers with hybrid [79,80] and surface-modified metal nanoparticles [73] have been studied for high surface area, excellent biocompatibility, and site-selective binding to biomolecules. We explore various nanostructures for biosensors, including pathogenic bacteria, heavy metals, antibodies, nucleic acids, cancer biomarkers, glucose and hydrogen peroxide detection [26,81–88], and many other targets [34,89–92].

However, there is no overall review in the research literature on fluorescent biosensors and various electrochemical fields using graphene-based nanocomposites. Graphene-based fluorescent biosensors need to be studied more, and not many reviews have yet appeared. Therefore, this chapter summarizes recent developments in graphene nanocomposite-based fluorescent biosensors and electrochemical fields such as graphene, GO, polymer hydrogel network, various 3D graphene integrated with metal/metal oxide nanoparticles, graphene/inorganic NP nanocomposite, rGO/polymer nanocomposites, and sensor effects increased by surface modification. We discuss cancer biomarkers, nucleic acids (NAs), uric acid (UA), cofactors nicotinamide adenine dinucleotide (NADH), adenosine triphosphate (ATP), ascorbic acid (AA), pathogens, cholesterol, dopamine (DA), hydrogen peroxide (H_2O_2), and glucose for the detection of pesticides. We also investigate the selectivity of graphene-based fluorescent biosensors and electrochemistry to optimize analysis in biological systems. This chapter will serve as a single reference for researchers in the fields of electrochemical, chemical, nanotechnology, material science, and physical properties of graphene and biosensors. Accordingly, this chapter will attract a wide audience from various research fields and further arouse interest in graphene-based biosensors in the sensor industry in the future. The biosensor system shown in Figure 2.2 shows a typical platform consisting of bioreceptors interfacing with the sensor. Bioreceptors should be able to efficiently sense and recognize biomolecule elements in research fields such as cells, RNA, DNA, antibodies, and enzymes. The varying amounts of biological signals that the receptor can detect in the combination should be transformed into observable visual information through quantitative, thermal, optical, chemical, or electrochemical actions and physical analysis.

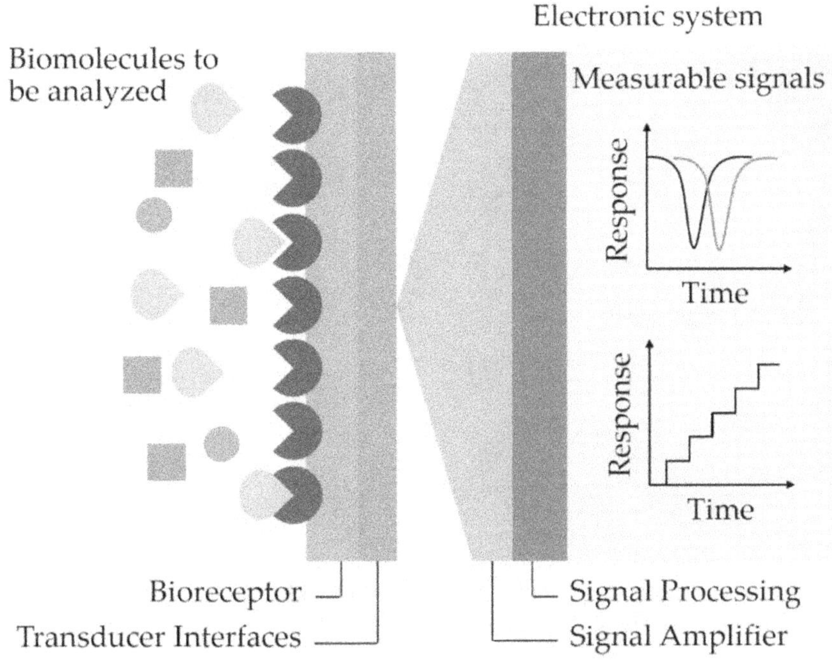

FIGURE 2.2 Schematic illustration of a typical biosensor system [93].

2.1.1 Graphene-Based Nanomaterials as Biosensors

Typically, a sensor consists of two components: the receiver part and the transducer part (see Figure 2.1). Receptor active materials are inorganic, organic, or organic–inorganic composite materials that specifically interact with target molecules. A transducer that converts chemical information such as a chemical reaction into a measurable signal is part of a sensor. We use graphene-based nanomaterials involved in converting interactions between target molecules and receptor active materials into detections and possible measurements for use as sensors in biosensors [94]. For these chemical reactions, biological receptors (e.g., molecules such as ssDNA, antibodies, and enzymes) must be attached to the surface of the transducer. Physical adsorption immobilization is most commonly used in enzymes, and EDC/NHS chemistry is the most common method used to fix ssDNA and antibodies to graphene and its derivatives (reduced graphene oxide, graphene oxide) (see Figure 2.3). We designed different biosensor schemes using graphene for various conversion modes, such as the ability to fix different heterogeneous molecules, high electron transfer rates, high electrical conductivity, and large surface areas [95]. For example, electron transfer between the bioreceptor and the sensor active material is facilitated by the conjugated structure of graphene, so its electrochemical sensors have high signal sensitivity [12,94,96,97]. Furthermore, a fluorescent biosensor can be made using graphene-based nanomaterials as active materials for converters. reduced graphene oxide, graphene oxide, and graphene (G) have been found to have very high fluorescence quenching efficiency

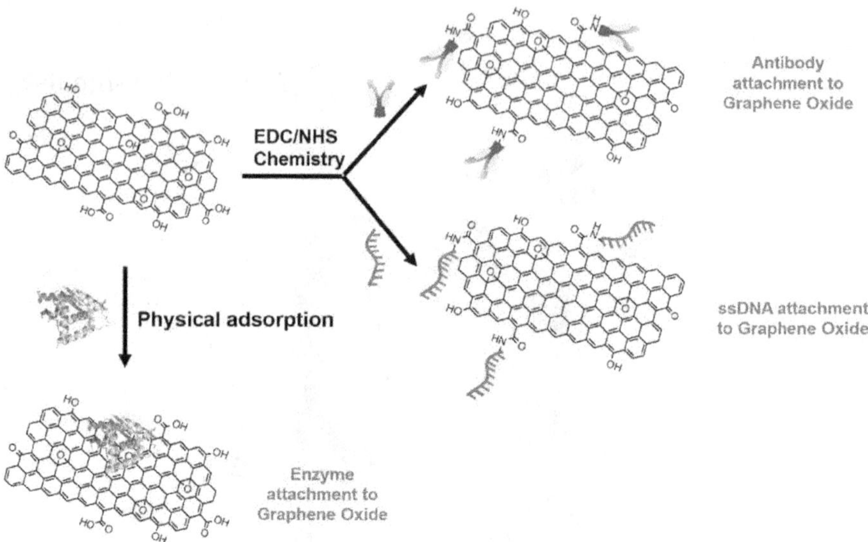

FIGURE 2.3 Schematic of the most common attachment methods of bioreceptors, such as enzymes, DNA, and antibodies onto graphene surfaces [103].

[98–100]. When designing a sensor using a graphene nano-sized active material, specific aspects of graphene characteristics that affect the detection limit of target molecules should be considered. For example, the different functions and properties of graphene-based nanomaterials in biosensors correlate with different synthesis methods and different synthesis arrangements of different graphene and their derivatives. The sensitivity and selectivity of the biosensor may be directly affected by the direction and orientation between the biosensor and the G, GO, or rGO layer. In addition, factors such as the oxidation state, functional group, and number of layers of graphene and its derivatives not only have a significant impact on the difference in sensing performance between sensors but also have a significant impact on the combination of sensors and bioreceptors. The detection limit of the target molecule and the interaction with the active material are also affected by the number of functional groups and active groups of the nanomaterial. In such cases, it is often required to block non-specific adsorption sites for nanomaterials to prevent non-specific binding of biomolecules other than target molecules. Applying blocking reagents can achieve the same effect as sensor treatment such as casein or superblock, bovine serum albumin (BSA), or surfactant [101–103]. Among these applied reagents, graphene-based nanomaterial biosensors have fast response times and high stability/sensitivity, potentially promoting the development of healthcare and biomedical diagnosis. In this chapter, the recent development of graphene and graphene-based nanomaterial biosensor technology is briefly summarized. In addition, we will focus progressively on DNA, enzyme, and antibody-based biosensors for applications in biological clinical environments and life sciences. Furthermore, our goal is to show progress in the application and synthesis of biosensors for real-time molecular detection and clinical diagnosis.

2.1.2 Nano-Graphene–Based Nanomaterials for Antibodies

Immunosensors measure antibody and antigen-specific binding reactions and are analytical detection platforms. In several fields, high-affinity binding and biocompatibility of antigens to antibodies make this molecule attractive, especially diagnostics. The antibody (Ab) structure consists of four polypeptide chains with a characteristic "Y" shape (Figure 2.4). We use a single disulfide bond to link these chains. The Ab structure contains two distinct parts: the Ab "arm" consists of two domains, a variable domain and a constant domain. The selectivity of an antibody for a particular antigen is conferred by the variable domains. The Ab "body" portion contains two distinct fragments, the antigen-binding fragment (Fab) and the crystallizable fragment (Fc). Fab and Fc consist of amino (-NH_2) and carboxyl (-COOH) groups that combine with a high affinity for target molecules [105,106]. This specific antibody–antigen response's high-affinity recognition is mainly due to the properties, reactivity, and structure of antibodies, making them excellent candidates for sensing applications. The diversity of functional groups on the GO surface allows for different Ab attachment strategies. Most strategies for functionalizing GO using antibodies include the addition of a 1-pyrenebutyric acid succinimidyl ester (PASE) linker, electrostatic bonding or carbodiimide hydrochloride (EDC)/N-hydroxysuccinimide (NHS) (EDC/NHS) or are via 1-ethyl-3-(3-dimethylaminopropyl). EDC/NHS chemical functionalization for producing biochemical conjugates is the most versatile and popular method. EDC is a water-soluble crosslinker that enables direct bioconjugation between amine groups and carboxyl. Amide bonds are formed by nucleophilic attack from antibody primary amino groups with carboxyl groups on the surface of GO in this reaction. Conjugates are allowed to form between two different amide groups' molecules in the process [107]. Different approaches enable the detection of target molecules.

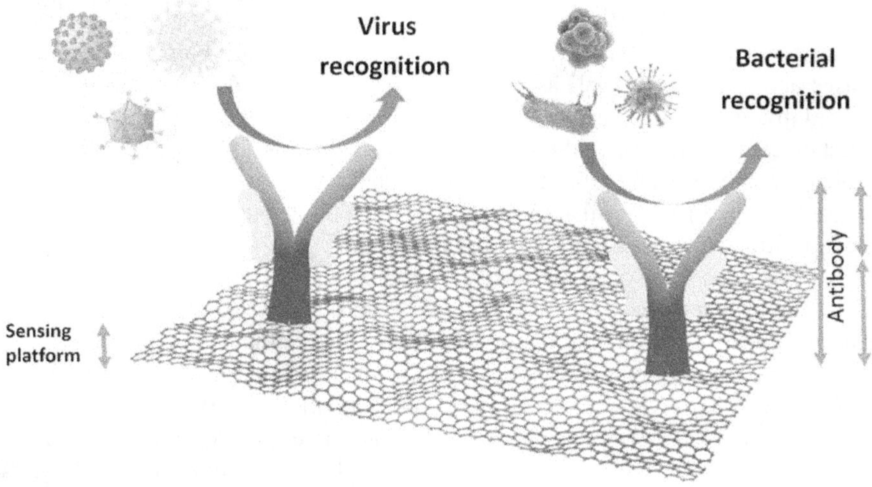

FIGURE 2.4 Scheme of graphene modified with antibodies for the recognition of pathogens [104].

Electrochemical methods are most often described. One of the methods to measure any chemical or electrical change at the electrode/electrolyte interface is electrochemistry. This method produces biological recognition between antigen and antibody, which is a conformational change. These nanosensors contain a reference electrode (which allows current to flow between the two electrodes and is connected to an electrolyte) and a working electrode (where the reaction occurs). Electrochemical sensors contain measurements of resistance, potential, or current, where electrode sensors are able to detect changes in electrical signals caused by binding reactions [108]. Because it is sensitive, fast, and simple; has good selectivity; and uses small sample volumes compared to other immunosensor methods, this method is better [105]. However, this approach has some limitations, for example, the irreversible antigen–antibody interactions and binding affinity [104]. In pathogen detection, graphene-based nanomaterials show broad versatility in antibody biosensors. Recently, in order to cope with the early detection of diseases, we have developed a variety of graphene antibody biosensors with clinical applications. Antibody nanosensors with G were developed to detect Zika virus and *E. coli* [109–111]. In other words, GO has been used to detect cardiovascular disease, rotavirus, and dengue virus [112–114]. In different samples, we have used rGO to detect *E. coli* [115] but had higher detection limits compared to G [109,110] and G modified with polymethylmethacrylate (PMMA) [116]. More advanced studies have shown that the sensing properties of transducers can be improved by modifying G with nanoparticles. In this case, we modified G with silver nanoparticles for the detection of hepatitis C virus (HCV) and *Salmonella typhimurium* [117,118]. We have used gold nanoparticles attached to the surface of G to detect avian influenza virus H7 [119] and for prediction, prognosis, and diagnosis of cancer treatment recurrence and efficacy [120,121]. Modification of G with magnetic nanoparticles enables early detection of Alzheimer's disease [122], as well as cancer diagnosis [123]. More complex biosensors modifying the G surface with dendrimers [124], polymers [125,126], or cyclodextrins [127] have been developed for the detection of cancer, cholera toxin, HIV, and celiac disease. We have developed immunosensors for different types of microorganisms, such as viruses and bacteria, as well as for diseases. In bacterial detection, compared to reduced graphene oxide, graphene and graphene oxide as sensor platforms had the lowest detection limit (10-fold lower). For viruses, antibodies covalently attached to graphene with silver and gold nanoparticles can detect virus concentrations down to picograms per mL (pg/mL). In the case of detecting cancer cells, femtogram (fg) detection limits can be achieved by functionalizing graphene oxide with magnetic Fe_3O_4. A total comparison of all currently usable sensing platforms shows that functionalizing graphene or graphene oxide with gold, silver, or other metal nanoparticles and covalently attaching antibodies often achieves the lowest detection limits. Using such sensors for early detection of these diseases can help diagnose, prevent, and manage individuals with "high risk" disease, which in turn can lead to patient survival and better management. In recent years, many biosensors based on graphene nanomaterials have been proposed for real-time monitoring and diagnosis of patients' health status. Although these types of sensors' limitations (irreversible antigen–antibody binding and binding affinity) have not been fully corrected, the proposed biosensor exhibits extremely low detection limits, selectivity, sensitivity, and speed, making them ideal candidates for medical diagnostic tests with these graphene-based biosensors.

2.1.3 Graphene-Based Nanomaterials for Sensing of DNA

Deoxyribonucleic acid is ideal for biosensor technology and has numerous biological, chemical, and physical properties. One of the most important properties of DNA for technology in biosensors is ease of synthesis, flexibility, ease of chemical binding to high specificity, simple regeneration, and formation of various platforms due to unique nucleotide sequences [129,130]. However, we have identified several disadvantages and advantages of DNA-based biosensors. Important advantages of DNA biosensors include multiple measurements that diversify targets, various designs as small measurement systems, real-time analysis functions, and high specificity [131,132]. However, since easy decomposition of DNA is one of the main disadvantages of DNA biosensors, difficult analytical conditions and specific reaction factors such as reactivity to specific media or specific buffer solutions are needed to maintain and maintain DNA connectivity to the sensor. Changes in pH as well as temperature are other factors that may affect the effectiveness of DNA-based sensors [133]. For example, experimental temperatures have a significant impact on determining the sensitivity of DNA biosensors because hybridization between target molecules and probes can solve problems at optimal temperatures determined prior to sensor placement. For pH, the amperemetric response showed the highest signal at pH 7.0, with little signal when the pH value was below 7.0. Therefore, in order to increase the efficiency of the sensor, a buffer solution containing sodium phosphate or potassium is required [134,135]. Despite these shortcomings, DNA is gaining increasing attention in the field of biosensors and bioanalysts because it is applied to pathogen detection in infectious diseases, genetics, and clinical settings [136]. There are two types of sensors, a fluorescent sensor and an electrochemical sensor, and graphene-based nanomaterials are used as sensors in DNA biosensors. The electrochemical sensor may explain a simple principle of action based on measuring a change in impedance, current, or voltage that may occur due to a change in electrochemical factors such as DNA hybridization or adenine oxidation (A), an electron loss of thymine (T) in DNA, or a change in capacitance (C). Electrochemical signals from these biosensors can be detected using electrochemical impedance spectroscopy (EIS), differential pulse voltage measurement (DPV), or cyclic voltammetry [96,137]. In the electrochemical method, DNA immobilization is accomplished through the graphene-based nanomaterial surface of π–π interactions (Figure 2.5). DNA can also covalently interact with rGO and GO or G edges and their functional groups (carboxyl, hydroxyl, and epoxy) [97,138]. EDC/NHS is the most common chemistry used to immobilize DNA on graphene-based nanomaterials, which is described in the antibody section. Our research mainly focuses on the improvement of transducers to increase the selectivity and sensitivity of electrochemical biosensors. For example, we can modify a pristine glassy carbon electrode (GCE) with GO for direct detection of dsDNA at pH 7.0 using C, G, T, and A of ssDNA or the DPV method [139]. In other types of studies, GCE was modified with DNA probes and rGO, hybridized to target DNA for detection by CV or EIS [103]. In this study, in the characteristics of rGO, high conductivity

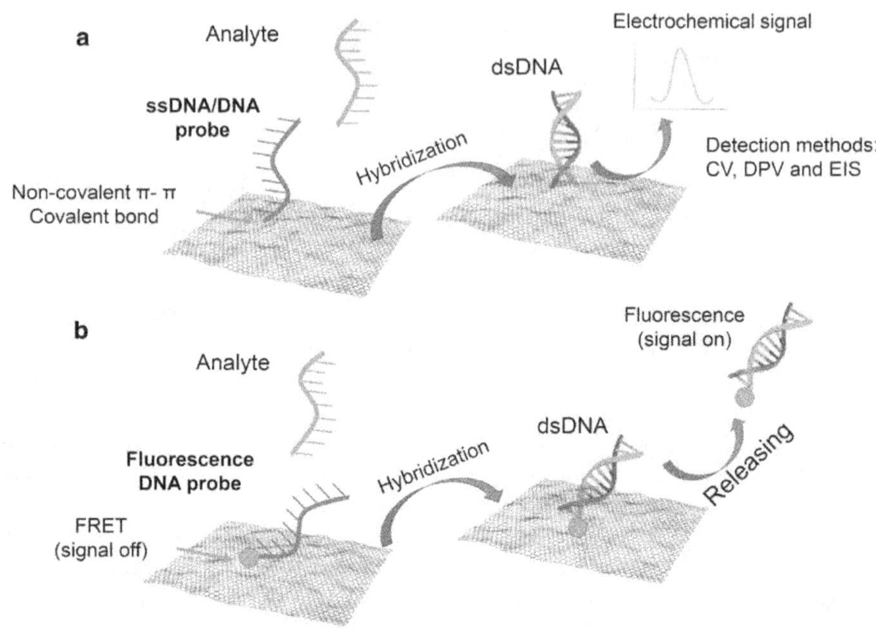

FIGURE 2.5 Structure of graphene-based nanomaterials as DNA biosensors. Electrochemical detection (a) and fluorescence detection (b) [128].

and a large surface area are important variables for sensor research, and using these characteristics, the utilization of sensors can be increased. Many other studies have proposed the design of DNA sensors using the active and reactive energies of reduced graphene nanowalls (RGNWs) to detect dsDNA with high sensitivity over a wide range between 0.1 fM and 10 mm. In these studies, the authors suggest that the active surface edge area of the RGNW sheet can improve the surface properties that more evenly distribute the electron transfer effect between the electrode and DNA in DPV [128]. In particular, the biosensor may expand the detection range according to wider sensitivity and detection between the target material and the active material and may increase the sensitivity. For example, graphene nanowalls, the best material presented in the literature, were identified as excellent materials capable of detecting up to 0.1 fM for dsDNA detection. In another active material test, the reduction-graphene oxide sensor measurement result of an active material using gold nanoparticles and ssDNA (ssDNA-AuNPs-ERGO) shows that sensitivity is improved and the detection limit is extended to 0.005 fM, which is a trace limit [140]. We studied nanographene-based DNA biosensors, focusing on the manufacturing process and biomedical applications of sensor active materials, and as another variable, we focused on fast measurement time and detection limitations. Therefore, These studies require optical detection and complementary DNA corresponding to the target DNA reacts to detect ssDNA labeled with fluorescent dye. Thus, the photo-hardening properties of graphene-based materials are utilized to improve visualization and detection of target ssDNA [12]. Through the π–π interaction between the DNA base ring structure and the graphene surface, the DNA probe is directly adsorbed on the graphene-based surface, and the immobilization of fluorescently labeled DNA can be achieved. GO-based sensors are an example of fluorescent biosensors that have been developed. The sensor is made of multicolor DNA probes to detect DNA with different sequence specificities. When hybridization occurs, this GO-based multiplex DNA sensor exhibits excellent emission signals from specific targets and low background fluorescence [141]. Fluorescence resonance energy transfer (Förster or FRET) is another widely used fluorescence sensing method that can also use graphene-based materials. In this detection method, first, a fluorescently labeled DNA probe is quenched to the surface of a graphene-based nanomaterial via FRET, turning off the fluorescent signal (Figure 2.5). After the active material probe reacts by hybridization with the target DNA, the injected fluorescent molecules exhibit fluorescent signals for optical detection and emit signals with dsDNA on the graphene surface, increasing sensor sensitivity [94]. For example, for HIV-1 detection, this method can be used to develop reliable biosensors by making use of the fact that they can be used to extend their range of use and that they have biocompatibility. In addition, fluorescent carbon dots (CDs) and DNA reaction probes can be considered, as well as active materials of needle-shaped nanocomposites, also known as nanocouples of GO and gold nanoparticles (AuNPs). Hybridization of the active material increases the effectiveness of the fluorescence signal generated in the presence of the target ssDNA. In the absence of target DNA, GO nanosheets containing AuNP may increase the effectiveness and reactivity of CD fluorescence reactions. In DNA biosensors, AuNPs/GO's excellent sensitivity and selectivity are important considerations [142]. The sensing limit of this sensor is capable of sensing a trace amount as much as 15fM and enables higher accuracy. Various nano-based metal particles and graphene-based material complexes were used to find the best sensor conditions to achieve the desired sensitivity. As another example, fluorescent graphene base sensors are useful for detecting ssDNA with sensitivity as low as 0.5 pM using target recycled nucleic acid hydrolyzase III, which will expand the use of nano-active materials. All together, both methods appear to be efficient and have low detection limits. However, the measurement method and immobilization ability of DNA in graphene-based nanomaterials mainly determine that each technique has its disadvantages and advantages. The conductivity and large surface area of nanomaterials are taken into account in the electrochemical detection method. Detection is based on the type and amount of bases present in the DNA, which cause a change in the electrical potential used for the measurement. Thus, uniform deposition of probes on graphene materials is crucial for precise measurements. Additionally, DNA length and electrostatic potential affect sensor efficiency. In the other word, regardless of the DNA length, dsDNA or ssDNA can perform fluorescent detection. The optical and quenching capabilities of graphene-based nanomaterials dictate this approach. Due to the presence of high background fluorescence signals, in some complex samples (e.g., serum samples), one of the main drawbacks of this approach can make errors that can overestimate the generated fluorescence signal. On the other hand, the fluorescent labeled probe has a disadvantage in that its strength decreases over time. The results of graphene-based DNA biosensor studies still require further research on the mechanism of interaction between modified DNA probes or DNA probes and graphene-based sensors to provide more accurate and reliable measurements. These studies can be considered a way to reduce the detection limit of current sensors and target sensitivity to overcome the shortcomings of current methods.

2.1.4 Chemical Detection

The graphene sensor of the first attempt at solution detection is a simple pH sensor using a single-layer and multi-layer graphene sheet grown on a SiC substrate [143]. The sensor provides super Nernstian pH sensitivity (98 mV per pH vs. 59.2 mV per pH) by monitoring changes in the Dirac (neutral) point. The authors propose a detection mechanism for the pH-dependent surface potential modulation (field effect) via ion-attached and adsorbed amphoteric OH groups. The authors also suggest that the high carrier mobility of epitaxial graphene leads to such a high sensitivity, which is an order of magnitude higher than that of hydrogen-terminated diamond or silicon. In the near future, Ohno et al. The detection limit of pH was found to be more than 26 times lower at 0.025 than carbon nanotube-based electronic pH sensors, a study of the pH-sensing capabilities of mechanically exfoliated graphene [144]. Zhang et al. demonstrated a heavy metal sensor with a detection limit of 10 ppm (about 5 mm) for Hg^{2+} using mechanically exfoliated graphene. Self-assembled 1-octadecanethiol, whose thiol group has a high binding affinity for heavy metal ions, modified 220 graphene. Recently, Chen and colleagues demonstrated a metal ion sensor based on centimeter-long and micrometer-wide RGO films fabricated by microfluidic patterning [145]. Ca^{2+} can be detected at a concentration of 1 mM by functionalizing a Ca^{2+} binding protein (calmodulin) onto RGO. One of the important variables, the field effect due to positively charged ions such as Ca^{2+}, plays a decisive role in determining detection. Tiny amounts of heavy metal ions (e.g., Cd^{2+}, Hg^{2+}) are functionalized heavy metal ion bonds in RGO, which can clearly act in detecting proteins (metallothionein type II proteins, MT-II). The authors suggest that this detection method is a field effect through negatively charged MT-II after protein binds to heavy metal ions. This sensor is a good example that can be applied to lake water samples, a complex water composed of various impurities, microorganisms, and ions, to prove practical use in environmental monitoring and pollutant control. Aromatic organic materials such as Myers et al., cyclohexane, xylene, ethylbenzene, toluene, and benzene were used as active materials for detecting RGO nanocomposites functionalized with octadecylamine (ODA) for electrochemical detection in ppm units [146]. In this chapter, the increase in electron tunneling barriers between RGO sheets is a problem, and we propose that this is because the conductivity is significantly reduced due to the adsorption characteristics of the target molecule.

2.1.5 Graphene-Based Nanomaterials for Bioenzymes

From a biological point of view, the design of enzymes in biosensors is of great interest due to their high stability and ease of operation. Furthermore, the metabolism of all living organisms and organisms contains these molecules. They act as highly selective and reusable catalysts, and can be distinguished by R and L in different molecules. The role of enzymes requires high selectivity, efficiency, and specificity essential for sensor design, and these factors can induce many catalytic reactions [152]. However, since the enzyme types vary, careful consideration and selective modification are required, which should be used when the enzyme design of biosensors has special uses due to their high stability and ease of operation. This problem was initially solved through the application of aerobic enzymes, but today aerobic enzymes are modified or produced by biotechnology to create more powerful organisms so they can improve their usefulness [153]. For example, one method for changing the properties of enzymes requires chemical modifications to increase enzyme stability, as well as the generation of site-oriented mutations [153]. Over-expression and replication of the desired enzyme gene during these modifications can rapidly transform enzymes with the development of recombinant DNA technology [153]. Several problems related to enzyme specificity and stability are solved by this approach, and the process of enzyme-based biosensor research improves stability while reducing reaction time and loss rate [154]. It depends on the stability of the enzyme as well as temperature, solvent polarity, chemical inhibitors, ionic strength, and pH. Because the graphene-based nanomaterial structure allows direct electron transfer between the electrode and the enzyme, it can be an effective sensor [97]. Moreover, one of the substrates that has proven to be excellent is the graphene-based material for enhanced enzyme immobilization, enzyme activity, and thermal stability [155–157]. Enzymes need to be immobilized on graphene surfaces in order to create enzyme-based biosensors, and several methods have been developed. Cyclic voltammetry, ultrasound, mixing, and sonication are some of the most common methods. These methods allow enzymes attachment by physical entrapment, covalent bonding, or adsorption. Enzymes' nonspecific binding to graphene via physisorption is by far the most common, as this immobilization technique is straightforward and chemical-free. EDC/NHS chemistry is another method applied to immobilize enzymes on nanomaterials. Due to its high stability and robustness, the method is still common for enzymes, as described earlier. Enzyme-based biosensors usually have electrochemical auxiliary properties. This method has an advantage over other methods because the electrodes can be used without damaging the system when detecting the presence of material in the body. Enzyme-based electrochemical biosensors are mainly derived from two mechanisms; one is based on the regulation/inhibition of enzyme activity, and the other is based on the enzymes' catalytic properties (enzyme catalytic detection from undetectable to detectable forms) [158]. Both mechanisms can produce a detectable change in the electrical signal across the sensor electrodes, allowing quantification for specific analyses, Especially because the activity of the enzyme will directly cause this electrical signal to be generated by the change of the substrate surface current. Redox reactions of enzymes catalyze the consumption or generation of electrons, which alters the current flow across the detection platform. Figure 2.6 shows the basic principle of how an enzymatic biosensor works. Although enzymes are expensive to use, when using enzymes, sensors can detect various compounds with high specificity, which of course are difficult to detect in complex mixtures. For instance, these sensors are especially useful for the detection of compounds such as bilirubin, glucose, 17β-estradiol, hydrogen peroxide, and phenol, which will be described in

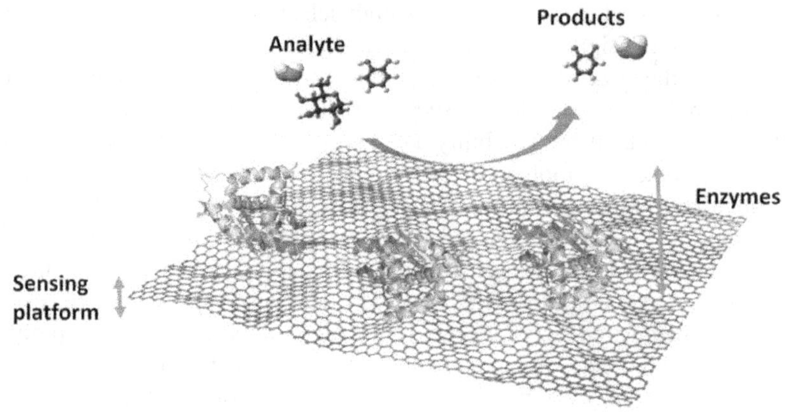

FIGURE 2.6 Example of an enzyme biosensor [147].

detail later in this chapter. The various compounds that can be detected by commonly used immobilized enzymes along with the detection range that each fabricated sensor possesses. We can already detect different molecules using enzyme-based nanosensors. When developing these sensors, horseradish and laccase peroxidase (HRP) are the most commonly used model enzymes [159]. These enzymes are less expensive, more common, and more versatile, applicable for the detection of lots of different compounds. Laccase is an oxygen reductase enzyme that has multiple uses. For instance, for detecting 17β-estradiol, we have developed a laccase-based electrochemical biosensor. We define it as a natural hormone of emerging pollutants affecting humans and aquatic life [160]. In addition, laccase can be used to detect phenols and catechols [159,161–163]. Another enzyme widely used in enzyme immobilization research is HRP, which can still help determine the concentration of hydrogen peroxide under complex test conditions [164]. We have immobilized HRP on graphene-coated porous calcium carbonate microspheres for hydrogen peroxide showing high selectivity. We can use this sensor platform stably for a long time as a biosensor for the immobilization of different enzymes [164]. Furthermore, HRP and laccase have been immobilized on rGO-Fe3O4 based substrates [159]. Utilizing the properties of the magnetism and rGO of iron oxide makes this hybrid nanomaterial an attractive substrate. While laccase and HRP are critical in enzymatic biosensor research, we can create highly specific biosensors by immobilizing other enzymes. For instance, we immobilized bilirubin oxidase on a GO-based surface [165,166]. Because of its ability to detect bilirubin, an important compound in the assessment of liver function, the biosensor could have a major impact in the field of medicine. Another enzyme is glucose oxidase (GOx) with medical applications. We have used this enzyme to develop a biosensor for measuring glucose levels due to its high specificity [167–175]. Figure 2.7 shows the installation of sensor head. Diabetic patients in particular need this type of biosensor. Therefore, in last few years, we have immobilized GOx on different sensing platforms, such as gold-palladium-modified

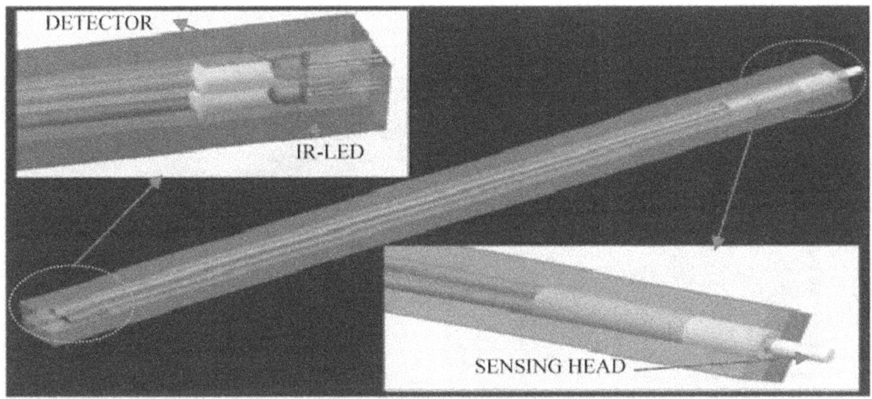

FIGURE 2.7 The schematic diagram of the sensor head [177].

polyimide/rGO films, organically modified inorganic support ligands (OISLs) for titanium dioxide and GO nanoparticles, polyaniline (PANI) and 3D GO composites, ZnS-decorated graphene, graphene field effect Silk fibroin films, three-dimensional graphene, and nanostructured graphene-conducting polyaniline composites on transistors, among others [167–175]. Among these platforms, the Pt/Nafion/chitosan nanoparticles/SGGT composite provided the largest linear range (up to 1 mm) and highest sensitivity (down to 0.5 μM) in glucose detection [171]. The versatility of graphene and its nanocomposites is demonstrated in these sensing platforms in detecting different substrate chemistries.

Biosensors analyze biological responses and translate them to electrical signals. Biosensors must be very reusable and specific, independent of physical conditions such as temperature and pH. Cammann [175] created the term "biosensor", and IUPAC defined it [176]. Figure 2.8 shows how to set up hybrid waveguide gas sensor. Material research, transducing devices, and immobilization procedures are required to fabricate biosensors. According to the mechanism of action, we can divide the materials used in biosensors into three categories: nucleic acids, antibodies and enzymes, and microbes. Figure 2.9 illustrates the DNA hybridization and DNAzyme-GO based fluorescence. Challenges like persistent population stress, variable climate conditions, and increased resource competition have all presented danger, necessitating the urgent need to ensure global food security. Figure 2.10 shows how graphene-based electrochemical sensor perform in various fields. Existing agricultural practices include unrestrained resource consumption, advanced technology, and the increase and misuse of agrochemicals. These activities have significantly harmed soil, air, and water resources, ultimately affecting human and animal health. Monitoring agroecosystems now includes gas chromatography, HPLC, mass spectroscopy, and more (Figure 2.11). The sensitivity, specificity, and repeatability of such measures are undeniable, but their implementation is constrained by time, cost, and the need for specialized equipment and expert staff [177]. Thus, simple, rapid,

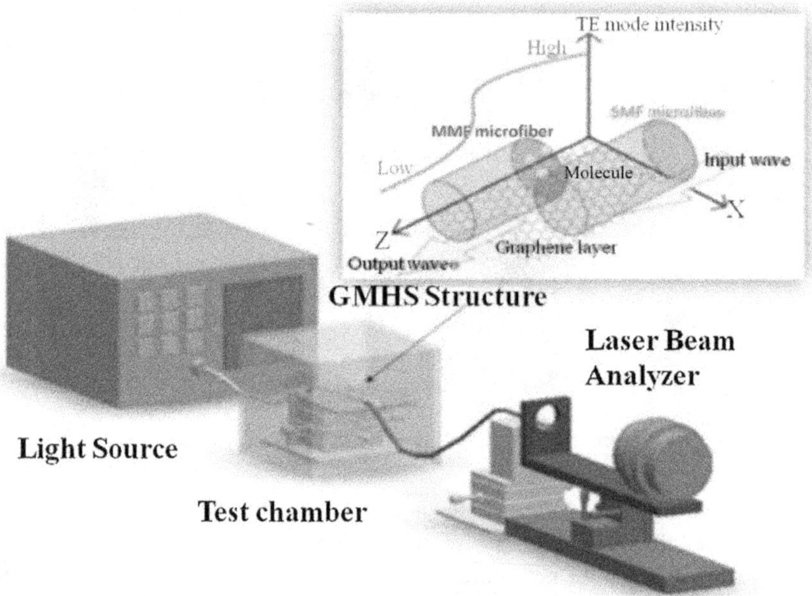

FIGURE 2.8 Experiment setup of hybrid waveguide gas sensor [178].

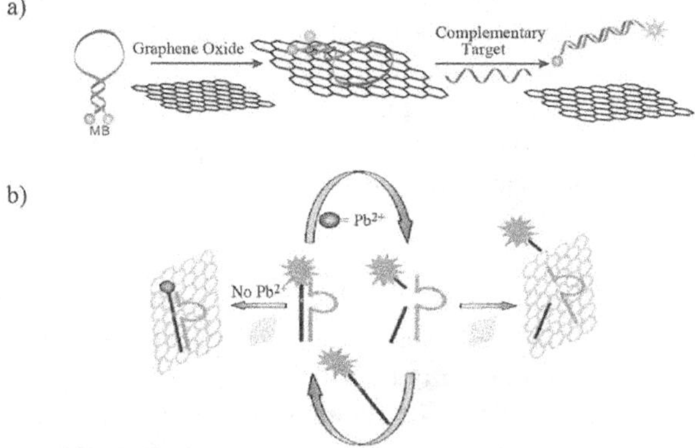

FIGURE 2.9 GO as the quencher in a FRET sensor. (a) Schematic illustration of DNA hybridization using a double-quenching system consisting of molecular beacon (MB) and GO. (b) Schematic illustration of the DNAzyme-GO based fluorescence sensor for detection of Pb^{2+} [179].

and cost-effective approaches for monitoring agricultural pollutants are required [178]. Nanosensors are tiny element devices designed to detect molecules, biological components, or environmental conditions. These sensors are small and portable and detect at a far lower level than their macroscale counterparts [179].

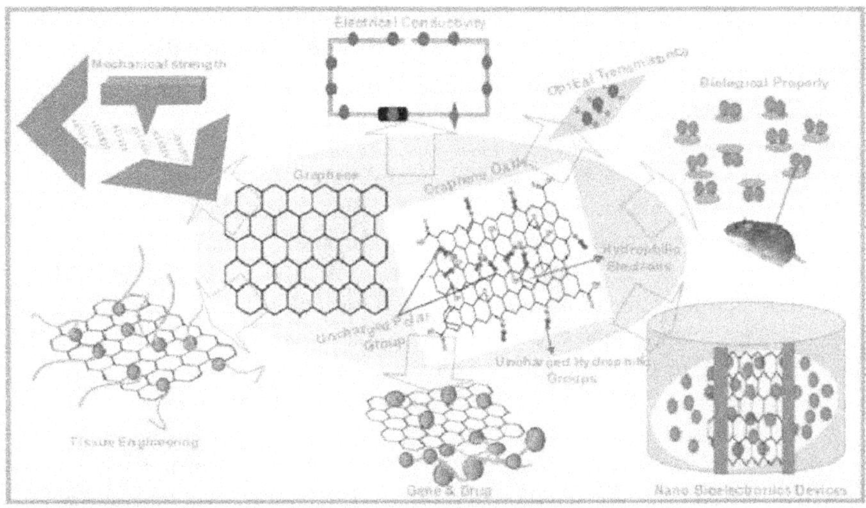

FIGURE 2.10 Representation of graphene-based electrochemical sensor in cancer cell, protein, and DNA detection [180].

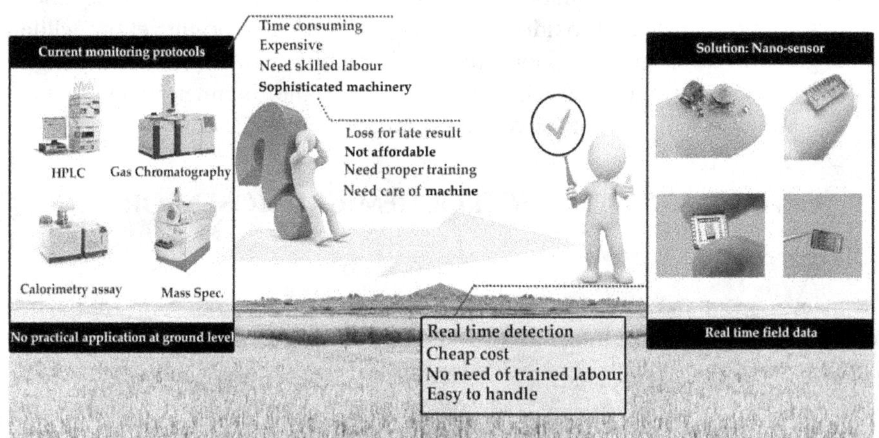

FIGURE 2.11 Schematic representation of advanced and traditional monitoring technologies [179].

2.2 CHEMICAL SYNTHESIS OF GRAPHITE OXIDE AND GRAPHENE OXIDE

(1) According to Brodie's method (1859), we use Ceylon as starting material to prepare the GpO, resulting in 99.96% carbon purification. There is an oxidizing agent called carbonic acid, which is a boiling mixture of sulfuric

acids and concentrated nitric. Elemental analysis shows that the C:O:H content of Ceylon graphite oxide is 67.79:30.37:1.84, and the C-O ratio is 2.23. The material, known as graphitic acid, is thus the first graphite oxide experimentally prepared sample.

(2) According to Staudenmaier's method (1898), we think it has a lot of similarities to Brody's approach. We mixed the fuming nitric acid and concentrated sulfuric acid to prepare the graphite oxide. In addition, we added potassium chlorate oxidizing agent and let it react for 4 days. The sulfonate ions were removed by dispersing in dilute hydrochloric acid and rinsing in water. Finally, at 60°C, we dried the graphite oxide for 2 days. We found that graphite oxide has an elemental composition C:O:H of 58.73:23.28:17.99 if we use this method to prepare it. The C-O ratio is 2.52, indicating the oxidation's lowest degree.

(3) According to the method of Offeman and Hummers (1958), we realized that using nitric acid would take a considerable amount of time to oxidize graphite and that there was a risk of explosion and release of highly corrosive vapors, so this method was developed. A less hazardous method of graphite oxidation is the method of Offeman and Hummers. To make an oxidizing agent, we need mix of concentrated potassium permanganate, sodium nitrate, and sulfuric acid. The whole process to complete the reaction takes 1–2 h. graphite oxide has a carbon to oxygen ratio between 2.1–2.9. The degree of oxidation refers to the color of the product in aqueous solution. For the most oxidized graphite, the product is bright yellow, while green to black indicate poorly oxidized graphite with an excessively high C-O ratio. The most commonly used method is the Hummers and Offeman method, which is often called the Hummers method [180].

2.3 GRAPHENE-BASED ELECTROCHEMICAL BIOSENSORS

2.3.1 Enzymatic Biosensors

The measurement and detection of biomolecules is of great significance for the treatment and diagnosis of various diseases in clinics. Enzymatic biosensors of electrochemical reactions (reduction or oxidation) for sensing biomolecules with enzymes immobilized on electrode surfaces is a well-accepted system. The analyte molecule's concentration corresponds to the electrochemical output signal. Electron transfer between enzyme metal active sites and electrode surfaces mainly determines the analytical performance of the biosensor. Figure 2.12 shows the schematic diagram of an enzymatic biosensor. In order to achieve direct electron transfer between the enzyme and the electrode, the enzyme can be directly immobilized on the surface of the electrode. However, it may cause enzyme denaturation, thereby affecting the response of the biosensor. Nanomaterials are widely used as immobilized substrates, that is, mediators between enzymes and electrodes, to enhance direct electron transfer, increase stability, and improve enzyme adsorption. Graphene and its oxidized derivative-based nanomaterials exhibit excellent electrochemical performance, as explained in the introduction, including excellent electron transfer rate, better electrocatalytic activity, large surface area, proximity to defect sites, and high electrical conductivity; these are expected to be used in the fabrication of enzymatic biosensors [182].

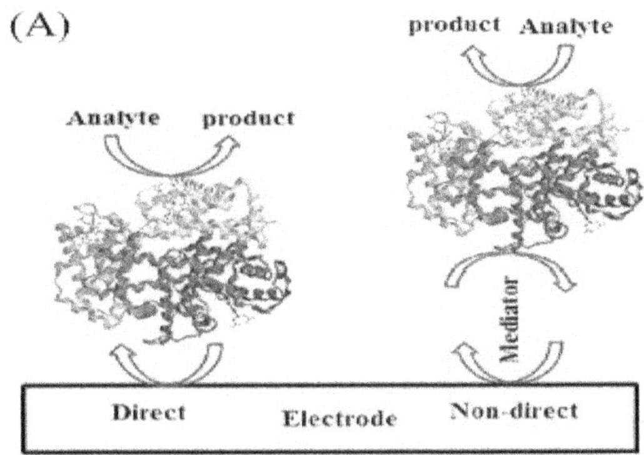

FIGURE 2.12 Schematic of the mediated direct electrochemical biosensor [181].

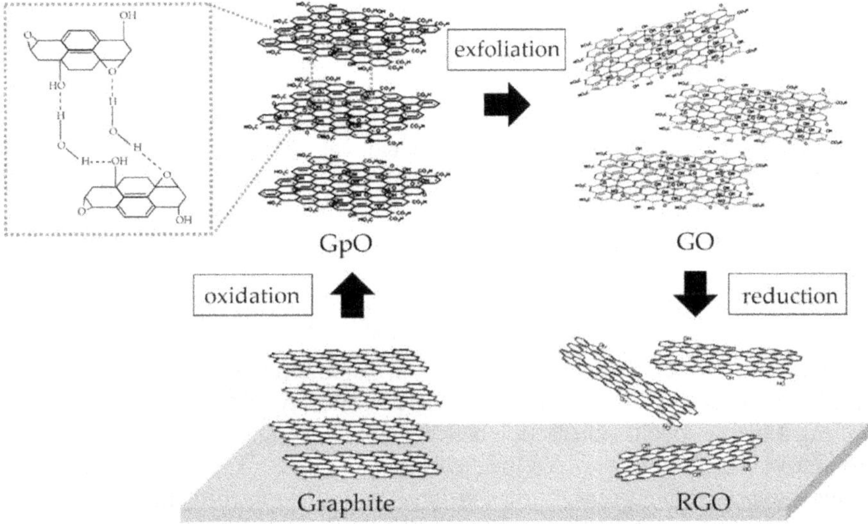

FIGURE 2.13 Schematic structure of chemical synthesis of reduced graphene oxide (RGO), graphene oxide (GO), and graphite oxide (GPO) [186].

2.3.2 Non-Enzymatic Biosensors

Another important application is using non-enzymatic electrodes made of graphene-based nanomaterials is to detect clinically important biomarkers' electrochemical parameters. Figure 2.13 show the schematic structure of graphene-based nanomaterials. A non-enzymatic sensor used ErGO to detect bilirubin, which is an important biomarker for jaundice. In screen-printed carbon electrodes, they found that ErGO electrodes far outperformed MWCNT electrodes in terms of sensitivity, detection range, and detection limit. This is due to the higher electrical conductivity and faster electron transfer rate of ErGO. Moreover, it is coated with Nafion membrane to ensure selectivity. A schematic diagram of using MWCNTs and ErGO to develop

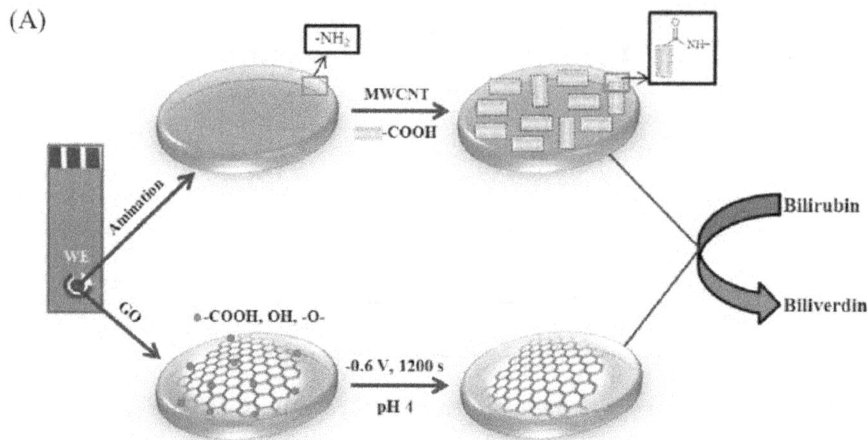

FIGURE 2.14 Schematic showing the electrochemically reduced graphene oxide (ErGO) based (bottom row) or preparation of the MWCNT based (top row) bilirubin sensors [182].

a non-enzymatic bilirubin sensor is shown in Figure 2.14. It provides a miniaturized, reliable, and low-cost point-of-care electrochemical sensor for bilirubin. Similarly, non-enzymatic electrodes based on graphene nanomaterials have been reported to detect uric acid, ascorbic acid, and dopamine. Because of overlapping electrochemical oxidation potentials, it is challenging to detect dopamine in the presence of AA [182].

2.3.3 Immunosensors

We have also used electrodes based on graphene nanomaterials to develop electrochemical immunosensors. The high selectivity and sensitivity of the immunosensor were confirmed by the specific interaction between antibody and antigen. Electrochemical immunoassays are widely known for their rapid analysis, miniaturized size, simplicity, high selectivity and sensitivity, and large-scale manufacturability. They can be label-free immunosensors or sandwich sensors. Several electrochemical immunosensors have been reported to detect well-known biomarkers, for example, prostate-specific antigen (PSA), human chorionic gonadotropin (hCG), interleukin 6 (IL-6), and carcinoembryonic antigen (CEA). Figure 2.15(a) shows a schematic of a sandwich immunosensor electrode with four different antibodies specific to its antigen immobilized. Mao et al. used graphene sheet-methylene blue-chitosan nanocomposites for PSA label-free electrochemical immunosensor detection. A schematic diagram of label-free electrochemical immunosensor fabrication is shown in Figure 2.15(b) [182].

2.3.4 Glucose Biosensors

The history of electrochemical glucose sensor development so far can be largely classified into three generations. First-generation electrodes for measuring glucose enzymes were made by relying on the measurement of oxygen consumption by enzyme catalysis. In particular, the GOx enzyme electrode and glucose undergo an enzymatic reaction in

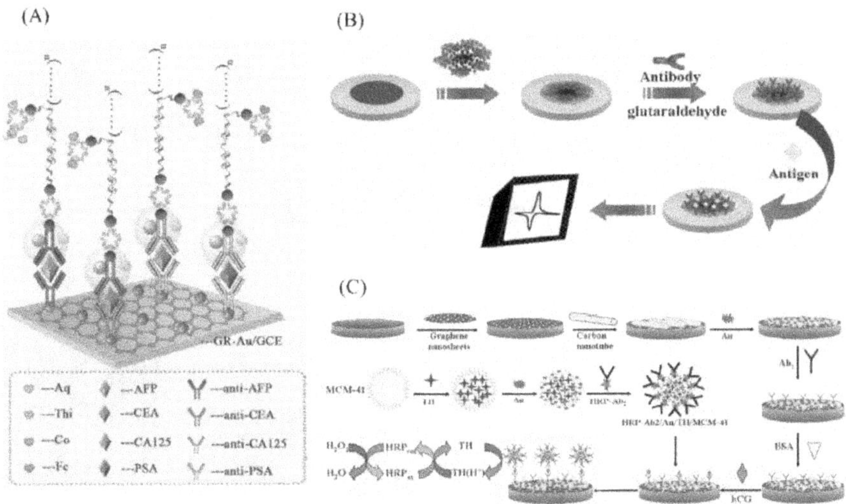

FIGURE 2.15 (a) Schematic illustration of the principle of sandwich-type simultaneous detection of four antigens. (b) Fabrication steps of the label-free electrochemical immunosensor. (c) The measurement protocol of the electrochemical immunosensor and fabrication process of Au/TH/MCM-41 nanomaterials [182].

the oxygen presence to generate hydrogen peroxide (H_2O_2) and monitor glucose levels by enzymatically producing the amount of H_2O_2. However, over the past 40 years, enzyme-based electrochemical glucose sensors have attracted attention for their high selectivity, simplicity, and sensitivity. The basis for developing a medium-free third-generation electrochemical glucose biosensor with high selectivity and sensitivity is to enhance the electrode surface contact with redox enzymes. The electrical conductivity and electrocatalytic activity may vary greatly depending on the graphene derivative. For instance, graphene's electrical properties can change upon chemical functionalization and reduction to rGO or GO. Graphene-based materials have abundant oxygen-containing functional groups, biocompatibility, excellent electrochemical performance, and large specific surface area, such as epoxy, carboxyl, carbonyl, and hydroxyl. Covalent binding efficiently immobilizes the oxidoreductase for subsequent glucose sensing. We have used graphene-based nanocomposites extensively to fabricate enzymatic glucose biosensors. Figure 2.16 shows a schematic fabrication of a glucose biosensor [183].

2.3.5 Biosensor Devices Using Graphene-Based Materials and Current Progress

Previously, since graphene is a semimetal, it has a large surface area, excellent electronic properties, ultra-high charge mobility, and the ability to be functionalized on its surface, graphene-based materials have been developed using different synthetic methods. Research has been able to take full advantage of the structure and excellent performance in biosensing applications. Receptors targeting biomolecules can be designed in many possible ways. In the field of biomedicine, pristine

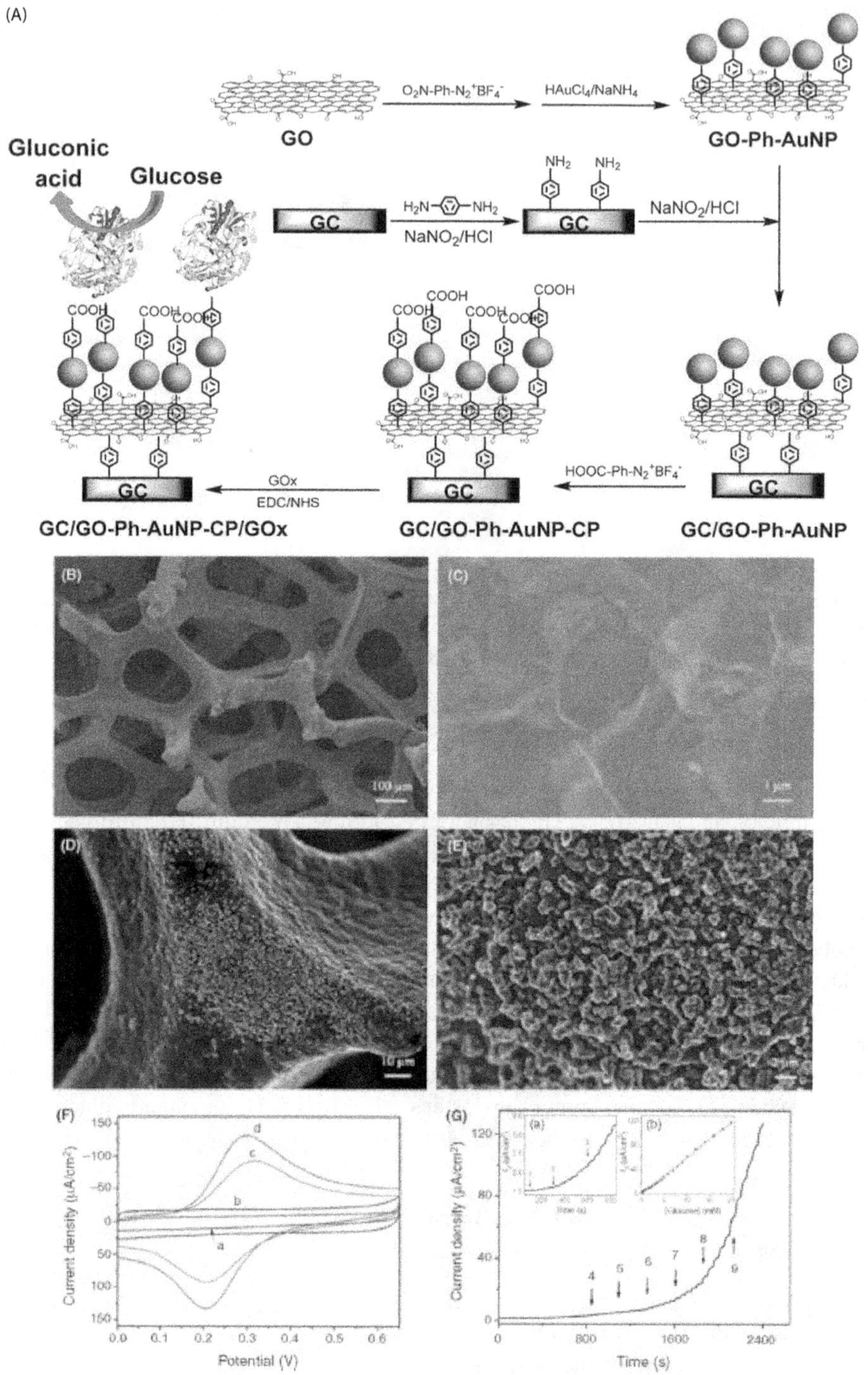

FIGURE 2.16 (a) Schematic diagram of glucose detection biosensors manufactured using GO nanosheets transformed into nanoparticle Au. Using aryldiazonium salt chemistry

FIGURE 2.16 *(Continued)*
(GO-Ph-AuNPs), AuNP is decorated on a GO nanosheet via a benzene bridge and then attached to a GC electrode modified with 4-aminophenyl. GC/GO-Ph-AuNPs were further functionalized with 4-carboxyphenyl (CP) and then covalently linked with GOx via amide binding to form a GC/GOph-AuNPS-CP/GOX-based glucose sensor [183]. Schematic of fabrication of 3D-GR-based enzyme glucose biosensors using high magnification 3D-GR foam (b, c), and low magnification SEM images (d and e). A high-magnification SEM image of the Fc-CS/SWCNTs/GOX composite film was electrodeposited on 3D graphene. After (f) CV curves of the (a) 3D-GR, (b) CS/GOx/3D-GR, (c) Fc-CS/GOx/3D-GR, and (d) Fc-CS/SWCNTs/GOx/3D-GR electrodes were continuously stirred at a scanning rate of 100 mV s-1 and then added at a pH (V-0). Insertion (a) shows an enlargement curve from 50 to 850 seconds. Insertion (b) is a calibration figure that plots the current as a function of glucose concentration [183].

graphene not only has an infinite surface at the molecular level but is also known as oxide-free graphene with high electrostatic force, non-covalent interactions, and stacking. Thus, Graphene provides high-probability active sites for charge–biomolecular interactions because of the functionalization required to support target biomolecules to enhance selectivity, and the large specific surface area of graphene leads to enhanced sensing. Figure 2.17 shows a view of the possible interactions of a graphene-based material system. Figure 2.18 shows a schematic diagram of a graphene-based biosensor. For example, pure graphene regions as shown can provide charged regions to absorb vacancy defects, as well as any charged molecular or metal ion interactions. Owing to the synthesis of abundant carboxyl, hydroxyl, and epoxide groups at the surface and edge positions, the functionalized graphene domains enable the direct detection of biomolecules through their own oxide components. Furthermore, functionalized graphene allows incorporation of antibodies, antigens, proteins, enzymes, DNA, quantum dots, nanoparticles (NPS), heteroatoms, and other specific molecules [180].

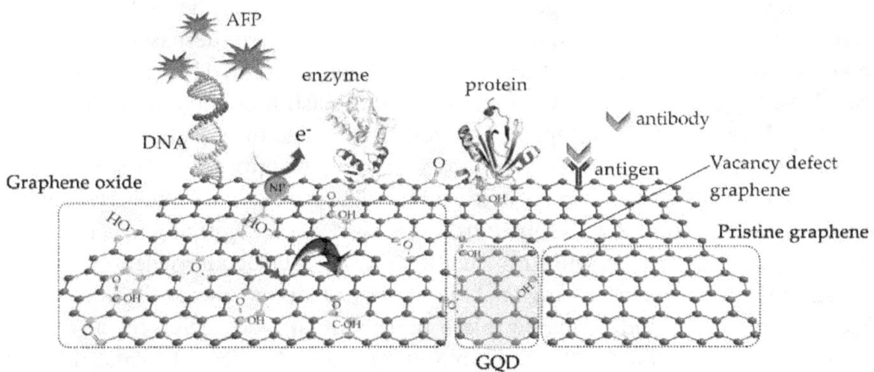

FIGURE 2.17 Schematic diagram of graphene-based materials that can immobilize biomolecules as receptors [180].

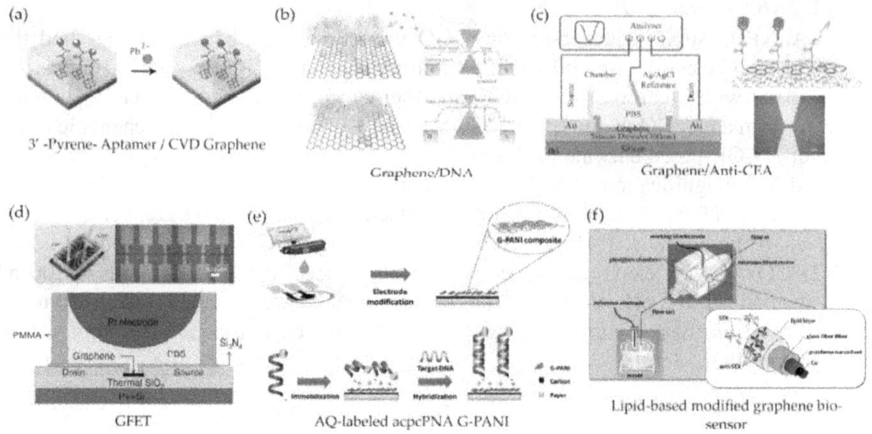

FIGURE 2.18 Schematic illustrations of graphene-based biosensors: (a) Pb2+ in blood biosensor based on GFET. (b) Pb2+ biosensor based on graphene/DNA. (c) CEA protein biosensor based on graphene/anti-CEA. (d) Affinity and real-time binding kinetics of GFET-based DNA hybridization. (e) Paper-based biosensor for human papillomavirus (HPV) detection. (f) a lipid-based modified graphene electrochemical biosensor [180].

2.4 REVIEW OF LITERATURE

2.4.1 Types of Biosensors

In the 1960s, Lyons and Clark developed biosensors, which were used piezoelectric biosensors, DNA-based thermal sensors, immunosensor, tissue and enzyme-based sensors. In 1967, the first enzyme sensor was reported by Hicks and Updike. The immobilization of enzymes via covalent bonding, van der Waals forces, or ionic bonding was developed. These include amino oxidases, peroxidases, polyphenol oxidases, and oxidoreductases (Figure 2.19) [184]. Diviès created the first cell-based or microbe sensor [181]. Plant and animal tissues are used in tissue-based sensors, and the analyte of interest may be a substrate or inhibitor. Rechnitz [185] first created the tissue-based arginine sensor. A membrane, chloroplast, mitochondrion, and microsome-based sensor was developed. As a result, the time of detection is prolonged, and the specificity is lowered. Antibodies have strong affinity for their antigens; that is, they selectively attach to poisons or infections, or they have interaction within immune system components. DNA biosensors work by recognizing and binding to the corresponding strands within a sample. The contact occurs when two nucleic acid strands create stable hydrogen bonds [186]. Through the magnetoresistance effect, nanoparticles and magnetic micro-particles can be detected in microfluidic channels in these miniature magnetic biosensors [93]. To create thermal or calorimetric biosensors, we need to combine biosensor components into physical sensors. These include quartz crystal microbalances and surface acoustic wave devices. They monitor variations in a piezoelectric crystal's resonance frequency owing to quality

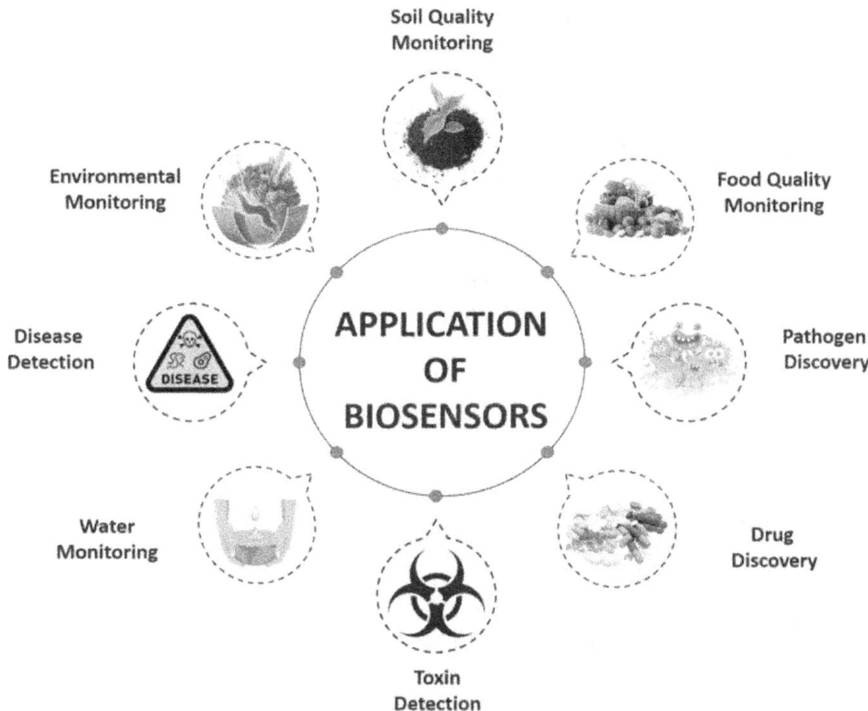

FIGURE 2.19 Different applications of biosensors [184].

change. Optical biosensors include many optical components, a photodetector, a customized sensing head, and a light source [187]. Genetic fusion reporters, AFP variations, and green fluorescent protein (GFP) have helped construct genetically encoded biosensors [188]. Biosensors that are easy to design, control, and implant into cells are of this form. Single-chain FRET biosensors are another type of sensor. They consist of two AFPs that may exchange fluorescence resonance energy when placed close to each other. Depending on the strength, ratio, or longevity of AFPs, FRET signals may be regulated. Synthetic peptide and protein biosensors are readily made by enzymatic labeling using synthetic fluorophores. They are appealing alternatives to genetically encoded AFPs because they are independent of AFPs and can improve the signal-to-noise ratio and response sensitivity by adding photoactivatable groups and chemical quenchers.

2.4.2 Applications of Biosensors

Compared with traditional methods, biosensors have better stability and sensitivity and have been used in many areas, such as the marine field, food industry, and medical field.

2.4.2.1 Sensor Quality and Safety in Terms of Food Processing, Monitoring, and Food Authenticity

The food processing sector has a difficult dilemma regarding quality, safety, and processing. Chemical experiments and spectroscopy are time consuming, costly, and labor intensive. The food sector seeks profitable alternatives for food inspection and monitoring. Therefore, the requirements for economical, selective, real-time and simple procedures seem to be conducive to the development of biosensors (Figure 2.20) [189]. Figure 2.20 shows the different methods to detect water samples and how biosensors use for detection. Ghasemi-Varnamkhasti [190] studied monitoring of beer aging with a cobalt phthalocyanine-based enzymatic biosensor. The aging of beer in storage can be effectively monitored by these biosensors. Infections in food are detected by biosensors. A biomarker of fecal contamination is *Escherichia coli* in vegetables [191]. Ammonia (produced by a urease-*E. coli* antibody conjugate) has been quantified using a potentiometric surrogate biosensing system, a liquid phase

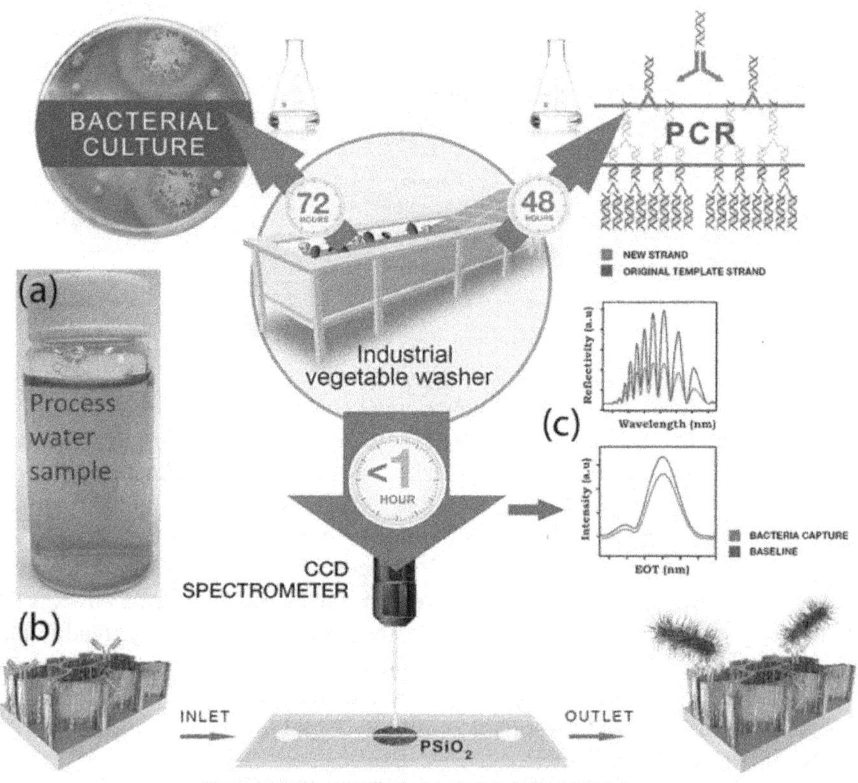

FIGURE 2.20 (a) Water samples from a Dutch fresh-cut fruit firm were tested for *E. coli* using three methods: label-free optical biosensors (bottom), PCR, and culturing. (b) The biosensor is activated by specific capture probes (antibodies) mounted on the $PSiO_2$ surface. (c) The observed optical signal comes from the porous nanostructure [196].

obtained by washing vegetables with peptone water. It is then sonicated to dissociate bacterial cells from food [192]. Dairy uses enzymatic biosensors, which are flow cells with screen-printed carbon electrode-based biosensors [193]. Enzymes were engulfed in a photo cross-linkable polymer to immobilize them. Automated flow-based biosensors might detect contaminants in milk. Sweeteners are among the most widely used food additives today, and they have been linked to problems such as type 2 diabetes, tooth decay, cardiovascular disease, and obesity. Artificial sweeteners are believed to be addictive, allowing us to eat more high-calorie foods without knowing it, leading to weight gain. So, their identification and measurement are crucial. Ion chromatographic techniques are used to differentiate two kinds of sweeteners. A multichannel biosensor measuring the electrophysiological activity of gustatory epithelial cells has been examined as a more effective tool for sensitive and rapid sweetener screening. Spatiotemporal analysis of data was performed using MATLAB, where sucrose and glucose are saccharin and natural sugars, and cyclamate is an artificial sweetener. They have multiple binding sites to recognize the sweet taste stimuli of various structures due to heterodimeric G protein-coupled receptors in type II cells of buds. The diacylglycerol and inositol triphosphate pathways use artificial sweeteners, while the cyclic AMP system uses natural sugars such as sucrose to transmit signals. The amino-terminal domain of taste receptors serves as the ligand-binding site for artificial sweeteners. Taste receptor cells respond differently to artificial and natural sweeteners. When glucose was administered, sparse positive waveforms were given by gustatory epithelial biosensors, but sucrose maintained negative spikes. Artificial sweeteners elicited stronger responses from the taste epithelium than natural sugars both in the frequency and temporal domains.

2.4.2.2 In Fermentation Processes

Product quality and process safety are critical in fermentation. Good fermentation monitoring is a must for the development, optimization, and maintenance of bioreactors. Biosensors can detect process conditions by monitoring the presence of by-products, antibodies, enzymes, biomass, or products. Because of their cheap cost, ease of automation, and high selectivity, biosensors have revolutionized the fermentation business. Modern commercial biosensors can detect biochemical parameters (alcohol, lysine, lactate, and glucose) and are extensively utilized in China (approximately 90% of the market). Glycation is examined using the classic Fehiing technique. This method's results were erroneous due to decreased sugar titration. Since the commercialization of glucose biosensors in 1975, the field of fermentation manufacturing has been very profitable. For production management in fermentation and saccharification workshops, firms now employ glucose biosensors. In ion exchange retrieval, biosensors detect changes in biological composition. For example, a glutamate biosensor was employed to study the ion exchange recovery of a glutamic acid isoelectric supernatant. Fermentation is a complex process with many variables, many of which are difficult to assess in real time. Controlling biological processes requires continuous key metabolite monitoring. Due to their quick response and simplicity, biosensors have gained popularity in online fermentation monitoring [194].

2.4.2.3 Biosensing Technology for Sustainable Food Safety

Food quality includes look, flavor, texture, and chemicals (Figure 2.21) [195]. In terms of food safety and quality, rapid detection and smart nutrient monitoring of pollutants are critical. Microfluidic systems, electromechanics, nanotechnology, and materials science are driving advances in sensing technology. Control mechanisms for food safety and quality as well as human health are being developed. Glucose monitoring is required because food composition might change during storage. German [196] investigated the glucose oxidase electrochemistry immobilized on a graphite rod and enhanced with gold nanoparticles. Glutamine is essential for vital processes, including precursor, transport, and signaling in biosynthesis of proteins, amino sugars, and nucleic acids. The immune system, digestive function, and bacterial translocation are improved when glutamine deficient patients are supplied. "Glutaminase-based microfluidic biosensor chip for flow injection analysis with electrochemical detection" [197]. Because biosensors can only react to the poisonous portions of metal ions, they are used to detect generality and toxicity. Pesticides endanger the environment. Organophosphates and carbamic insecticides are widely used pesticides. Immunosensors have shown their value in agri-food and environmental monitoring. Many chemicals have AChE and butyrylcholinesterase biosensors. Arduino and coworkers created Oxon using screen-printed electrodes. Pesticides in wine and

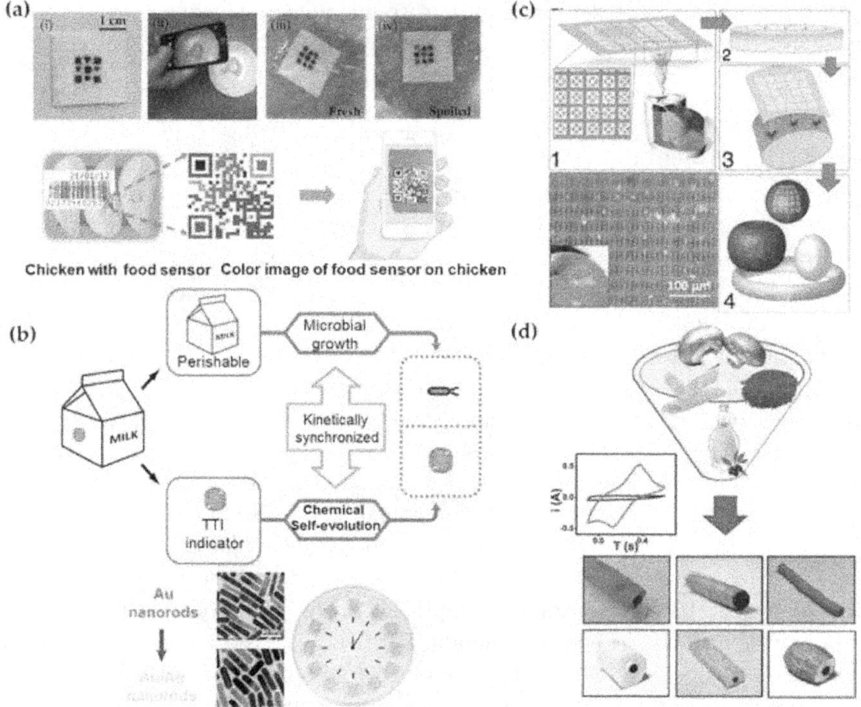

FIGURE 2.21 Various demonstrations of biosensors in the food industry [195].

orange juice were detected using a similar biosensor [198]. Bacterial bioassays can test for arsenic [199].

2.4.2.4 In the Medical Field

The use of biosensors in medicine is rapidly expanding. Diabetes mellitus needs careful management of blood glucose levels; therefore, clinical applications frequently use glucose biosensors (Figure 2.22) [200]. Eighty-five percent of the huge global market is occupied by household blood glucose biosensors. Biosensors are widely utilized in medicine to diagnosis infectious illnesses. A potential biosensor method for antibiotic susceptibility, UTI diagnosis, and pathogen identification is being studied. Identifying end-stage heart failure patients at risk for complications following left ventricular aided device placement is critical. A novel biosensor based on hafnium oxide (HfO_2) was used for early detection of IL-10 [202]. Early cytokine detection using recombinant human IL-10 and a monoclonal antibody has been explored. Fluorescence mode and electrochemical impedance spectroscopy are used to assess antibody–antigen interactions and protein biorecognition. Chen et al. used HfO_2 as a biofield effect transistor [203]. Electrochemical impedance spectroscopy was used by HfO_2 biosensors to detect human antigens. The main problem now is heart failure, which affects almost a million individuals every year. Cardiovascular disease detection methods contain enzyme-linked immunosorbent tests, fluorometrics, and immunoaffinity columns [204]. These are time intensive and need trained workers. Electrical biosensors use biomolecules to recognize the desired selectivity of specific biomarkers. In addition to this, other biosensor applications include fast and accurate

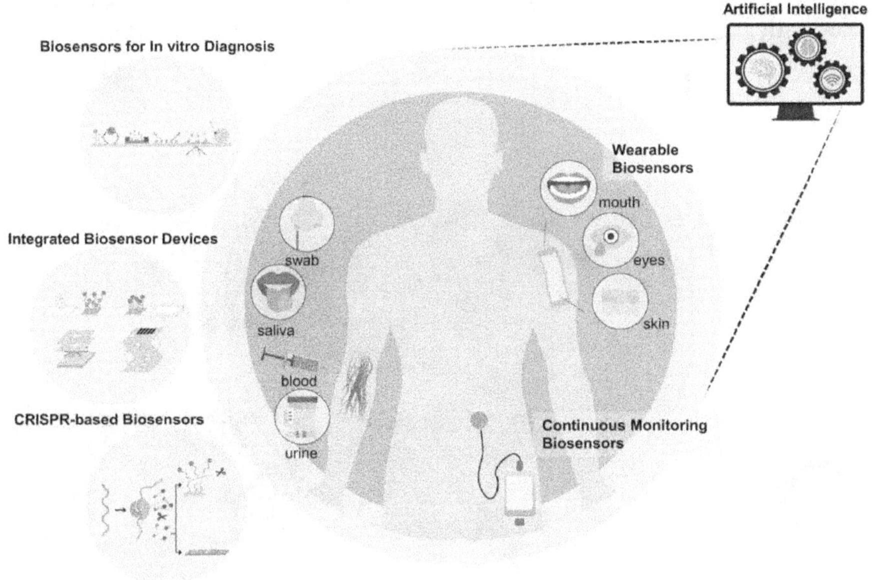

FIGURE 2.22 Some medical biosensors [200].

biochips, detection of histone deacetylase (HDAC) inhibitors from resonance energy transfer, malignant effects of oxaboridine on immobilized fructosyltransferases in dental diseases, use in acute leukemia immunosensor arrays for clinical immunophenotyping, microfluidic impedance assays for control of endothelin-induced cardiac hypertrophy, and quantitative measurement of cardiac markers in undiluted serum.

2.4.2.5 Fluorescent Biosensors

In cancer and drug discovery, fluorescent biosensors can image. They have shown biological function and control of enzymes. FRET biosensors with GFP have an important function. An enzyme, chemical, or genetically modified fluorescent probe is attached to a fluorescent biosensor through a receptor. The receptor detects and measures a particular target or analyte by transducing a fluorescence signal [201]. Figure 2.23 show the schematic diagram of fluorescent biosensor and different kinds of fluorescent biosensor. Fluorescent biosensors can detect ions, metabolites, and protein biomarkers in complicated solutions and indicate their status, activity, or presence (cell extracts, serum). Signal transmission, transcription, cell cycle, and apoptosis all use them to probe conformation, protein location, and gene expression. Their use enables detection of illnesses including cancer and metastases as well as arthritic conditions. Fluorescent biosensors to enable lead optimization, post-hit screening analysis, high-content screening, and high throughput are used in drug discovery programs. These are powerful techniques for assessing drug candidates' therapeutic potential, biodistribution, and pharmacokinetics [205]. Fluorescent biosensors are used for early biomarker detection in image-guided surgery, intravital imaging, disease progression monitoring, and genetic and clinical diagnostics [206]. Using a genetically encoded FRET biosensor to detect Bcr-Abl kinase activity on cancer patient cells helped demonstrate a link between Bcr-Abl kinase activity

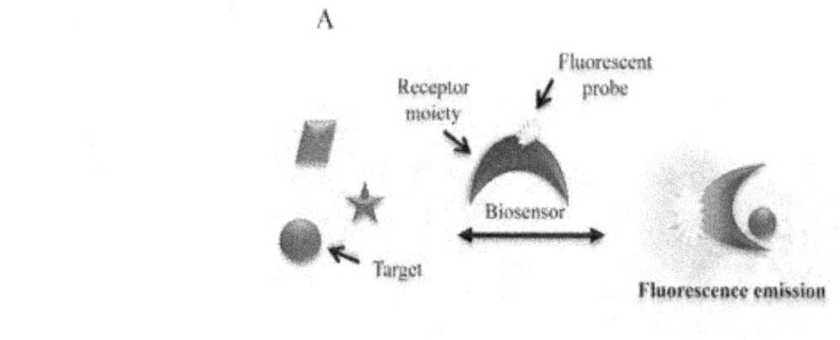

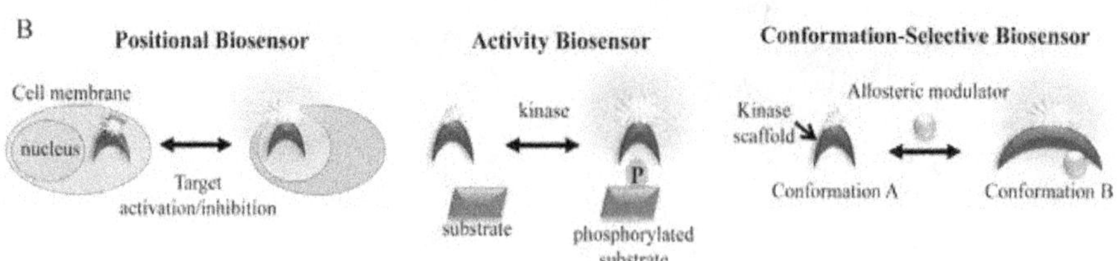

FIGURE 2.23 Fluorescent biosensors (a) Schematic representation of a fluorescent biosensor. (b) Different examples of fluorescent biosensors [211].

and chronic myeloid leukemia disease state. The probe was also used to control treatment response and detect drug-resistant cells, thereby predicting alternative treatments [207].

2.4.2.6 Biodefense Biosensing Applications

Biological assault biosensors may be employed for military objectives. These biosensors are designed to recognize viruses, poisons, and biowarfare agents (BWAs), which include bacteria (spores and vegetative). Several efforts have been made to design such biosensors utilizing molecular approaches that identify BWA chemical markers. Nucleic acid-based sensing systems are superior to antibody-based detection methods because they can provide gene-based specificity without the need for an amplification stage. There are two types of human papillomavirus (HPV): HPV 16 and 18. A sink-type surface acoustic wave peptide nucleic acid biosensor with a dual-port resonator was developed for the rapid detection of HPV. The probe can efficiently and accurately bind specific DNA sequences.

2.4.2.7 In Metabolic Engineering

A growing demand for microbial cell factories for chemical synthesis is being driven by environmental concerns and petroleum-derived product insufficiency. Metabolic engineering is seen as a key technique for a sustainable bioeconomy [208]. They also expect that instead of relying on petroleum refining or plant extraction, most commodity chemicals, pharmaceuticals, and fuels will be produced by microorganisms from renewable raw materials. The enormous variety of creation capacity necessitates effective screening procedures to choose the target phenotype. Earlier enzymatic assay analytics approaches used spectroscopy but had low throughput. To overcome this hurdle, researchers have created genetically encoded biosensors that can be used for high-throughput selection and screening using cell survival and fluorescence-activated cell sorting (FACS). In a FRET sensor, two acceptor and donor fluorophores sandwich the ligand-binding peptide. When a ligand of interest contacts the peptide, it changes conformation and therefore FRET [209]. Despite their high orthogonality, temporal resolution, and simplicity of assembly, FRET sensors cannot modulate downstream signals and can only report the abundance of metabolites [210]. Transcription factors are naturally occurring proteins through controlling gene expression in response to a change environment [211]. This is achieved by exploiting the host transcription system and driving a reporter gene with a synthetic condition-specific promoter. These have weak orthogonality and noise [212]. The third family of biosensors is riboswitches, which are mRNA regulatory domains that may selectively bind ligands and modify their structure, hence controlling protein production. They are quicker than TF-based biosensors since the RNA is already transcribed, and they do not depend on protein–protein or protein–metabolite interactions. Ribosomes have recently been widely developed in bacterial systems [213].

2.4.2.8 Biosensors in Plant Biology

Plant research has benefited from revolutionary new tools in DNA sequencing and molecular imaging. However, traditional mass spectroscopy approaches did not provide information on the dynamics and position of transporters, receptors, and enzyme substrates. Using biosensors, these data may be readily accessed. To quantify dynamic processes in physiological contexts, we need to visualize them, such metabolite conversion or signaling events. Sensors that react dynamically may accomplish this depiction. Roger Tsien's lab was the first to manipulate calcium levels and quantify caspase activity in live cells [214]. These sensors used FRET between two spectrum GFP versions [215]. Using chameleon sensors, researchers can monitor calcium oscillations in real time. Biosensors can discover missing components in analyte transport, regulation, or metabolism. The sucrose FRET sensor transports sucrose from the mesophyll to the phloem. Fluorimeter-based sugar transporter detection detects sugar transporters that work soon after glucose administration in deprived yeast cells [216]. Other studies have shown that biosensors may be used in genetic screening if high-throughput imaging tools are available. Using nanoelectrodes, nanofilms, and single-molecule multifunctional nanocomposites [217] remains difficult. Nanomaterial customization, availability of high-quality nanomaterials, interfaces, characterization, processing issues, and processes affecting the behavior of these nanoscale composites on electrode surfaces are all major concerns. Improving signal-to-noise ratios, transduction, and signal amplification are other key obstacles. The next step should be to understand how biomolecules and nanomaterials interact with electrodes or nanofilm and create a new biosensor generation with innovative functions. Despite this, biosensors based on nanomaterials have broad prospects in environmental monitoring, process control, food analysis, and clinical diagnosis.

2.5 NANOSENSORS FOR DETECTING PLANT PATHOGENS

Pathogen detection, identification, and evaluation are critical for food security, ecological monitoring, and scientific research. Investigators must ensure that the delicate ingredients of biomimetic components or biological origin interact with the analyte in the test. Many reliable and quick identification components, such as cell receptors, bacterial imprints, phages, antibodies, aptamers, or lectin, have been discovered for bacterial exposure [218]. Bacterial receptors, antibodies, and lectins are the most often employed biosensing components. Because they may be combined into biosensors, these constituents are widely used to detect infections [219]. Aptamers, single-stranded nucleic acids, are cheaper and more stable than recognition elements based on antibodies for detecting bacteria [220]. However, they have drawbacks such as preparation difficulty, sturdiness in complicated materials, and batch-to-batch variability. The "chemical nose" is a relatively new tool for identifying infections. It assigns different discriminative receptors to each target, allowing sequencing. It works like our brain when we smell something [221]. To create a reference database, sensors are trained with competent bacterial samples. The reference catalog is used to identify bacterial pathogens [222].

2.6 CONCLUSION

In this review, we report on the possibilities of recent research on the application of graphene and its associated biosensors in life sciences and clinical settings. We present results on each sensor's analytical performance and illustrate their applications in medicine and life sciences. Enzyme-based biosensors, antibodies, and DNA all have their pros and cons. In general, the type of application determines the type of sensor chosen. For instance, the use of DNA in biosensing could be a cost-effective way to rapidly detect markers of cancer, viruses, or microbes. However, the use of antibodies or enzymes in biosensors can detect or monitor certain diseases more effectively because of the wide variety of molecules present in the human body. For example, we can use antibodies to specifically detect viruses such as influenza virus, HIV, and Zika virus A. On the other hand, enzymes hold great promise for detecting glucose levels with small samples. Overall, graphene and graphene-based nanomaterials have shown great promise for applications in biosensor technologies due to their electron transfer rate, electrical conductivity, high specific surface area, and ability to immobilize various different biomolecules. The clinical promise lies in development of biosensors that are specific, sensitive, stable, and rapidly processable for target molecules. However, more research is needed to obtain reliable and uniform results and to produce biosensors that can be used in the medical field to check the reliability and safety of the sensors. One of the excellent sensing electrode materials in the medical field is graphene, so we need to pay more attention to new methods for well-controlled graphene processing and synthesis, which need to be investigated in future research. Current chemical strategies for modifying graphene surfaces with biomolecules are effective in targeting specific analyses. However, this sensing platform can be further refined to improve the orientation of biomolecules on the graphene platform and avoid adsorption of unwanted molecules on the graphene. Therefore, in order to realize the most important role in nanomaterial graphene-based nanosensors, the chemical and physical properties of graphene surfaces and the interaction with biomolecules at the interface should be better understood, and sensor mechanisms should be implemented. In addition, the manufacture and miniaturization of small diagnostic biosensors is important in sensor technology, and the development of sensors with high specificity, sensitivity, precision, cost efficiency, reproducibility, and reliability is required. Therefore, these are urgent requirements for sensor technology. In order to improve availability in remote locations, it is necessary to reduce the cost of sensors in economic terms. In addition, miniaturization of sensors enables rapid detection of bacterial pathogens and viruses. It can also be used in bioimplants by self-monitoring to detect serious health conditions. The applications described previously in the life sciences will help improve people's health and protect human life in the fields of treatment and examination. However, it is important to ensure that there are no health risks in long-term use. Furthermore, many studies still need to be conducted in terms of demonstrating, guaranteeing, and ensuring non-toxicity and biocompatibility of graphene-based nanomaterials.

In this review, we also discuss the intrinsic properties of graphene and its surface functionalization as a perspective on the transduction mechanism for biomedical applications. Several well-known techniques for the synthesis of graphene-based materials and their properties have been explained. Functionalization of various graphene-based materials is also presented, including reduced graphene oxide and graphene quantum dots, pristine graphene, and graphene oxide. The mechanisms discussed in this chapter involve the latest biosensing devices for bioimaging, healthcare sensors, biosensors, drug delivery, and other new technologies. One of the most famous two-dimensional materials is currently graphene. The main properties and characteristics of graphene are highlighted, namely high thermal conductivity and tensile strength, proximity induction capability, nontoxicity, large surface area, transparency, ultrahigh charge mobility, relativistic charge velocity, Dirac point linear semiconductor, and zero band gap semiconductor. However, graphene must be very suitable to maintain these properties. Here we also present alternative techniques for chemical synthesis suitable for the surrounding environment. Graphite is functionalized and modified with various oxygen-containing groups during synthesis. Furthermore, functionalized graphene is always polluted by disorders, defects, and impurities. As a result, the structure of graphene will undergo important changes, especially distorting the electronic properties. In other words, very important for biosensing applications is the ability to functionalize them into graphene structures and the presence of oxygen-containing groups, such as detection of biomolecules in imaging applications, increased number of active regions and sites, enhanced sensing signals, and electrochemical labeling of biomolecules for identification. The last section introduces a novel graphene-based biosensor. We discuss the technologies, mechanisms, and engineering principles of the biosensing platform. This chapter discusses many of the new methods in detail. We have now explored other 2D materials since graphene is very well established. This plays a vital role in biosensors and sensor applications utilizing the maximum surface area with endless possibilities. In the near future, these new 2D materials will be further customized for the specificity of biological receptors. We can integrate and use these materials with ultrahigh sensitivity in different biosensors and sensor platforms, and they may provide solutions to some problems such as early cancer detection.

2.7 FUTURE SCOPE

In vivo or ex vivo genetically modified proteins injected into cells form cell-based biosensors. They use biophotonics or other physical concepts to detect levels of medicines, hormones, or poisons noninvasively and continually. This might be useful in ageing studies. Marine biosensors detect eutrophication using nitrate and nitrite sensors. The Monterey Bay Aquarium Research Institute is developing an environmental sample processor that will use ribosomal RNA probes to automatically identify harmful algae in situ at moorings. Another goal is to use biosensors to detect pesticides, heavy metals, and pollution. Nanomaterials in biosensors may help develop a new generation of biosensors. Nanomaterials are required to realize single-molecule biosensors with high-throughput biosensor arrays with mechanical, electrochemical, optical, and magnetic capabilities. Harnessing the structures and functions of nanomaterials and biomolecules to fabricate nanoelectrodes, nanofilms, and single-molecule multifunctional nanocomposites remains a difficult problem. Availability, interfacial issues, characterization, processing of nanomaterial customization, high-quality nanomaterials, and processes that affect the behavior of these nanoscale composites on electrode surfaces are all major issues. Improving signal-to-noise ratios, transduction, and signal amplification are other key obstacles. The next step should be to understand how biomolecules and nanomaterials interact with nanofilms or electrodes and use innovative functions to create a new generation of biosensors. Despite the challenges, nanomaterial-based biosensors have bright prospects in environmental monitoring, process control, food analysis, and clinical diagnosis.

REFERENCES

[1] Haque R, Kress K, Wood S, Jackson TF, Lyerly D, Wilkins T, Diseases I, et al. Diagnosis of pathogenic *Entamoeba histolytica* infection using a stool ELISA based on monoclonal antibodies to the galactose-specific adhesin. *J Infect Dis.* 2018;167:247–249.

[2] Lazcka O, Del Campo FJ, Muñoz FX. Pathogen detection: a perspective of traditional methods and biosensors. *Biosens Bioelectron.* 2007;22:1205–1217.

[3] Josephson KL, Gerba CP, Pepper IL. Polymerase chain reaction detection of nonviable bacterial pathogens. *Appl Environ Microbiol.* 1993;59:3513–3515.

[4] Wilson WJ, Strout CL, DeSantis TZ, Stilwell JL, Carrano AV, Andersen GL. Sequence-specific identification of 18 pathogenic microorganisms using microarray technology. *Mol Cell Probes.* 2002;16:119–127.

[5] Lindsley MD, Mekha N, Baggett HC, Surinthong Y, Autthateinchai R, Sawatwong P, et al. Evaluation of a newly developed lateral flow immunoassay for the diagnosis of cryptococcosis. *Clin Infect Dis.* 2011;53:321–325.

[6] Schulze H, Rubtsova M, Bachmann TT. DNA microarrays for pathogen detection. In *Modern Techniques for Pathogen Detection.* New York: Wiley; 2015. p. 113–220.

[7] Wilson CB. Sensors in medicine. *West J Med.* 1999;171:322.

[8] Sapsford KE, Bradburne C, Delehanty JB, Medintz IL. Sensors for detecting biological agents. *Mater Today.* 2008;11:38–49.

[9] Patolsky F, Zheng G, Lieber CM. Nanowire sensor for medicine and the life science. *Nanomedicine.* 2006;1:51–65.

[10] Zhu Y, Murali S, Cai W, Li X, Suk JW, Potts JR, et al. Graphene and graphene oxide: synthesis, properties, and applications. *Adv Mater.* 2010;22:3906–3924.

[11] Morales-Narváez E, Baptista-Pires L, Zamora-Gálvez A, Merkoçi A. Graphene-based biosensors: going simple. *Adv Mater.* 2017;29:1604905.

[12] Chauhan N, Maekawa T, Kumar DNS. Graphene based biosensors—accelerating medical diagnostics to new-dimensions. *J Mater Res.* 2017;32:2860–2882.

[13] Janegitz BC, Silva TA, Wong A, Ribovski L, Vicentini FC, Taboada Sotomayor MDP, et al. The application of graphene for in vitro and in vivo electrochemical biosensing. *Biosens Bioelectron.* 2017;89:224–233.

[14] Wang Y, Li Z, Wang J, Li J, Lin Y. Graphene and graphene oxide: biofunctionalization and applications in biotechnology. *Trends Biotechnol.* 2011;29:205–212.

[15] Liu J, Tang J, Gooding JJ. Strategies for chemical modification of graphene and applications of chemically modified graphene. *J Mater Chem.* 2012;22:12435.

[16] Gubala V, Harris LF, Ricco AJ, Tan MX, Williams DE. Point of care diagnostics: status and future. *Anal Chem.* 2012;84:487–515.

[17] Malhotra BD, Chaubey A. Biosensors for clinical diagnostics industry. *Sens Actuators B.* 2003;91:117–127.

[18] Rodriguez-Mozaz S, Alda M, Marco M, Barcelo D. Biosensors for environmental monitoring a global perspective. *Talanta* 2005;65:291–297.

[19] Su S, Chen S, Fan C. Recent advances in two-dimensional nanomaterials-based electrochemical sensors for environmental analysis. *Green Energy Environ.* 2018;3:97–106.

[20] Thakur MS, Ragavan KV. Biosensors in food processing. *J Food Sci Technol.* 2013;50:625–641.

[21] Velasco-Garcia MN, Mottram T. Biosensor technology addressing agricultural problems. *Biosyst Eng.* 2003;84:1–12.

[22] Singh E, Nalwa HS. *Nanomaterial-Based Flexible and Multifunctional Sensors.* Los Angeles, CA: American Scientific Publishers; 2019.
[23] Kelley SO, Mirkin CA, Walt DR, Ismagilov RF, Toner M, Sargent EH. Advancing the speed, sensitivity and accuracy of biomolecular detection using multi-length-scale engineering. *Nat Nanotechnol.* 2014;9:969–980.
[24] Liu Q, Wu C, Cai H, Hu N, Zhou J, Wang P. Cell-based biosensors and their application in biomedicine. *Chem Rev.* 2014;114:6423–6461.
[25] Rong G, Corrie SR, Clark HA. In vivo biosensing: progress and perspectives. *ACS Sens.* 2017;2:327–338.
[26] Du Y, Dong S. Nucleic acid biosensors: recent advances and perspectives. *Anal Chem.* 2017;89:189–215.
[27] Huang X, Liu Y, Yung B, Xiong Y, Chen X. Nanotechnology-enhanced no-wash biosensors for in vitro diagnostics of cancer. *ACS Nano.* 2017;11:5238–5292.
[28] Bereza-Malcolm LT, Mann G, Franks AE. Environmental sensing of heavy metals through whole cell microbial biosensors: a synthetic biology approach. *ACS Synth Biol.* 2014;4:535–546.
[29] Gaddes D, Reeves WB, Tadigadapa S. Calorimetric biosensing system for quantification of urinary creatinine. *ACS Sens.* 2017;2:796–802.
[30] De Wael K, Daems D, Van Camp G, Nagels LJ. Use of potentiometric sensors to study (bio)molecular interactions. *Anal Chem.* 2012;84:4921–4927.
[31] Heller A, Feldman B. Electrochemical glucose sensors and their applications in diabetes management. *Chem Rev.* 2008;108:2482–2505.
[32] Tamura T, Hamachi I. Recent progress in design of protein-based fluorescent biosensors and their cellular applications. *ACS Chem Biol.* 2014;9:2708–2717.
[33] Wang Z, Zong S, Wu L, Zhu D, Cui Y. SERS-activated platforms for immunoassay: probes, encoding methods, and applications. *Chem Rev.* 2017;117:7910–7963.
[34] Tiwari JN, Vij V, Kemp KC, Kim KS. Engineered carbon-nanomaterial-based electrochemical sensors for biomolecules. *ACS Nano.* 2016;10:46–80.
[35] Maduraiveeran G, Sasidharan M, Ganesan V. Electrochemical sensor and biosensor platforms based on advanced nanomaterials for biological and biomedical applications. *Biosens Bioelectron.* 2018;103:113–129.
[36] Zhang H, Zhang H, Aldalbahi A, Zuo X, Fan C, Mi X. Fluorescent biosensors enabled by graphene and graphene oxide. *Biosens Bioelectron.* 2017;89:96–106.
[37] Zhu C, Yang G, Li H, Du D, Lin Y. Electrochemical sensors and biosensors based on nanomaterials and nanostructures. *Anal Chem.* 2015;87:230–249.
[38] (a) *Handbook of Nanostructured Materials and Nanotechnology*, ed. HS Nalwa. San Diego: Academic Press; 2000, vol. 1–5; (b) *Encyclopedia of Nanoscience and Nanotechnology*, ed. HS Nalwa. Los Angeles: American Scientific Publishers; 2004, vol. 1–10; (c) *Handbook of Nanostructured Biomaterials and Their Applications in Nanobiotechnology*, ed. HS Nalwa. Los Angeles: American Scientific Publishers; 2005, vol. 1–2; (d) Georgakilas V, Perman JA, Tucek J, Zboril R. Broad family of carbon nanoallotropes: classification, chemistry, and applications of fullerenes, carbon dots, nanotubes, graphene, nanodiamonds, and combined superstructures. *Chem Rev.* 2015;115:4744–4822.
[39] Cobley CM, Chen J, Cho EC, Wang LV, Xia Y. Gold nanostructures: a class of multifunctional materials for biomedical applications. *Chem Soc Rev.* 2011;40:44–56.
[40] Sun S-K, Wang H-F, Yan X-P. Engineering persistent luminescence nanoparticles for biological applications: from bio sensing/bio imaging to theranostics. *Acc Chem Res.* 2018;51:1131–1143.
[41] Nandini S, Nalini S, Reddy MBM, Suresh GS, Melo JS, Niranjana P, Sanetuntikul J, Shanmugam S. Synthesis of one-dimensional gold nanostructures and the electrochemical application of the nanohybrid containing functionalized graphene oxide for cholesterol biosensing. *Bioelectrochemistry*, 2016;110:79–90.

[42] Tan C, Cao X, Wu X-J, He Q, Yang J, Zhang X, Chen J, Zhao W, Han S, Nam G-H, Sindoro M, Zhang H. Recent advances in ultrathin two-dimensional nanomaterials. *Chem Rev.* 2017;117:6225–6331.

[43] (a) Singh E, Kim KS, Yeom GY, Nalwa HS. Atomically thin-layered molybdenum disulfide (MoS_2) for bulk-heterojunction solar cells. *ACS Appl Mater Interfaces*, 2017;9:3223–3245; (b) Singh E, Kim KS, Yeom GY, Nalwa HS. Two-dimensional transition metal dichalcogenide-based counter electrodes for dye-sensitized solar cells. *RSC Adv.* 2017;7:28234–28290; (c) Singh E, Singh P, Kim KS, Yeom GY, Nalwa HS. Flexible molybdenum disulfide (MoS2) atomic layers for wearable electronics and optoelectronics. *ACS Appl Mater Interfaces.* 2019. https://doi.org/10.1021/acsami.8b19859; (d) Singh R, Singh E, Nalwa HS. Inkjet printer nanomaterial based flexible radio frequency identification (RFID) tag sensors for the internet of Nano things. *RSC Adv.* 2017;7:48597–48630.

[44] Dasgupta NP, Sun J, Liu C, Brittman S, Andrews SC, Lim J, Gao H, Yan R, Yang P. 25th anniversary article: semiconductor nanowires—synthesis characterization, and applications. *Adv Mater.* 2014;26:2137–2184.

[45] Sapsford KE, Algar WR, Berti L, Gemmill KB, Casey BJ, Oh E, Stewart MH, Medintz IL. Functionalizing nanoparticles with biological molecules: developing chemistries that facilitate nanotechnology. *Chem Rev.* 2013;113:1904–2074.

[46] Fadel TR, Farrell DF, Friedersdorf LE, Griep MH, Hoover MD, Meador MA, Meyyappan M. Toward the responsible development and commercialization of sensor nanotechnologies. *ACS Sens.* 2016;1:207–216.

[47] Novoselov KS, Geim AK, Morozov SV, Jiang D, Zhang Y, Dubonos SV, Grigorieva IV, Firsov AA. Electric field effect in atomically thin carbon films. *Science.* 2004;306:666–669.

[48] Cheng C, Li S, Thomas A, Kotov NA, Haag R. Functional graphene nanomaterials based architectures: bio interactions, fabrications, and emerging biological applications. *Chem Rev.* 2017;117:1826–1914.

[49] Geim AK, Novoselov KS. The rise of graphene. *Nat Mater.* 2007;6:183–191.

[50] Zhu Y, Murali S, Cai W, Li X, Suk JW, Potts JR, Ruoff RS. Graphene and graphene oxide: synthesis, properties, and applications. *Adv Mater.* 2010;22:3906–3924.

[51] Avouris P, Dimitrakopoulos C. Graphene: synthesis and applications. *Mater Today.* 2012;15:86–97.

[52] Mao HY, Laurent S, Chen W, Akhavan O, Imani M, Ashkarran AA, Mahmoudi M. Graphene: promises, facts, opportunities, and challenges in nanomedicine. *Chem Rev.* 2013;113:3407–3424.

[53] Meyer JC, Geim AK, Katsnelson MI, Novoselov KS, Booth TJ, Roth S. The structure of suspended graphene sheets. *Nature.* 2007;446:60–63.

[54] Xu Y, Bai H, Lu G, Li C, Shi G. Flexible graphene films via the filtration of water-soluble noncovalent functionalized graphene sheets. *J Am Chem Soc.* 2008;130:5856–5857.

[55] Chen H, Müller MB, Gilmore KJ, Wallace GG, Li D. Mechanically strong, electrically conductive, and biocompatible graphene paper. *Adv Mater.* 2008;20:3557–3561.

[56] Li X, Wang X, Zhang L, Lee S, Dai H. Chemically derived, ultrasmooth graphene nanoribbon semiconductors. *Science.* 2008;319:1229–1232.

[57] Xi Q, Chen X, Evans DG, Yang W. Gold nanoparticle-embedded porous graphene thin films fabricated via layer-by-layer self-assembly and subsequent thermal annealing for electrochemical sensing. *Langmuir.* 2012;28:9885–9892.

[58] Hu C, Xue J, Dong L, Jiang Y, Wang X, Qu L, Dai L. Scalable preparation of multifunctional fire-retardant ultralight graphene foams. *ACS Nano.* 2016;10:1325–1332.

[59] Balandin AA, Ghosh S, Bao W, Calizo I, Teweldebrhan D, Miao F, Lau CN. Superior thermal conductivity of single-layer graphene. *Nano Lett.* 2008;8:902–907.

[60] Morozov SV, Novoselov KS, Katsnelson MI, Schedin F, Elias DC, Jaszczak JA, Geim AK. Giant intrinsic carrier mobilities in graphene and its bilayer. *Phys Rev Lett.* 2008;100:11–14.
[61] Nair RR, Blake P, Grigorenko AN, Novoselov KS, Booth TJ, Stauber T, Peres NMR, Geim AK. Fine structure constant defines visual transparency of graphene. *Science.* 2008;320:1308.
[62] Stoller MD, Park S, Zhu Y, An J, Ruoff RS. Graphene-based ultracapacitors. *Nano Lett.* 2008;8:3498–3502.
[63] Singh E, Nalwa HS. Graphene-based bulk-heterojunction solar cells: a review. *J Nanosci Nanotechnol.* 2015;15:6237–6278.
[64] Singh E, Nalwa HS. Graphene-based dye-sensitized solar cells: a review. *Sci Adv Mater.* 2015;7:1863–1912.
[65] Lv W, Li Z, Deng Y, Yang Q-H, Kang F. Graphene-based materials for electrochemical energy storage devices: opportunities and challenges. *Energy Storage Mater.* 2016;2:107–138.
[66] Khan U, Kim T-H, Ryu H, Seung W, Kim S-W. Graphene tribotronics for electronic skin and touch screen applications. *Adv Mater.* 2017;29:2–8.
[67] Schwierz F. Graphene transistors. *Nat Nanotechnol.* 2010;5:487–496.
[68] Han T-H, Lee Y, Choi M-R, Woo S-H, Bae S-H, Hong BH, Ahn J-H and Lee T-W. Extremely efficient flexible organic light-emitting diodes with modified graphene anode. *Nat Photonics.* 2012;6:105–110.
[69] Singh E, Meyyappan M, Nalwa HS. Flexible graphene-based wearable gas and chemical sensors. *ACS Appl Mater Interfaces.* 2017;9:34544–34586.
[70] Iannazzo D, Pistone A, Salam'o M, Galvagno S, Romeo R, Giofré SV, Branca C, Visalli G, Di Pietro, A. Graphene quantum dots for cancer targeted drug delivery. *Int J Pharm.* 2017;518:185–192.
[71] Kumar V, Kim K-H, Park J-W, Hong J, Kumar S. Graphene and its nanocomposites as a platform for environmental applications. *Chem Eng J.* 2017;315:210–232.
[72] Chen D, Feng H, Li J. Graphene oxide: preparation, functionalization, and electrochemical applications. *Chem Rev.* 2012;112:6027–6053.
[73] Yin PT, Shah S, Chhowalla M, Lee K-B. Design, synthesis, and characterization of graphene—nanoparticle hybrid materials for bio applications. *Chem Rev.* 2015;115:2483–2531.
[74] Zhao X, Zhang P, Chen Y, Su Z, Wei G. Recent advances in the fabrication and structure-specific applications of graphene-based inorganic hybrid membranes. Nanoscale. 2015;7:5080–5093.
[75] Qiu L, Li D, Cheng H-M. Structural control of graphene-based materials for unprecedented performance. *ACS Nano.* 2018;12:5085–5092.
[76] Nardecchia S, Carriazo D, Ferrer ML, Gutiérrez MC, del Monte F. Three dimensional macro porous architectures and aerogels built of carbon nanotubes and/or graphene: synthesis and applications. *Chem Soc Rev.* 2013;42:794–830.
[77] Liu F, Piao Y, Choi JS, Seo TS. Three-dimensional graphene micropillar based electrochemical sensor for phenol detection. *Biosens Bioelectron.* 2013;50:387–392.
[78] Liu J, Fu S, Yuan B, Li Y, Deng Z. Toward a universal "adhesive nanosheet" for the assembly of multiple nanoparticles based on a protein-induced reduction/decoration of graphene oxide. *J Am Chem Soc.* 2010;132:7279–7281.
[79] Wang M, Duan X, Xu Y, Duan X. Functional three-dimensional graphene/polymer composites. *ACS Nano.* 2016;10:7231–7247.
[80] Chee WK, Lim HN, Huang NM, Harrison I. Nanocomposites of graphene/polymers: a review. *RSC Adv.* 2015;5:68014–68051.
[81] Zhang R, Chen W. Recent advances in graphene-based nanomaterials for fabricating electrochemical hydrogen peroxide sensors. *Biosens Bioelectron.* 2017;89:249–268.

[82] Bollella P, Fusco G, Tortolini C, Sanz'o G, Favero G, Gorton L, Antiochia R. Beyond graphene: electrochemical sensors and biosensors for biomarkers detection. *Biosens Bioelectron.* 2017;89:152–166.
[83] Wang L, Xiong Q, Xiao F, Duan H. 2D nanomaterials based electrochemical biosensors for cancer diagnosis. *Biosens Bioelectron.* 2017;89:136–151.
[84] Chikkaveeraiah BV, Bhirde AA, Morgan NY, Eden HS, Chen X. Electrochemical immunosensor for detection of cancer protein biomarkers. *ACS Nano.* 2012;6:6546–6561.
[85] Farka Z, Juřík T, Kovář D, Trnková L, Skládal P. Nanoparticle-based immunochemical biosensors and assays: recent advances and challenges. *Chem Rev.* 2017;117:9973–10042.
[86] Li M, Gou H, Al-Ogaidi I, Wu N. Nanostructured sensors for detection of heavy metals: a review. *ACS Sustainable Chem Eng.* 2013;1:713–723.
[87] Saidur MR, Aziz ARA, Basirun WJ. Recent advances in DNA-based electrochemical biosensors for heavy metal ion detection: a review. *Biosens Bioelectron.* 2017;90:125–139.
[88] Reta N, Saint CP, Michelmore A, Prieto-Simon B, Voelcker NH. Nanostructured electrochemical biosensors for label-free detection of water- and food-borne pathogens. *ACS Appl Mater Interfaces*, 2018;10:6055–6072.
[89] Zhang Y, Shen J, Li H, Wang L, Cao D, Feng X, Liu Y, Ma Y, Wang L. Recent progress on graphene-based electrochemical biosensors. *Chem Rec.* 2016;16:273–294.
[90] Lawal AT. Progress in utilisation of graphene for electrochemical biosensors. *Biosens Bio Electron.* 2018;106:149–178.
[91] Liu Y, Dong X, Chen P. Biological and chemical sensors based on graphene materials. *Chem Soc Rev.* 2012;41:2283–2307.
[92] Rasheed PA, Sandhyarani N. Carbon nanostructures as immobilization platform for DNA: a review on current progress in electrochemical DNA sensors. *Biosensor Bio Electron.* 2017;97:226–237.
[93] Wang J. DNA biosensors based on peptide nucleic acid (PNA) recognition layers. A review. *Biosens Bioelectron.* 1998;13:757–762.
[94] Pumera M. Graphene in biosensing. *Mater Today.* 2011;14:308–315.
[95] Rao CNR, Sood AK, Subrahmanyam KS, Govindaraj A. Graphene: the new two-dimensional nanomaterial. *Angew Chemie.* 2009;48:7752–7777.
[96] Park CS, Yoon H, Kwon OS. Graphene-based nanoelectronic biosensors. *J Ind Eng Chem.* 2016;38:13–22.
[97] Kuila T, Bose S, Khanra P, Mishra AK, Kim NH, Lee JH. Recent advances in graphene-based biosensors. *Biosens Bioelectron.* 2011;26:4637–4648.
[98] Kasry A, Ardakani AA, Tulevski GS, Menges B, Copel M, Vyklicky L. Highly efficient fluorescence quenching with graphene. *J Phys Chem C.* 2012;116:2858–2862.
[99] Batır GG, Arık M, Caldıran Z, Turut A, Aydogan S. Synthesis and characterization of reduced graphene oxide/rhodamine 101 (rGO-Rh101) nanocomposites and their heterojunction performance in rGORh101/p-Si device configuration. *J Electron Mater.* 2018;47:329–336.
[100] Wu X, Xing Y, Zeng K, Huber K, Zhao JX. Study of fluorescence quenching ability of graphene oxide with a layer of rigid and tunable silica spacer. *Langmuir.* 2018;34:603–611.
[101] Cheng S, Hideshima S, Kuroiwa S, Nakanishi T, Osaka T. Label-free detection of tumor markers using field effect transistor (FET)-based biosensors for lung cancer diagnosis. *Sens Actuators B.* 2015;212:329–334.
[102] Wang F, Horikawa S, Hu J, Wikle H, Chen I-H, Du S, et al. Detection of *Salmonella typhimurium* on spinach using phage-based magnetoelastic biosensors. *Sensors.* 2017;17:386.
[103] Benvidi A, Rajabzadeh N, Mazloum-Ardakani M, Heidari MM, Mulchandani A. Simple and label-free electrochemical impedance Amelogenin gene hybridization biosensing based on reduced graphene oxide. *Biosens Bioelectron.* 2014;58:145–152.

[104] Singh P, Pandey SK, Singh J, Srivastava S, Sachan S, Singh SK. Biomedical perspective of electrochemical nanobiosensor. *Nano-Micro Lett.* 2016;8:193–203.
[105] Wan Y, Su Y, Zhu X, Liu G, Fan C. Development of electrochemical immunosensors towards point of care diagnostics. *Biosens Bioelectron.* 2013;47:1–11.
[106] Wujcik EK, Wei H, Zhang X, Guo J, Yan X, Sutrave N, et al. Antibody nanosensors: a detailed review. *RSC Adv.* 2014;4:43725–43745.
[107] Fischer MJE. Amine coupling through EDC/NHS: a practical approach. In: Mol NJ, Fischer MJE, editors. *Surf Plasmon Reson Methods Protoc.* Totowa: Humana Press; 2010. p. 55–73.
[108] Cho IH, Lee J, Kim J, Kang MS, Paik JK, Ku S, et al. Current technologies of electrochemical immunosensors: perspective on signal amplification. *Sensors.* 2018;18:207.
[109] Huang Y, Dong X, Liu Y, Li L-J, Chen P. Graphene-based biosensors for detection of bacteria and their metabolic activities. *J Mater Chem.* 2011;21:12358.
[110] Pandey A, Gurbuz Y, Ozguz V, Niazi JH, Qureshi A. Graphene-interfaced electrical biosensor for label-free and sensitive detection of foodborne pathogenic *E. coli* O157:H7. *Biosens Bioelectron.* 2017;91:225–231.
[111] Afsahi S, Lerner MB, Goldstein JM, Lee J, Tang X, Bagarozzi DA, et al. Novel graphene-based biosensor for early detection of Zika virus infection. *Biosens Bioelectron.* 2018;100:85–88.
[112] Navakul K, Warakulwit C, Yenchitsomanus PT, Panya A, Lieberzeit PA, Sangma C. A novel method for dengue virus detection and antibody screening using a graphene-polymer based electrochemical biosensor. *Nanomed Nanotechnol Biol Med.* 2017;13:549–557.
[113] Jung JH, Cheon DS, Liu F, Lee KB, Seo TS. A graphene oxide based immuno-biosensor for pathogen detection. *Angew Chemie.* 2010;49:5708–5711.
[114] Kailashiya J, Singh N, Singh SK, Agrawal V, Dash D. Graphene oxide-based biosensor for detection of platelet-derived microparticles: a potential tool for thrombus risk identification. *Biosens Bioelectron.* 2015;65:274–280.
[115] Thakur B, Zhou G, Chang J, Pu H, Jin B, Sui X, et al. Biosensors and bioelectronics rapid detection of single *E coli* bacteria using a graphene-based field-effect transistor device. *Biosens Bioelectron.* 2018;110:16–22.
[116] Akbari E, Nikoukar A, Buntat Z, Afroozeh A, Zeinalinezhad A. *Escherichia coli* bacteria detection by using graphene-based biosensor. *IET Nanobiotechnol.* 2015;9:273–279.
[117] Sign C, Sumana G. Antibody conjugated graphene nanocomposites for pathogen detection. *J Phys.* 2016;704:7.
[118] Valipour A, Roushani M. Using silver nanoparticle and thiol graphene quantum dots nanocomposite as a substratum to load antibody for detection of hepatitis C virus core antigen: electrochemical oxidation of riboflavin was used as redox probe. *Biosens Bioelectron.* 2017;89:946–951.
[119] Huang J, Xie Z, Xie Z, Luo S, Xie L, Huang L, et al. Silver nanoparticles coated graphene electrochemical sensor for the ultrasensitive analysis of avian influenza virus H7. *Anal Chim Acta.* 2016;913:121–127.
[120] Dharuman V, Hahn JH, Jayakumar K, Teng W. Electrochemically reduced graphene—gold nano particle composite on indium tin oxide for label free immuno sensing of estradiol. *Electrochim Acta.* 2013;114:590–597.
[121] Elshafey R, Siaj M, Tavares AC. Au nanoparticle decorated graphene nanosheets for electrochemical immunosensing of p53 antibodies for cancer prognosis. *Analyst.* 2016;141:2733–2740.
[122] Demeritte T, Viraka Nellore BP, Kanchanapally R, Sinha SS, Pramanik A, Chavva SR, et al. Hybrid graphene oxide based plasmonic-magnetic multifunctional nanoplatform for selective separation and label-free identification of Alzheimer's disease biomarkers. *ACS Appl Mater Interfaces.* 2015;7:13693–13700.

[123] Sharafeldin M, Bishop GW, Bhakta S, El-Sawy A, Suib SL, Rusling JF. Fe3O4 nanoparticles on graphene oxide sheets for isolation and ultrasensitive amperometric detection of cancer biomarker proteins. *Biosens Bioelectron.* 2017;91:359–366.
[124] Gupta S, Kaushal A, Kumar A, Kumar D. Ultrasensitive transglutaminase based nanosensor for early detection of celiac disease in human. *Int J Biol Macromol.* 2017;105:905–911.
[125] Wu Y-M, Cen Y, Huang L-J, Yu R-Q, Chu X. Upconversion fluorescence resonance energy transfer biosensor for sensitive detection of human immunodeficiency virus antibodies in human serum. *Chem Commun.* 2014;50:4759–4762.
[126] Veerapandian M, Hunter R, Neethirajan S. Dual immunosensor based on methylene blue-electroadsorbed graphene oxide for rapid detection of the influenza A virus antigen. *Talanta.* 2016;155:250–257.
[127] Gao J, Guo Z, Su F, Gao L, Pang X, Cao W, et al. Ultrasensitive electrochemical immunoassay for CEA through host—guest interaction of β-cyclodextrin functionalized graphene and Cu@Ag core-shell nanoparticles with adamantine-modified antibody. *Biosens Bioelectron.* 2015;63:465471.
[128] Song Y, Luo Y, Zhu C, Li H, Du D, Lin Y. Recent advances in electrochemical biosensors based on graphene two-dimensional nanomaterials. *Biosens Bioelectron.* 2016;76:195–212.
[129] Premkumar T, Geckeler KE. Graphene–DNA hybrid materials: assembly, applications, and prospects. *Prog Polym Sci.* 2012;37:515–529.
[130] Pedersen R, Marchi AN, Majikes J, Nash JA, Estrich NA, Courson DS, et al. Properties of DNA. In: *Handbook of Nanomaterials Properties*; 2014. p. 1125–1157.
[131] Recent KR. Advances and applications of biosensors in novel technology. *Biosens J.* 2017;6:1–12.
[132] Bora U, Sett A, Singh D. Nucleic acid based biosensors for clinical applications. *Biosens J.* 2013;2:1–8.
[133] Koyun A, Ahlatcolu E, Koca Y. *Biosensors and Their Principles. A Roadmap of Biomedical Engineers and Milestones.* New York: InTech; 2012.
[134] Kavita V. DNA biosensors—a review. *J Bioeng Biomed Sci.* 2017;7:222.
[135] Rahman M, Heng LY, Futra D, Chiang CP, Rashid ZA, Ling TL. A highly sensitive electrochemical DNA biosensor from acrylic-gold nanocomposite for the determination of arowana fish gender. *Nanoscale Res Lett.* 2017;12:484.
[136] Ray M, Ray A, Dash S, Mishra A, Achary KG, Nayak S, et al. Fungal disease detection in plants: traditional assays, novel diagnostic techniques and biosensors. *Biosens Bioelectron.* 2017;87:708–723.
[137] Liu A, Wang K, Weng S, Lei Y, Lin L, Chen W, et al. Development of electrochemical DNA biosensors. *Trends Anal Chem.* 2012;37:101–111.
[138] Zhou M, Zhai Y, Dong S. Electrochemical sensing and biosensing platform based on chemically reduced graphene oxide. *Anal Chem.* 2009;81:5603–5613.
[139] Akhavan O, Ghaderi E, Rahighi R. Toward single-DNA electrochemical biosensing by graphene nanowalls. *ACS Nano.* 2012;6:2904–2916.
[140] Dong H, Zhu Z, Ju H, Yan F. Triplex signal amplification for electrochemical DNA biosensing by coupling probe-gold nanoparticles—graphene modified electrode with enzyme functionalized carbon sphere as tracer. *Biosens Bioelectron.* 2012;33:228–232.
[141] He S, Song B, Li D, Zhu C, Qi W, Wen Y, et al. A graphene nanoprobe for rapid, sensitive, and multicolor fluorescent DNA analysis. *Adv Funct Mater.* 2010;20:453–459.
[142] Qaddare SH, Salimi A. Amplified fluorescent sensing of DNA using luminescent carbon dots and AuNPs/GO as a sensing platform: a novel coupling of FRET and DNA hybridization for homogeneous HIV-1 gene detection at femtomolar level. *Biosens Bioelectronics.* 2017;15(89):773–780.

[143] Guo Y, Guo Y, Dong C. Ultrasensitive and label-free electrochemical DNA biosensor based on water-soluble electroactive dye azophloxine-functionalized graphene nanosheets. *Electrochim Acta*. 2013;113:69–76.
[144] Han X, Fang X, Shi A, Wang J, Zhang Y. An electrochemical DNA biosensor based on gold nanorods decorated graphene oxide sheets for sensing platform. *Anal Biochem*. 2013;443:117–123.
[145] Mihrican M, Surbhi S, Arzum E, Pagona P. Electrochemical monitoring of nucleic acid hybridization by single-use graphene oxide-based sensor. *Electroanalysis*. 2011;23:272–279.
[146] Singh A, Sinsinbar G, Choudhary M, Kumar V, Pasricha R, Verma HN, et al. Graphene oxide-chitosan nanocomposite based electrochemical DNA biosensor for detection of typhoid. *Sens Actuators B*. 2013;185:675–684.
[147] Shen X, Xia X, Du Y, Ye W, Wang C. Amperometric glucose biosensor based on AuPd modified reduced graphene oxide/polyimide film with glucose oxidase. *J Electrochem Soc*. 2017;164:285–291.
[148] Mansouri N, Babadi AA, Bagheri S, Hamid SBA. Immobilization of glucose oxidase on 3D graphene thin film: novel glucose bioanalytical sensing platform. *Int J Hydrogen Energy*. 2017;42:1337–1343.
[149] Kishore P, Dinakar D, Shankar M, Srimannarayana K, et al. Non-contact vibration sensor using bifurcated bundle glass fiber for real time monitoring. *Proc SPIE*. 2011;8311. https://doi.org/10.1117/12.904566.
[150] Yao BC, Wu Y, Chen Y, et al. Graphene-based microfiber gas sensor, *Proc SPIE*. 2012;8421:8421CD-8421CD-4.
[151] Li F, Huang Y, Yang Q, Zhong ZT, Li D, Wang LH, Song SP, Fan CH. A graphene-enhanced molecular beacon for homogeneous DNA detection. *Nanoscale*. 2010;2:1021–1026.
[152] Staiano M, Pennacchio A, Varriale A, Capo A, Majoli A, Capacchione C, D'Auria S. *Enzymes as Sensors. Methods in Enzymology*. New York: Academic Press; 2017.
[153] Staiano M, Pennacchio A, Varriale A, Capo A, Majoli A, Capacchione C, et al. Enzymes as sensors. *Methods Enzymol*. 2017;589:115–131.
[154] Amine A, Mohammadi H, Bourais I, Palleschi G. Enzyme inhibition-based biosensors for food safety and environmental monitoring. *Biosens Bioelectron*. 2006;21:1405–1423.
[155] Zhang J, Zhang F, Yang H, Huang X, Liu H, Zhang J, et al. Graphene oxide as a matrix for enzyme immobilization. *Langmuir*. 2010;26:6083–6085.
[156] Hermanová S, Zarevúcká M, Bouša D, Pumera M, Sofer Z. Graphene oxide immobilized enzymes show high thermal and solvent stability. *Nanoscale*. 2015;7:5852–5858.
[157] Skoronski E, Souza DH, Ely C, Broilo F, Fernandes M, Fúrigo A, et al. Immobilization of laccase from *Aspergillus oryzae* on graphene nanosheets. *Int J Biol Macromol*. 2017;99:121–127.
[158] Gaudin V. Advances in biosensor development for the screening of antibiotic residues in food products of animal origin—a comprehensive review. *Biosens Bioelectronics*. 2017;15(90):363–377.
[159] Patel SK, Choi SH, Kang YC, Lee JK. Eco-friendly composite of Fe_3O_4-reduced graphene oxide particles for efficient enzyme immobilization. *ACS Appl Mater Interfaces*. 2017;9:2213–2222.
[160] Povedano E, Cincotto FH, Parrado C, Díez P, Sánchez A, Canevari TC, et al. Decoration of reduced graphene oxide with rhodium nanoparticles for the design of a sensitive electrochemical enzyme biosensor for 17 β -estradiol. *Biosens Bioelectron*. 2017;89:343–351.
[161] Mei LP, Feng JJ, Wu L, Zhou JY, Chen JR, Wang AJ. Novel phenol biosensor based on laccase immobilized on reduced graphene oxide supported palladium–copper alloyed nanocages. *Biosens Bioelectron*. 2015;74:347–352.

[162] Patel SKS, Anwar MZ, Kumar A, Otari SV, Pagolu RT, Kim S-Y, et al. Fe_2O_3 yolk-shell particle-based laccase biosensor for efficient detection of 2,6-dimethoxyphenol. *Biochem Eng J.* 2018;132:1–8.
[163] Palanisamy S, Ramaraj SK, Chen S-M, Yang TCK, Yi-Fan P, Chen T-W, et al. A novel laccase biosensor based on laccase immobilized graphene–cellulose microfiber composite modified screen-printed carbon electrode for sensitive determination of catechol. *Sci Rep.* 2017;7:41214.
[164] Fan Z, Lin Q, Gong P, Liu B, Wang J, Yang S. A new enzymatic immobilization carrier based on graphene capsule for hydrogen peroxide biosensors. *Electrochim Acta.* 2015;151:186–194.
[165] Chauhan N, Rawal R, Hooda V, Jain U. Electrochemical biosensor with graphene oxide nanoparticles and polypyrrole interface for the detection of bilirubin. *RSC Adv.* 2016;6(68):63624–63633.
[166] Filip J, Andicsová-Eckstein A, Vikartovská A, Tkac J. Immobilization of bilirubin oxidase on graphene oxide flakes with different negative charge density for oxygen reduction the effect of GO charge density on enzyme coverage, electron transfer rate and current density. *Biosens Bioelectron.* 2017;89:384–389.
[167] Suganthi G, Arockiadoss T, Uma TS. ZnS nanoparticles decorated graphene nanoplatelets as immobilisation matrix for glucose biosensor. *Nanosyst Phys Chem Math.* 2016;7:637–642.
[168] You X, Pak JJ. Graphene-based field effect transistor enzymatic glucose biosensor using silk protein for enzyme immobilization and device substrate. *Sens Actuators B.* 2014;202:1357–1365.
[169] Feng X, Cheng H, Pan Y, Zheng H. Development of glucose biosensors based on nanostructured graphene-conducting polyaniline composite. *Biosens Bioelectron.* 2015;70:411–417.
[170] Haghighi N, Hallaj R, Salimi A. Immobilization of glucose oxidase onto a novel platform based on modified TiO2 and graphene oxide, direct electrochemistry, catalytic and photocatalytic activity. *Mater Sci Eng C.* 2017;73:417–424.
[171] Zhang M, Liao C, Mak CH, You P, Mak CL, Yan F. Highly sensitive glucose sensors based on enzyme-modified whole-graphene solution-gated transistors. *Sci Rep.* 2015;5:8311.
[172] Novak MJ, Pattammattel A, Koshmerl B, Puglia M, Williams C, Kumar CV. "Stable-on-the-table" enzymes: engineering the enzyme-graphene oxide interface for unprecedented kinetic stability of the biocatalyst. *ACS Catal.* 2016;6:339–347.
[173] Kang Z, Jiao K, Xu X, Peng R, Jiao S, Hu Z. Graphene oxide-supported carbon nanofiber-like network derived from polyaniline: a novel composite for enhanced glucose oxidase bioelectrode performance. *Biosens Bioelectron.* 2017;96:367–372.
[174] Pimenta MA, Dresselhaus G, Dresselhaus MS, Cancado LG, Jorio A, et al. Studying disorder in graphite-based systems by Raman spectroscopy. *Phys Chem Chem Phys.* 2007;9(11):1276–1290. https://doi.org/10.1039/B613962K
[175] Cammann K. Biosensors based on ion-selective electrodes. *Fresenius' Z Anal Chem.* 1977;287(1):1–9.
[176] Thevenot DR, Toth K, Durst RA, Wilson GS. Electrochemical biosensors: recommended definitions and classification. *Pure Appl Chem.* 1999;71:2333–2348.
[177] Thevenot DR, Toth K, Durst RA, Wilson GS. Electrochemical biosensors: recommended definitions and classification. *Biosens Bioelectron.* 2001;16:121–131.
[178] Thevenot DR, Toth K, Durst RA, Wilson GS. Electrochemical biosensors: recommended definitions and classification. *Anal Lett.* 2001;34:635–659.
[179] Sharma P, Pandey V, Sharma MMM, Patra A, Singh B, Mehta S, Husen A. A review on biosensors and nanosensors application in agroecosystems. *Nanoscale Res Lett.* 2021;16(1):1–24.

[180] Suvarnaphaet P, Pechprasarn S. Graphene-based materials for biosensors: a review. *Sensors*. 2017;17(10):2161.

[181] Singh S, Kumar V, Dhanjal DS, Datta S, Prasad R, Singh J. Biological biosensors for monitoring and diagnosis. In: *Microbial Biotechnology: Basic Research and Applications*. Singapore: Springer; 2020. p. 317–335.

[182] Thangamuthu M, Hsieh KY, Kumar PV, Chen GY. Graphene- and graphene oxide-based nanocomposite platforms for electrochemical biosensing applications. *Int J Mol Sci*. 2019;20(12):2975.

[183] Krishnan SK, Singh E, Singh P, Meyyappan M, Nalwa HS. A review on graphene-based nanocomposites for electrochemical and fluorescent biosensors. *RSC Adv*. 2019;9(16):8778–8881.

[184] Venugopal V. Biosensors in fish production and quality control. *Biosens Bioelectron*. 2002;17:147–157.

[185] Diviès C. Remarques sur l'oxydation de l'éthanol par une electrode microbienne d'acetobacter zylinum. *Ann Microbiol*. 1975;126A:175–186.

[186] Rechnitz GA. Biochemical electrodes uses tissues slice. *Chem Eng News*. 1978;56:16–21.

[187] Scognamiglio V, Arduini F, Palleschi G, Rea G. Biosensing technology for sustainable food safety. *Trends Anal Chem*. 2014;62:1–10.

[188] Leatherbarrow RJ, Edwards PR. Analysis of molecular recognition using optical biosensors. *Curr Opin Chem Biol*. 1999;3:544–547.

[189] Massad-Ivanir N, Shtenberg G, Raz N, Gazenbeek C, Budding D, Bos MP, Segal E. Porous silicon-based biosensors: towards real-time optical detection of target bacteria in the food industry. *Sci Rep*. 2016;6(1):1–12.

[190] Lippincott-Schwartz J, Patterson GH. Development and use of fluorescent protein markers in living cells. *Science*. 2003;300:87–91.

[191] Shaner NC, Steinbach PA, Tsien RY. A guide to choosing fluorescent proteins. *Nat Methods*. 2005;2:905–909.

[192] Tsien RY. Breeding and building molecules to spy on cells and tumors. *FEBS Lett*. 2005;579:927–932.

[193] Giepmans BN, Adams SR, Ellisman MH, Tsien RY. The fluorescent toolbox for assessing protein location and function. *Science*. 2006;312:217–224.

[194] Ibraheem A, Campbell RE. Designs and applications of fluorescent protein-based biosensors. *Curr Opin Chem Biol*. 2010;14:30–36.

[195] Wang T, Ramnarayanan A, Cheng H. Real time analysis of bioanalytes in healthcare, food, zoology and botany. *Sensors*. 2018;18(1):5.

[196] Aye-Han NN, Qiang N, Zhang J. Fluorescent biosensors for real-time tracking of post-translational modification dynamics. *Curr Opin Chem Biol*. 2009;13:392–397.

[197] Scognamiglio V, Arduini F, Palleschi G, Rea G. Bio sensing technology for sustainable food safety. *Trends Anal Chem*. 2014;62:1–10.

[198] Ghasemi-Varnamkhasti M, Rodriguez-Mendez ML, Mohtasebi SS, et al. Monitoring the aging of beers using a bioelectronic tongue. *Food Control*. 2012;25:216–224.

[199] Arora P, Sindhu A, Dilbaghi N, Chaudhury A. Biosensors as innovative tools for the detection of food borne pathogens. *Biosens Bioelectron*. 2011;28:1–12.

[200] Kim ER et al. Biosensors for healthcare: current and future perspectives. *Trends Biotechnol*. 2023;41(3):374–395.

[201] Prével C, Kurzawa L, Van TNN, Morris MC. Fluorescent biosensors for drug discovery new tools for old targets—screening for inhibitors of cyclin-dependent kinases. *Eur J Med Chem*. 2014;88:74–88.

[202] Torun O, Boyaci I, Temur E, Tamer U. Comparison of sensing strategies in SPR biosensor for rapid and sensitive enumeration of bacteria. *Biosens Bioelectron*. 2012;37:53–60.

[203] Mishra R, Dominguez R, Bhand S, Munoz R, Marty J. A novel automated flow-based biosensor for the determination of organophosphate pesticides in milk. *Biosens Bioelectron*. 2012;32:56–61.
[204] Yan C, Dong F, Chun-yuan B, Si-rong Z, Jian-guo S. Recent progress of commercially available biosensors in China and their applications in fermentation processes. *J Northeast Agric Univ*. 2014;21:73–85.
[205] German N, Ramanaviciene A, Voronovic J, Ramanavicius A. Glucose biosensor based on graphite electrodes modified with glucose oxidase and colloidal gold nanoparticles. *Mikrochim Acta*. 2010;168:221–229.
[206] Chen QH, Yang Y, He HL, et al. The effect of glutamine therapy on outcomes in critically ill patients: a meta-analysis of randomized controlled trials. *Crit Care*. 2014;18:R8.
[207] Backer D, Rakowski M, Poghossiana A, Biselli M, Wagner P, Schoning MJ. Chip-based amperometric enzyme sensor system for monitoring of bioprocesses by flow-injection analysis. *J Biotechnol*. 2013;163:371–376.
[208] Amaro F, Turkewitz AP, Martin-Gonzalez A, Gutierrez JC. Whole-cell biosensors for detection of heavy metal ions in environmental samples based on metallothionein promoters from Tetrahymena thermophila. *Microb Biotechnol*. 2011;4:513–522.
[209] Arduini F, Ricci F, Tuta CS, Moscone D, Amine A, Palleschi G. Detection of carbamic and organophosphorus pesticides in water samples using cholinesterase biosensor based on Prussian blue modified screen printed electrode. *Anal Chim Acta*. 2006;58:155–162.
[210] Ivanov I, Younusov RR, Evtugyn GA, Arduini F, Moscone D, Palleschi G. Cholinesterase sensors based on screen-printed electrodes for detection of organophosphorus and carbamic pesticides. *Anal Bioanal Chem*. 2003;377:624–631.
[211] Suprun E, Evtugyn G, Budnikov H, Ricci F, Moscone D, Palleschi G. Acetylcholinesterase sensor based on screen-printed carbon electrode modified with Prussian blue. *Anal Bioanal Chem*. 2005;383:597–604.
[212] Diesel E, Schreiber M, van der Meer JR. Development of bacteria-based bioassays for arsenic detection in natural waters. *Anal Bioanal Chem*. 2009;394.
[213] Scognamiglio V, Pezzotti G, Pezzotti I, et al. Biosensors for effective environmental and agrifood protection and commercialization: from research to market. *Mikrochim Acta*. 2010;170:215–225.
[214] Rea G, Polticelli F, Antonacci A, et al. Structure-based design of novel *Chlamydomonas reinhardtii* D1–D2 photosynthetic proteins for herbicide monitoring. *Protein Sci*. 2009;18:2139–2151.
[215] Lee M, Zine N, Baraket A, et al. A novel biosensor based on hafnium oxide: application for early-stage detection of human interleukin-10. *Sens Actuators B*. 2012;175:201–207.
[216] Chen YW, Liu M, Kaneko T, McIntyre PC. Atomic layer deposited hafnium oxide gate dielectrics for charge-based biosensors. *Electrochem Solid State Lett*. 2010;13:G29–G32.
[217] Caruso R, Trunfio S, Milazzo F, et al. Early expression of pro- and anti-inflammatory cytokines in left ventricular assist device recipients with multiple organ failure syndrome. *Am Soc Art Int Org J*. 2010;56:313–318.
[218] Caruso R, Verde A, Cabiati M, et al. Association of pre-operative interleukin-6 levels with interagency registry for mechanically assisted circulatory support profiles and intensive care unit stay in left ventricular assist device patients. *J Heart Lung Transplant*. 2012;31(6):625–633.
[219] Watson CJ, Ledwidge MT, Phelan D, et al. Proteomic analysis of coronary sinus serum reveals leucine-rich 2-glycoprotein as a novel biomarker of ventricular dysfunction and heart failure. *Circ Heart Fail*. 2011;4:188–197.
[220] Maurer M, Burri S, de Marchi S, et al. Plasma homocysteine and cardiovascular risk in heart failure with and without cardiorenal syndrome. *Int J Cardiol*. 2010;141:32–38.

[221] Wang H, Nakata E, Hamachi I. Recent progress in strategies for the creation of protein-based fluorescent biosensors. *Chem BioChem*. 2009;10:2560–2577.
[222] Morris MC. Fluorescent biosensors of intracellular targets from genetically encoded reporters to modular polypeptide probes. *Cell Biochem Biophys*. 2010;56:19–37.

3 An Effective Study on Graphene-Based Biosensors and Their Electrochemical Working Principles

Chang-Min Yoon and Won-Chun Oh

3.1 INTRODUCTION

To investigate major symptoms, diagnose patients, and improve critical patient care, sensors are used in medicine and life sciences [1–10]. Many novel sensors have been developed to detect diseases in the early stage, and minimally invasive detection methods have been used. To synthesize nanosensors, there is a need to focus on miniaturization by using nanomaterials. Biosensors require minimal hygroscopicity and extreme sensitivity. The nanoscale of hybrid nanomaterials and their unique physical, chemical, and electrochemical properties can improve the treatment range of patients [10]. The excellent effects and detection limits of biosensors can affect positive and negative values. In general, many published studies describe a linear range of biosensors that can identify the detection limits of sensors. However, many new active material materials have been proposed, and detailed working principles or statistics on the precision, accuracy, and positive and negative prediction values of these parameters have not been described. In future studies, it is necessary to clearly present these parameters.

Graphene or graphene-based nanomaterials are the most valuable due to their good signal response in terms of sensing applications among nanomaterials used for the fabrication of nanosensors [11–13]. In addition, graphene-based hybrid nanomaterials have offered excellent biocompatibility and a large surface area with a various kind of biomolecules such as DNA, cells, antibodies, enzymes, and proteins [13]. The high detection of biomolecules in graphene or hybrid graphene-based nanomaterials (Figure 3.1) has enabled the development of advanced biosensors. These biosensors can detect different kinds of biomolecules and even various cells [14, 15]

Human serum albumin (HSA) is a major experimental protein in blood flow analysis. It adjusts the osmotic pressure required for fluid distribution between body tissues and intravascular compartments and is manufactured and secreted by the liver for this purpose. A healthy person releases a certain amount of albumin (up to

DOI: 10.1201/9781003425427-4

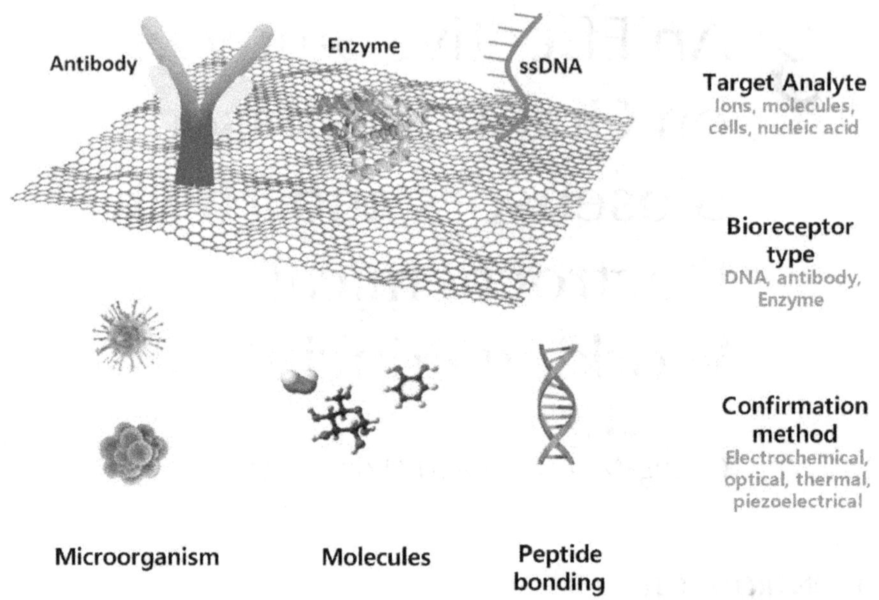

FIGURE 3.1 Biological reactivity and designs of biosensors with graphene-based platforms.

30 mg a day) into the urine [1]. In the case of antagonistic conditions, renal excretion of albumin may be increased, leading to elevated urinary albumin levels. Mini albuminuria refers to urinary albumin excretion of 30–300 mg/day from 24-hour urine collection or 30–300 mg/l from random or morning puncture collection [2]. For renal and cardiovascular infections, albumin in urine is an important clinical and medical marker [3]. Diabetic nephropathy and terminal renal disease are the main symptoms of 20–40% of diabetic patients. In the convergence of albumin in urine, semiquantitative or quantitative immunochemical strategies such as immunoturbidimetry (IT), immunonephelometry (IN), protein-linked immunosorbent assay (ELISA), radioimmunoassay (RIA), chemiluminescence immunoassay (CLIA), and fluorescent immunoassay (FIA) have been clinically estimated [4–9]. Between these techniques, RIA and ELISA are the most effective and require fewer antisera, but some of the reagents used in these two strategies can cause harm to human beings. IT is simple and fast but requires an enormous number of antisera. CLIA has effective potential but requires some expensive chemicals. In addition, this method requires long observation, developed machines and chemicals, many major steps, and skilled personnel. Biosensors could be an ideal choice for analyte determination in organic samples when skilled investigators and machines are limited [10–15].

One way to quickly test the amount of albumin and determine suitability is that biosensors can be considered attractive and easily used by the general public in the hospital setting. To date, several biosensors have been proposed to detect albumin in urine [16–18, 15]. As part of these studies, we propose nanobiosensors based on laser-induced surface plasmon reflection (LSPR) for microalbumin localization. To distinguish urine albumin fixation in the range of 1 ng/ml to 1 mg/ml to achieve fine sensitivity, researchers built a sensor that detects albumin in urine and tests the immune system by transferring human albumin neutralizers to silver nanoparticles [15] and screen printing modified with colloidal gold nanoparticles (AuNPs) and polyvinyl alkali (PVA). An independent and practical SPR biosensor is an easy analytical method for identifying HSA in urine by detecting albumin in the range of 2.5–200 mg/ml using a voltmeter and square wave voltmeter to measure differential heart beats. In one study, monoclonal immunity relative to HSA was immobilized in the gold detection region, and detection of HSA was achieved in the range of 1–100 mg/ml [16–19].

The active material nanographene has a two-dimensional layer of a one-molecule-thick carbon hexagonal ring composed of a hexagonal system with a large explicit part, exhibits excellent heat transfer and electrical conductivity, and thus has been used in high electron transfer rate, excellent electrocatalytic motion, and friendly bio-detection applications. In addition, due to the usefulness and high capacity of semiconductors, graphene-based sensors may have a higher influence than conventional electrochemical evaluation technologies [20–25]. As a sensing tool, a sensor is generally based on the connection between the material to be analyzed and the active material channel [26, 27, 29]. Due to their high electrical catalytic motion, excellent electrical conductivity, and abundant and inexpensive costs, many metal-based mixtures have recently been used as suitable candidates for glucose detection [30]. Metal-based sensors have an adjustable redox dynamic reaction site of N-trimethylchitosan (TMC), thereby improving electrochemical conductivity. In addition, metal-based active materials exhibit high electrical conductivity due to their thin band gap and additional covalency [31]. Among some TMC species, Ni-based nanomaterials form two redox pairs of Ni^{2+}/Ni^{3+}, which are widely used in basic electrolytes due to their attractive electrocatalytic activity, low toxicity, and low cost [32]. In this spinel structure, the Ni particles are periodically in a more chemically dynamic Ni^{+3} oxidation state than Ni^{2+}, improving the unique synergy effect [33]. Ni cations have exposed targets, and cations are suitable for a fully filled layer. Some chemically dynamic sites provide different paths for charge transport that may be introduced into the electrolyte [34]. Unlike the pulse structure, another factor that improves electrical catalyst efficiency is increasing the electrical conductivity of the reaction composite. In this regard, various carbon-based additive agents have also been used. High sensitivity is required in all biosensors, and their localization capabilities are not easily able to cover the entire range of microalbuminuria (30–300 mg/mL) without complex sample attenuation. Thus, in addition to increasing the possible sensitivity, extending the detection range to the full range of microalbuminuria without test attenuation is a key problem in the development of microalbuminuria biosensors. As an example of another study, a BiZnSbV-G-SiO_2@NF cathode was used to improve albumin detection. Here, we present a simple and sensitive BiZnSbV-G-SiO_2@NF-based electrochemical sensor for identifying urine albumin associated with microalbuminuria. BiZnSbV-G-SiO_2 was effectively mixed with graphene fixed on a Ni foam substrate (BiZnSbV-G-SiO_2@NF) by one-step electrodeposition on the composite cathode, and microalbumin in urine was screened. The produced synergistic composite had a direct range of 0.01 to 0.1 μmol/L for microalbumin identification, which was a very low working capacity of 0.20 V against Ag/AgCl. Their detection limit was low, such as 0.01 μmol. It showed particularly high characteristic evaluation performance in glucose oxidation. The electrochemical sensing performance of the BiZnSbV-G-SiO_2@NF cathode for microalbumin detection is a characteristic method studied using an amperemetric method.

On the other hand, the long persistence of antibiotics despite low concentrations has raised concerns about antibiotic resistance [16–21, 33]. Infections caused by bacteria such as streptococcus, pneumonia, typhoid, and bronchitis can be treated or prevented with the new macrolide antibiotic azithromycin [34, 35]. In the human body, azithromycin is excreted in a molecular sieve that hardly changes in urine. Overuse of antibiotics can cause problems with anti-bacterial resistance, excessive allergic reactions, organ damage, tooth discoloration, and gastrointestinal disorders. This affects genetic factors and is transmitted to offspring, increasing the incidence and increasing the risk of cancers such as malignant tumors [36]. Liquid chromatography-mass spectroscopy (LC-MS), spectroscopic absorption photometry, HPLC, atomic absorption spectroscopy, electrochemical emission, electrophoresis, receptor measurement, and electrochemical measurement are useful methods of analysis described so far for the detection of antibiotics and drugs [37]. Although this analysis technology has sufficient accuracy and sensitivity to identify drug molecules in the drug analysis process, it has several disadvantages. For example, these analysis methods take a long time to process and analyze samples. In addition, materials for sample preparation are very expensive. Also, a large number of samples are needed for one analysis. The amount of analyte (about ng/L) is an important factor for analysis, and if the amount of material for analysis is small, pre-concentration and purification are required prior to each analysis. Sensor analysis by electrochemical analysis has high sensitivity and can be manufactured quickly and easily, so it is inexpensive. In addition, interest is increasing because of its excellent mobility and simplicity in the sensing field [38]. Unlike various methods, the electrochemical analysis method has recently attracted a lot of attention by combining simple analytical chemistry and bioengineering analysis methods. Selectivity and reproducibility are important in analysis, so if these characteristics, speed, and convenience are combined, good analysis results can be derived. Electrodes coated with $MgAl_2O_4$, which is a sensor active material, have many advantages, such as being easy to manufacture and forming a thin film on the surface of the electrode. The surface stability of $MgAl_2O_4$ deposition on SiO_2 substrates is one of the important properties in terms of the geometry of thin film formation [11–13]. For better catalytic activity, catalyst and electrochemical sensing capabilities may be improved by using a dopant [30] in the active material. Since graphene is a two-dimensional layer to be used as an active material, its effects may be maximized by using the characteristics of the surface and the atomic carbon arrangement. Many sensor systems require the advantages of unique properties such as strong electron conductivity, high charge carrier mobility, and high specific surface areas. Graphene sensors satisfy these requirements and may sensitively detect environmental changes. Graphene-based biotechnology is ideal for sensitive diagnostic and immunological applications [14–20].

SiO_2-based nanocomposites can provide hydrophilic surfaces, which provide excellent adhesion properties to biomolecules because they have high conductivity, unique catalytic activity, and excellent biocompatibility. Mesoporous materials can be used to detect analytes with high precision, including rationally designed and highly detectable enzymatic and non-enzymatic biosensors. Mesoporous active materials with high surface area have many active chemical reaction sites and many interconnected pore structures, and these properties can provide strong biosensing and biocatalytic performance for non-enzyme biosensors. Due to the large surface area, characterized pore structure and size, and excellent catalyst characteristics and good stability, the use of a non-enabling biosensor shows excellent sensing characteristics, and demand for such characteristics is increasing day by day [21–23]. In addition to in the blood, bacteria can be found everywhere in the presence of water, including food, and are also found everywhere in nature. Even a small number of various bacteria in nature can seriously affect public health, their severity is increasing day by day, and new species such as pathogenic bacteria are emerging. For example, *E. coli* O157 shows the following characteristics. In particular, H7 can cause serious diseases in the human body when infected, such as hemolytic uremic syndrome and hemorrhagic inflammation in the large intestine. In particular, *E. coli* O157:H7 is the most dangerous type of pathogen for patients with weak immunity [36]. As a conventional method of detecting pathogenic bacteria, the most familiar methods are culture technology and polymerase chain reaction (PCR). Some expensive equipments require skilled personnel to operate for time constraints and detection [37]. Therefore, there is a need to develop new methods that can detect *E. coli* O157:H7 more sensitively and accurately while requiring fewer resources [38]. Electrochemical biosensing has both simple theoretical electrochemical practices and high sensitivity due to high chemical affinity, making it an important topic in analytical technology research when approached from a biological and medical perspective [39]. $MgAl_2O_4$, an active material of a manufactured electrode, has many advantages, considering the simplification of the electrode, the design of tunability, and the thin film coating on the electrode surface in the manufacturing process. The active material is manufactured in a thin film form, and in particular, the formation of a thin film of $MgAl_2O_4$ requires strong durability against a Si-SiO_2 substrate and maintains stable adhesive force in the long term. As mentioned, doping can significantly improve catalyst and electrochemical efficiency and may be effective in developing better active materials for sensors. Graphene, which is an effective carbon-based nanomaterial due to its unique chemical behavior, has high carrier mobility. In addition, the zero-band gap has excellent transistor characteristics. Mesoporous-based materials, including graphene, can be reasonably used to manufacture active materials for enzyme and non-enzyme biosensors, and they show excellent sensing performance for various analytes in fields of selectivity and reproducibility. For non-enzyme biosensors, reasonable use of mesoporous materials with interconnected pore structures can produce promising biosensor technologies with strong sensing and catalytic performance. In the design of non-hydroelectric biosensors, it is highly desirable to have an orderly pore structure, a high specific surface area, numerous biocatalytic chemical active sites, and excellent stability [38–42]. Commercially required sensors have major characteristics, which must form a solid and uniform thin film, and have been manufactured by the facility's own assembly to satisfy this. For example, Ca-attached $MgAl_2O_3$-G-SiO_2 (CMAGS) satisfies these characteristics and shows high selectivity for target analytes and *E. coli.* In particular, the proposed method for reducing viruses such as O157:H7 should also be useful for measuring other bacteria.

Diabetes is also emerging as a major challenge worldwide and could offer a breakthrough alternative to treating non-communicable complications and diseases such as stroke, cardiovascular disease, and persistent renal failure [43]. The measurement of microalbumin or glucose expands treatment options and can suggest significant improvements in handling methods and strategies for preventive measures. Simply controlling the amount of glucose or measuring the amount of glucose in the urine can be a suitable marker for effectively detecting and removing excess glucose in the body due to endocrine and hyperglycemia problems [40–45]. Recent methods of identifying non-enzyme glucose suggest that these improved methods can overcome the limitations of traditional chemical glucose sensors. Conventional chemical glucose sensors have many disadvantages due to their low reproducibility, mucosal decomposition, pH control, proteolysis, and high production costs [46]. According to Fletcher [47–50], the non-enzymatic glucose sensor mechanism is involved in the chemical adsorption of hydroxyl bundles on the kinetic momentum of metals. In these processes, the bond between the d-transition metal and atoms in biomolecules can be developed. In the glucose-metal bond, glucose particles on the surface of the metal affect the rate of desorption of glucose atoms reduced due to different oxidation states, indicating different metal-glucose links [51–58]. Due to simple manufacturing processes, excellent mechanical strength, and excellent sensing properties by synergy, some metal, including metal oxides such as Au, Cu, Pt, and Ni [59–62], exhibits excellent detection effects by direct electrochemical electron transfer of polysaccharides such as glucose. Among these metals or metal oxides, ZrO_2-based active material has been evaluated as having the best pulse. ZrO_2, which corresponds to an n-type semiconductor with a large band gap of 5.12 eV, is a metal oxide with excellent electron transfer effects by the redox layer [63]. Thus, combining ZrO_2 with p-type semiconductors such as graphene oxide or nanocarbon is a promising way to solve sensing tests, especially by adjusting the synergistic kinetic properties. Materials that combine metal oxides using nanocarbon-based materials such as graphene have applicability in various test fields, and their excellent performance has already been proven. Also, graphene oxide is rich in oxygen and is stably located on the surface of metal nanoparticles, which serves to improve the electrochemical properties of metal particles, thereby preventing aggregation of media. However, despite the excellent electron transfer of high concentrations of reactants, activity under acid inhibition conditions is not useful in most physiological examples. Nanocarbon has a high electron donating ability, a characteristic of improving the reaction effect with carbon atoms by improving a uniform charge distribution [64], and a large surface area that can be used as an adsorbent in the catalytic field [65]. Since a large specific surface area lowers the resistance of ZrO_2, which is a metal oxide, by high reactant motion, and a high surface area of mesoporous SiO_2 increases the adsorption rate, it is determined that the characteristics of the active material are increased by combining with nanocarbon such as graphene [50–55, 66]. Ascorbic acid (AA) is used in several applications to promote wound healing [67] and to prevent or treat various conditions such as burns [68], colds [69], and hyperemia [70]. Therefore, detection strategies that allow rapid, unobtrusive, and stable backup of AA are important in various fields, especially in the pharmaceutical and food industries.

3.2 SIMPLE STUDY OF METHODOLOGY AND SAMPLE PREPARATION

Graphite (20 g) can be used with sulfuric acid (450 mL) and purified water, and an amount of sonic (using VCX 750 sonic indicator) can be washed several times with hot water for 2 h and dried in a dry oven for 6 h to obtain graphene oxide powder [48]. When manufacturing a hybrid nanocomposite active material, a BiV lead was made on the graphene surface using self-assembly technology. Mesoporous BiZnSbV was synthesized by a self-assembly method for porous addition [36]. Water was added to the mixture (H_2O/metal molar ratio 20), and the precipitation was prevented by aging through an acidic reaction to induce an additional reaction, and 0.333 g of graphene oxide (GO) was added to 250 ml of water to perform ultrasonic treatment for several tens of minutes. Ultrasonic treatment plays a useful role in forming a graphene oxide-based BiZnSbV compound. The prepared composite may be mixed at 100°C for 2 h to obtain a BiZnSbV-G composite complex compounded with BiZnSbV [44–47]. Next, in order to complex porous SiO_2 using BiZnSbV-G as a base material, the mixture was mixed with 15 mL of deionized water containing 2 M HCl until complete decomposition at 40°C, and 1.1 g of Pluronic F-127 triblock copolymer was coagulated. The composite was put in a fixture and heated in an oven at 100°C for 20 h. At this time, a filtration process was required using a sieve, washed several times using water and ethanol, and dried quickly for a short time at 65°C. The BiZnSbV-G hybrid nanocomposite was dropped into a flask containing 0.3 g of silica powder, mixed at 100°C for 24 h, and then slowly subjected to ultrasonic treatment for 1.5 h. The precipitate was filtered through a sieve, washed several times with 1.5 ml of methanol, and dried at 65°C for a short time. For electrochemical evaluation by electrochemical equipment, BZSV, BZSVG, and BZSVGS were used as working anodes, respectively, and Pt wire and Ag/AgCl anodes were selected and used as counter and reference electrodes, respectively. As the working anode, nickel foam was used as a current collector, and the previous nanocomposite active material was coated and used as an electrode [46–48].

Prior to the electrochemical statement, Ni foam (current collector) was cut into 10 × 30–mm pieces, washed using 3 M HCl, DI water, and total ethanol, and then ultrasonic treatment was carried out in stages. The prepared anode may exchange particles with deionized water for 1 h and perform washing and drying processes in a vacuum dry oven at 60°C for 10 h. The weight of BiZnSbV-G-SiO_2 coated on the Ni foam is preferably about 0.8 mg [48]. As another example for synthesizing an active material, a $MgAl_2O_3$ (MACO) composite may be added to a graphene solution by mixing cobalt acetate tetrahydrate $Co(CH_3CO_2)_2$ and $4H_2O$, magnesium acetate $Mg(CH_3COO)_2 \cdot 4H_2O$, aluminum isopropoxide

($C_9H_{21}O_3Al$), and 4.16 g of a Pluronic F127 block copolymer with an ethanol solution. When 3.20 g of tetraethyl orthosilicate (TEOS) is added to it, stirred at 40°C for 20 h, washed with ethanol and water, and dried at 65°C, the $MgAl_2O_3$-G-SiO_2 (MACGS) composite can finally be synthesized [49]. The working electrode was coated using MACGS (0.8 mg) doped with Ca in the previous method as an active material. Ethyl cellulose (EC) was mixed with MAC, MACG, and MACGS to form a paste and pulverized in gate mortar. All samples were coated to a thickness of 5 mm [49]. To evaluate high-temperature strength, unformatted samples were fired at internal temperatures from ambient to 600°C and held below 600°C for 4 h [50, 51].

There are several nanocarbon composite (NC) synthesis methods, but the first direct mixing method is most commonly used [59]. Simply put, the variable ratios GO, PPy, PANI, and ZO of the NC components additionally use ultrasonic dispersion at appropriate temperature, reaction conditions, and time (1 M sulfuric acid for 2 h at 25°C), sometimes a low temperature and long time. The prepared colloidal suspension is centrifuged under the conditions of time and rotational speed to remove aggregates. The prepared slurry is washed several times with deionized (DI) water and dried in a vacuum oven. The final product (dried state) is stored at room temperature (RT) for electrode manufacturing. The electrodes were manufactured by directly coating the electrode GCE and washing and polishing with an alumina kit before modification. The prepared NC is dropped to a concentration of 1 mg/ml and dried at RT. Likewise, enzyme immobilization is performed by a physical adsorption method. It simply used 0.5 mg/ml cholesterol oxidase and bilirubin oxidase combined with peroxidase in 7.4% phosphate-buffered saline (PBS). The enzyme is dropped on a surface-modified electrode with Nafion, dried, and stored at 4°C. High-quality single-layer graphene (SLG) synthesis is typically manufactured using low-pressure chemical vapor deposition (LP-CVD) systems [37]. SLG prepared on Cu foil is coated with PMMA/acetone with an average molecular weight (AMW) of less than 15,000 (by GPC) by spin coating. Since PMMA/acetone has a low AMW, the poly-methylmetacrylate (PMMA) solution has a low doping amounts, and high carrier mobility is improved. It also provides a clean surface with little residue [38]. The dried sample is immersed in a diluted [$Fe(NO_3)_{39}H_2O$] solution to etch the Cu foil. PMMA/SLG film can be obtained from these processes. The film is washed with deionized water (DIW), and the $Fe(NO_3)_3$ PMMA/SLG plate is washed in a dilute hydrochloric acid solution. It is then transferred to an indium oxide (ITO) substrate. The ITO substrate is immersed in ethanol, isopropanol, and DI water for about 13 minutes to remove residues, dust, and contaminants from the surface layer and then transferred [39, 40]. The generated PMMA/SLG@ITO stack layer is dried. The step of removing the PMMA layer from the SLG surface after heating in an oven at 80°C for 15 minutes is essential. SLG@ITO film can be successfully obtained by immersing it in isopropyl alcohol and acetone. Scheme 1 shows a schematic diagram of sensor electrodes using Ab-N, S-GQDs@AuNP-PNP-PNI nanocomposites and HEV pulse-induced impedance sensing using them. Scheme 3.1 shows the simple preparation procedure of polymer nanocomposite loaded sensor electrode.

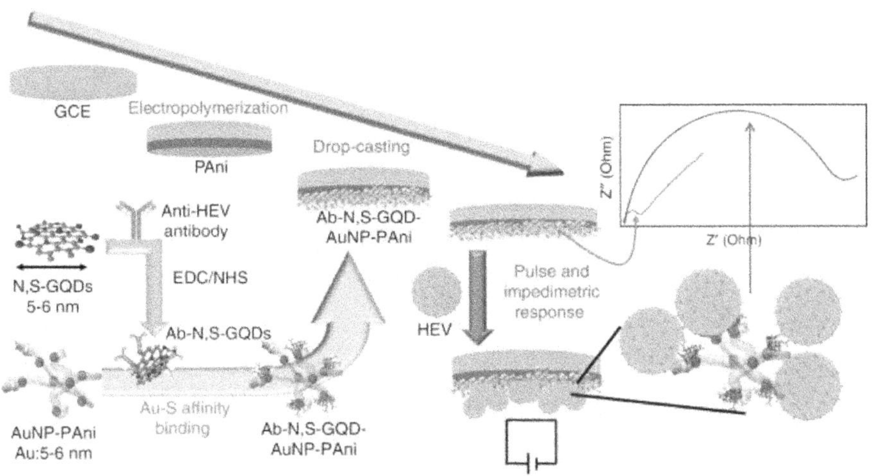

SCHEME 3.1 Schematic diagram of the Ab-N,S-GQDs@AuNP-PAni nanocomposite-loaded sensor electrode and its pulse-induced impedimetric sensing of HEV [60].

3.3 CHARACTERIZATION OF ACTIVE MATERIALS FOR SENSORS

3.3.1 Basic Analyses of Samples

Figure 3.2(a) shows an example XRD of BZSV, BZSVG, and BZSVGS. The peaks are deposited at (101), (012), (104), (110), (020), (202), (024), (208), and (220) tetragonal BiZnSbV-G-SiO_2. In comparison with the staged crystal structure BZSVGS [40–42], after modification with G, SiO_2, and hybrid BiZnSbV-G-SiO_2 nanocomposite, all diffraction peaks are associated with BZSVG and BZSVGS, confirming the glassy appearance of the material. Mixed forms of BZSV, BZSVG, and BZSVGS were investigated. XRD patterns [Figure 3.2(b)] showed evidence for organized meso-structural evolution via MAC, MACG, and MACGS. In the MACGS phase, MACGS, (002), (012), (009), (015), (018), (110), (101), (111), and (103) peaks were detected. Compared with the XRD patterns of MAC, MACG, and MACGS, the (015), (110), and (103) indexed diffraction patterns had significantly increased mesoscopic order, as indicated by the greater intensity of the diffraction peaks. With increasing d_{100} distance, Ca species were incorporated into the mesoporous framework of MACGS. Similar results were observed when additional heteroatoms were added to the 2D hexagonal mesostructured material [45]. As examples of several composite active materials, we investigated the crystal structure and development forms of the newly synthesized MAC, MACG, and MACGS. It may be seen that the active materials of CMA, CMAG, and CMAGS formed in an orderly crystal structure, and conclusive evidence thereof [Figure 3.2(c)] was presented. The results of analyzing the index of peaks for these samples were identified as (002), (012), (009), (015), (018), (110, (101), (111), and (103) planes, and the presence of $MgAl_2O_3$ uniformly doped with Ca was confirmed in these samples. In particular, the presence of graphene and silicon

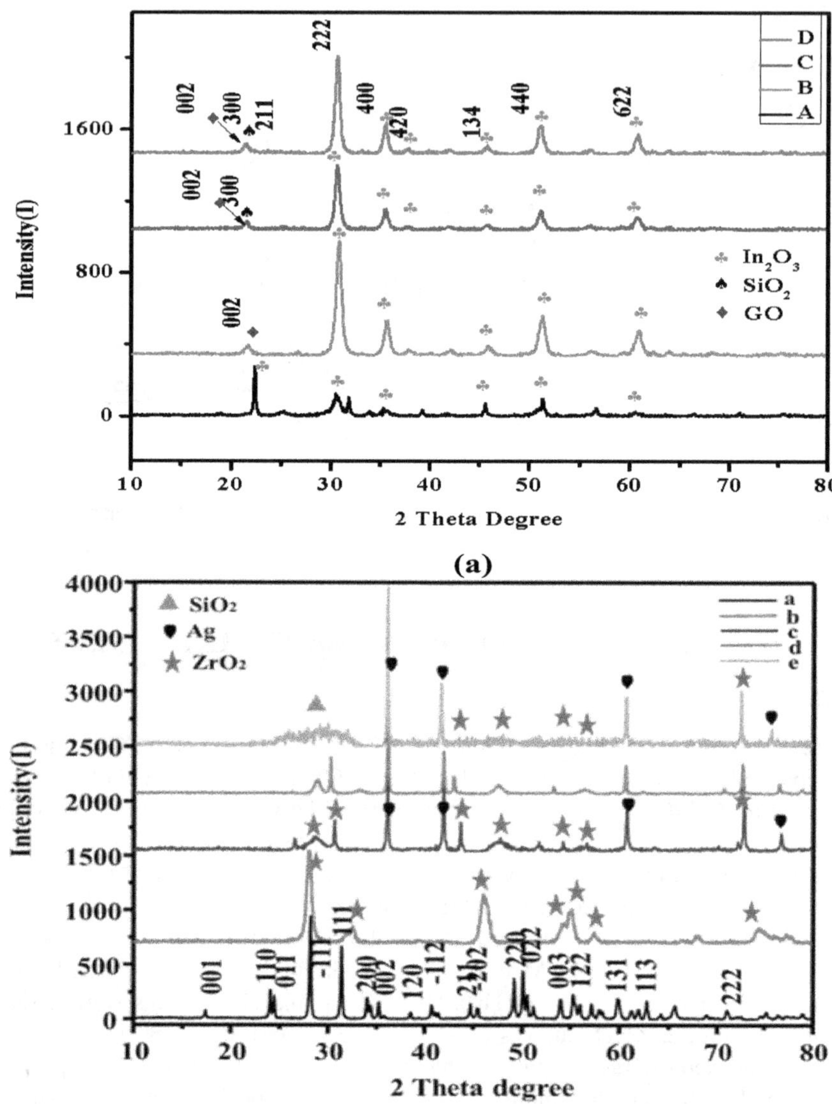

FIGURE 3.2 (a) XRD patterns of In_2O_3 (A), In_2O_3–G (B), In_2O_3-G-SiO_2-10% (C), and In_2O_3-G-SiO_2-20% (D), and commercial ZrO_2 **a**, synthesized ZrO_2 **b**, ZrO_2–Ag **c**, ZrO_2-Ag-G **d**, and ZrO_2-Ag-G-SiO_2 **e** [48].

oxide with a 2-dimensional hexagonal meso structure [24] by CMAGS was confirmed. Comparing the XRD data of CMA, CMAG, and CMAGS, (015), (110), and (103) diffraction peaks are relatively strong, showing a clear calcium doping effect. It was finally confirmed that the Ca species were doped into the mesoporous skeleton of the Ca-doped $MgAl_2O_3$-G-SiO_2. Other heteroatoms form regular 2D hexagonal meso structures, and similar phenomena were observed when these mesoporous materials were bonded to graphene [24]. After reforming the active material with G and SiO_2, $MgAl_2O_3$-G-SiO_2 secondarily doped with Ca could show all XRD peaks corresponding to CMAG and CMAGS, and it can be confirmed from these results that successful crystals of the material were formed. Next, we investigated the forms of synthesized CMA, CMAG, and CMAGS.

As another example of the treatment of zirconium oxide, Figure 3.2(d) shows XRD examples of ZrO_2-GO, GO-SiO_2, and hybrid ZrO_2-GO-SiO_2 nanocomposites. According to the measurement results, the large peaks (green) are related to ZrO_2 (=5.1623) (reference number JCPDS, No. 78~0047, 88~1007 [18]), and the plane indices of (002), (102), and (103) formed at the two theta positions of 30.63, 35.49, 50.84, and 60.46°. Another peak was the XRD peak derived from ZrO_2 in the tetragonal system, and it was confirmed that this component corresponds to 7.3% as a percentage. The subsequent peak represents the XRD arrangement of the GO–SiO_2 nanocomposite. This was a sharp and wide peak located in an area of 26.62°. The final peak shows the post-XRD effect of the ZrO_2-GO-SiO_2 ternary complex. All peaks corresponding to ZrO_2, SiO_2, and graphene are shown. It may be seen that a new peak was also shown in the XRD example of the ternary composite, ZrO_2 and mesoporous SiO_2 were slightly structural agglomerated, and many other XRD peaks derived therefrom were present. As a result of the XRD of these composites, the presence of graphite peaks was very low, so diffraction peaks were not shown. In addition, the XRD results confirmed that SiO_2 effectively adhered to the surface of the GO-ZrO_2 nanocomposite. One important fact was that the binding method and order of the compound proved to form the best material merging and active material conditions. The crystallite sizes of the ZrO_2 and SiO_2 particles were measured using the Debye-Scheer method, and the results were 72 and 15 nm. Surface potential and average particle size distribution are key parameters for controlling the biological, electrical, and optical functions of nanocomposites. Among the other variables, the component ratio greatly affects the NC particle size and the function of zeta potential. As one example, PPY:PANI:ZrO_2 with a NC ratio of 2:2:4 compared to other compositions shows the most suitable size arrangement, showing the best potential in sensor characteristics. Therefore, it has been shown that the much smaller particle size and improved stability of NCs are correlated with the sensor properties of the active material. While colloids with strong positive or negative potentials have electrical stability, colloids with low zeta potential tend to coagulate and thus exhibit the opposite sensor characteristics. As a result, NCs are very stable, and their strong negativity is suitable for electrostatic biological enzyme immobilization.

3.3.2 Surface Morphology

In order to investigate the form of NCs in the GO/PPy/PANI/ZO complex, a field emission scanning electron microscope (FESEM) was used to characterize them. Research on dispersibility has shown an effect on active material stability, and Figure 3.3(c and d) shows the uniform distribution of GO and ZO in a polymer matrix. The pores of the NCs can be observed from the shape photograph and have an high porosity with a mean pore size of 2 mm [Figure 3.3(e)]. The porosity of NCs improves the contact reaction by electron transfer rate and affinity with enzymes [60]. As illustrated in Figure

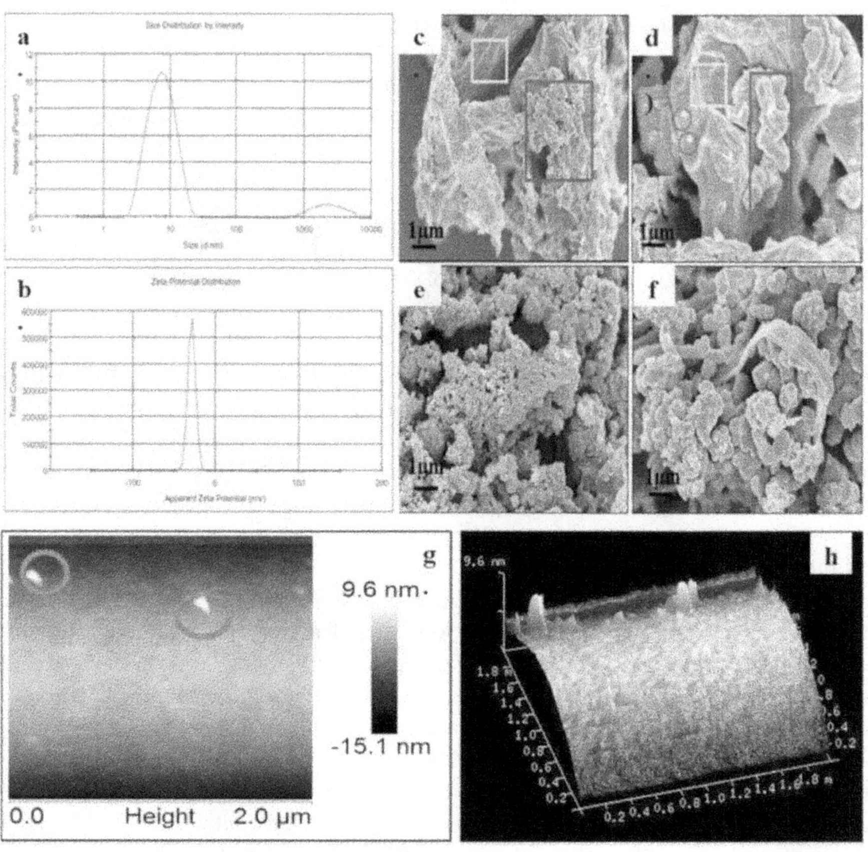

FIGURE 3.3 (a) NC particle size curve with an average size of 7.5 nm, (b) total number of particles and surface zeta potential of NCs, and (c and d) surface morphology of NCs (SEM). Distribution of GO and ZO in the polymer matrix (red: ZO, yellow square: GO, brown square: polymer), (e) SEM shape of porous NCs structure, (f) uniform distribution of filler through polymer matrix. AFM image, (g) GO (2D), red circle: graphene oxide, and (h) GO sheet height 3D is about 9.6 nm [60].

3.3(f), the uniform distribution of graphene oxide (GO) and zinc oxide (ZO) nanoparticles in the polymer matrix forms a geometric structure suitable for sensor design. In these photographs, the region between the nanoparticles and polymer is represented by bright areas, and the trace of GO is represented by dark areas. The dispersibility of GO, the shape of the NCs, and the height of plate-like GO particles can all be observed using atomic force microscopy (AFM), and Figure 3.3(g) shows a 2D AFM image of NCs. The NCs had a maximum particle diameter of 9.6 nm and a shape of 67.4 nm [Figure 3.3(h)], showing a plane structure of GO (red circle). Therefore, irregular shape, uniform and regular particle distribution, and proper particle height in NC are complex elements suitable for sensor application and are one indicator of sensor active material production. The shape of NCs appeared to be larger than the reported GO-based nanocomposites, and this effect has an effect on improving enzyme immobilization. The maximum reported size of the GO-enhanced nanocomposite was ~26 nm [60].

Microscopic (SEM) analysis is a useful method for observing samples and analysis has been performed to analyze the appearance and shape of various other samples. Mb/Au-PTY-f-MWCNT/GCE was prepared according to the synthesis process of Figure 3.4 and is shown in Figure 3.4 as a structurally modified feature for each step. SEM images of Mb/Au-PTY-f-MWCNT/GCE clearly show features with well-formed appearance and defined structure, curved surfaces, smooth surfaces and uniform features. In addition, the usefulness of graphene was presented by mixing graphene with Mb/Au-PTY-f-MWCNT/GCE, and the surface interaction by TEOS sol-gel formation in harmony with graphene was confirmed. The UV-vis spectrum and EIS data [49] of the Mb/Au-PTY-f-MWCNT samples in Figure 3.5 were used to evaluate the chemical properties of the generated samples. AgNP exhibits excellent electrochemical properties and is also good in terms of catalytic activity. Electrochemical sensors using nanomaterials containing AgNP as active materials have been proven to be able to provide higher responses with low response times and low detection limits. Interdigital electrodes containing gold by electrodeposition of AgNPs show higher sensitivity to hydrogen peroxide and show excellent potential because of this. Electrodes made of metal nanoparticles exhibit a larger surface-to-volume ratio compared to electrodes made of bulk metal. This means more potential sites where reactions can occur. Figure 3.6 shows sensor characteristics by coating the surface roughness of cured AgNP on a flexible substrate. It was confirmed that the RMS value for the surface roughness of the AgNP electrode was 313.7 nm,

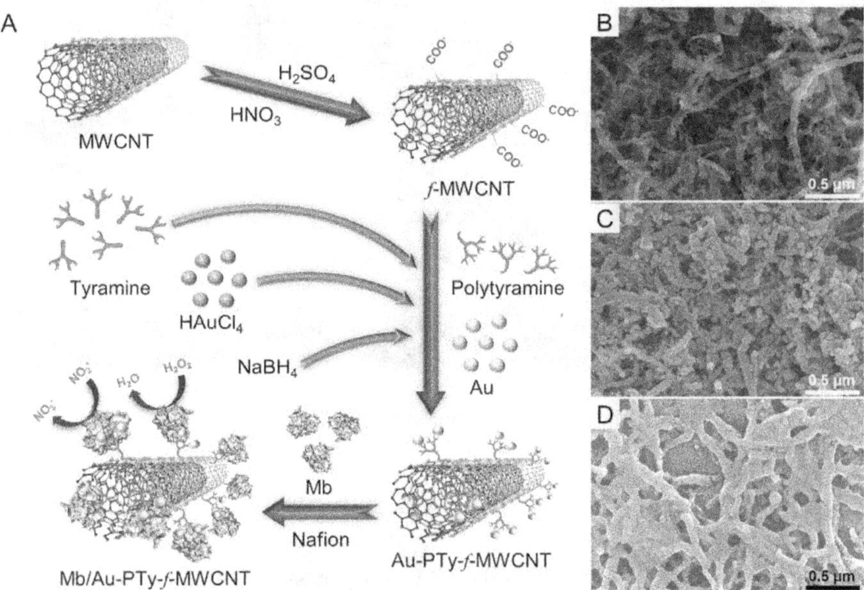

FIGURE 3.4 (a) Schematic representation of the procedure used to produce the Mb/Au-PTy-*f*-MWCNT/GCE composite; SEM images of (b) *f*-MWCNT, (c) Au-PTy-*f*-MWCNT, and (d) Mb/Au-PTy-*f*-MWCNT biocomposite [49].

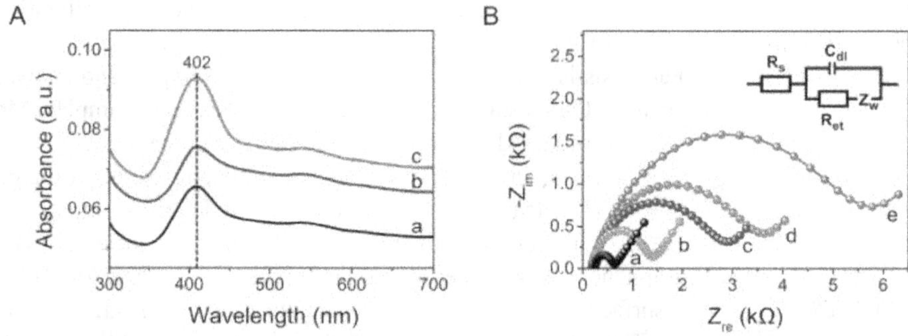

FIGURE 3.5 **A**. UV-vis spectra of the **a**. Mb, **b**. Mb/PTy-*f*-MWCNT, and **c**. Mb/Au-PTy-*f*-MWCNT films. **B**. EIS results for **a**. bare electrode, **b**. Au-PTy-*f*-MWCNT/GCE, **c**. PTy-*f*-MWCNT/GCE, **d**. Mb/Au-PTy-*f*-MWCNT/GCE, and **e**. Mb/GCE in the 5 mM $Fe(CN)_6^{4-/3-}$ containing 0.05 M PBS buffer solution. Inset: Randles equivalent circuit model [49].

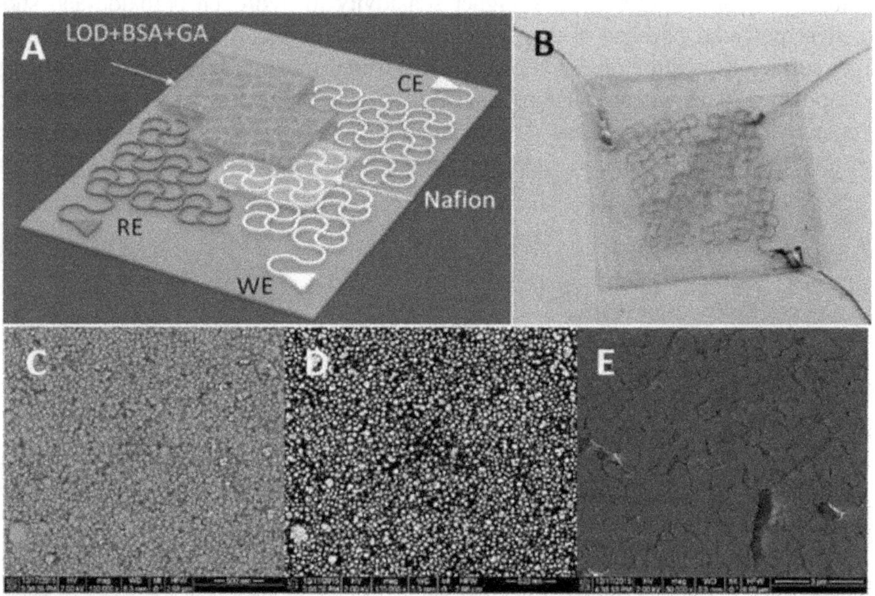

FIGURE 3.6 Images of fabricated and modified electrodes: (a) schematic illustration of bendable electro-chemical sensor for monitoring sweat lactate, (b) printed sweat lactate sensor, (c) surface morphology of working electrode, (d) B/W modified picture of (c), (e) reference electrode surface with chloridized AgNPs for 5 min [50].

which was significantly higher than the RMS roughness (164.018 nm for a modified electrode with a length of 5 minutes) of the Ag/AgCl reference electrode.

Figure 3.7 shows the overall particle shape of the ternary complex synthesized by the self-assembly method examined by SEM measurements. According to Figure 3.7(a–e), SEM studies of the aftermath of GO-SiO_2 and ZrO_2-GO double complexes show

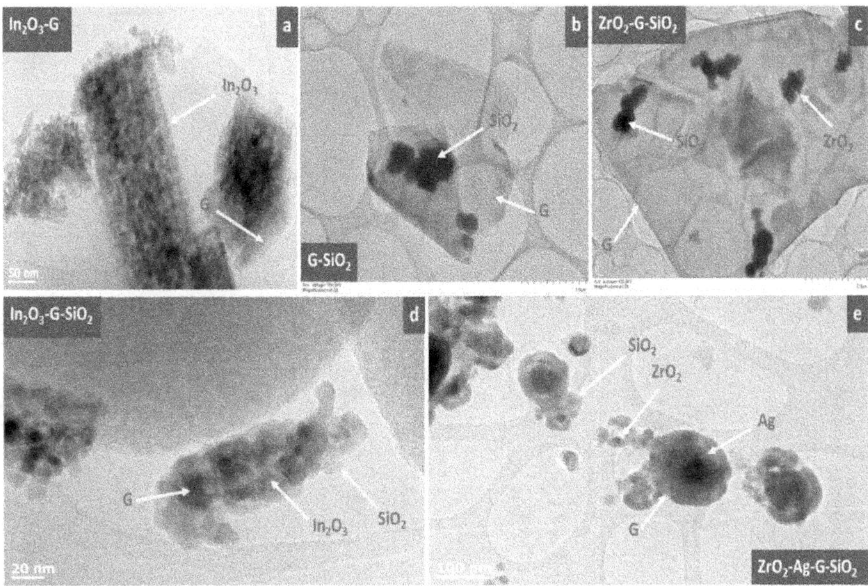

FIGURE 3.7 SEM images of (a) In_2O_3-G, (b) G-SiO_2, (c) ZrO_2-G-SiO_2, (d) In_2O_3-G-SiO_2, and (e) ZrO_2-Ag-G-SiO_2 samples [48].

that rectangular ZrO_2 and circular SiO_2 of an unpredictable three-dimensional structure are uniformly distributed on the graphene surface. The normal molecular size of ZrO_2 is 55.04 nm, the circular width of SiO_2 is 280.21 nm [19], and Figure 3.7(e) shows that metal oxide (MO) and silica oxide (SO) completely cover the entire graphene surface. The mixture in the form of the primary oxide has stability and has not changed. However, the size of silica in the graphene plane was reduced to nanosize. The main function of graphene provides a matrix region of suitable nanocomposites that can be unfolded on the surface during continuous complexation. In addition, graphene binds to the precursor reagent through oxygen-containing aggregates. SEM analysis showed the difference between binary and ternary complexes.

3.3.3 Morphological Analysis of AgNP/SLG-Based Electrodes

Figure 3.8(a) shows the XRD pattern of the resulting Fe_3O_4/CG nanocomposite. Strong diffraction peaks were shown at positions of 30.1°, 36.8°, 43.2°, 54.1°, 58.9°, and 63.7°, which correspond to indexing at (220), (311), (400), (422), (511), and (440), respectively. The peak position and relative strength were clearly represented, and the presence of Fe_3O_4 was clearly confirmed, as it was well matched with the reverse spinel structure of magnetite (JCPDS: 19–0629). No clear diffraction peak due to graphite was observed, and it may be seen that the graphene sheets of the Fe_3O_4/CG nanocomposites were distributed in a disorderly way. The Raman spectra of both CG and Fe_3O_4/CG nanocomposites clearly showed the characteristic D, G, and 2D bands for graphene [Figure 3.8(b)]. The D band of 1350 cm^{-1} is commonly seen in the

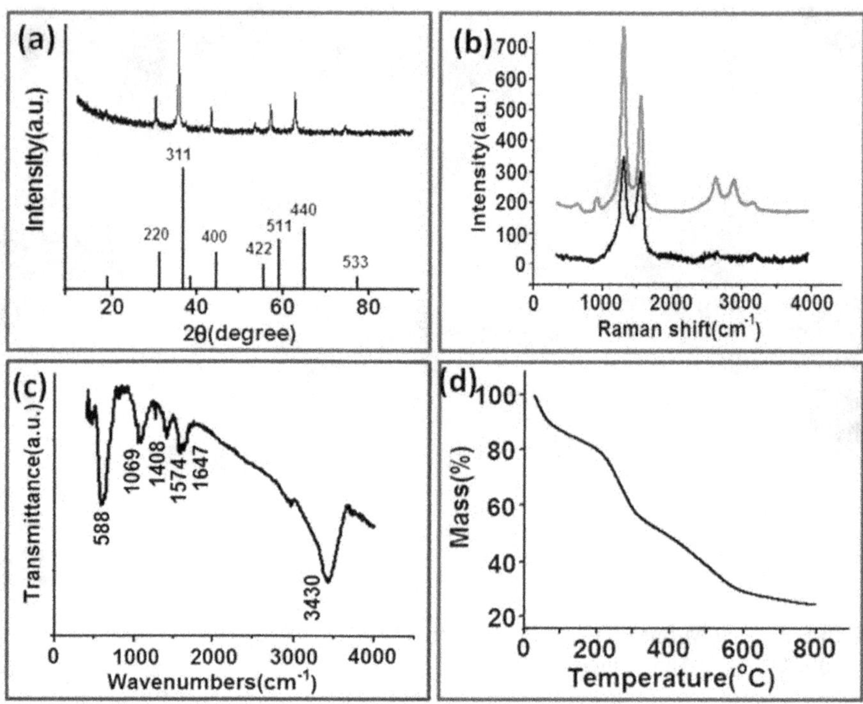

FIGURE 3.8 Physicochemical characterization of the Fe_3O_4/CG nanocomposites. (a) XRD patterns of the Fe_3O_4/CG. (b) Raman spectra of the CG and the Fe_3O_4/CG. (c) The FTIR spectrum of the Fe_3O_4/CG. (d) The TGA curve of the Fe_3O_4/CG [60].

vibrational motion of sp^3 carbon atoms, whereas the G band of 1595 cm^{-1} represents the E_{2g} mode of sp^2 carbon atoms in a 2D hexagonal lattice. The relative strength ratio (ID/IG ratio) of the D band to the G band represents the ratio of graphite carbon to defects. The ID/IG of the prepared CG was 1.15, which was much lower than nitrogen-doped graphene reported to be manufactured using several methods, including physicochemical treatment, electrochemical microwave, and plasma. Edge functionalization is performed using a ball milling technique, which is in good agreement with previous studies on nitrogen edge functionalized graphene nanoplatelets manufactured from dry ball milling graphite with N_2 and can be used to manufacture graphene nanosheets with low defects. The introduction of Fe_3O_4 nanoparticles to CG nanosheets significantly increases the ID/IG ratio to 1.48, introducing more defects and indicating the interaction between nanoparticles and nanosheets. Figure 3.8(c) shows the FTIR spectrum of the Fe_3O_4/CG nanocomposite. The strong band of 3430 cm^{-1} may be seen as the stretching vibration of the N-H bond and may be additionally confirmed by the peak of 1574 cm^{-1} generated from the bending vibration of N-H. The peak obtained at 1647 cm^{-1} is related to C = O stretching vibration, and the peak obtained at 1408 cm^{-1} is probably related to the cutting and bending of C-H. The peak at 1069 cm^{-1} is due to the increasing vibration of the C-N bond. A band of

588 cm^{-1} is associated with an Fe-O functional group that has been demonstrated by the characteristic peak of Fe_3O_4. The FTIR spectrum confirmed that Fe_3O_4 nanoparticles were successfully deposited on CG nanosheets. TGA was used to evaluate the mass ratio of Fe_3O_4 in Fe_3O_4/CG hybrids. As shown in Figure 3.8(d), the stage of weight reduction (10%) between 50 and 150°C may be due to the loss of residual water and adsorbed organic matter in the sample. Weight loss of 150 to 600°C (62%) is associated with loss of CG nanosheets. The content of residue Fe_3O_4 was found to be about 28%. So the mass ratio of Fe_3O_4 to CG is about 1:2.

Figure 3.9(a) schematically illustrates the shape of chitosan-graphite through ball milling and the reaction mechanism. Chitosan, an organic chemical, may act as a functional group at the edge of the graphite sheet in the initial stage. Milling weakens the shear force between chitosan chains, which breaks the chains between graphene sheets to promote the peeling of graphene nanosheets. The chitosan-graphite thus prepared can be further modified by plasma-treating acetic acid to introduce many active carboxyl groups for loading Fe_3O_4 nanoparticles [Figure 3.9(c)]. Figure 3.9(b and d) shows AFM images of chitosan-graphite and Fe_3O_4/chitosan-graphite composite nanomaterials. The size of the synthesized graphene nanosheet is confirmed to be about 1.641 nm [Figure 3.9(b)],

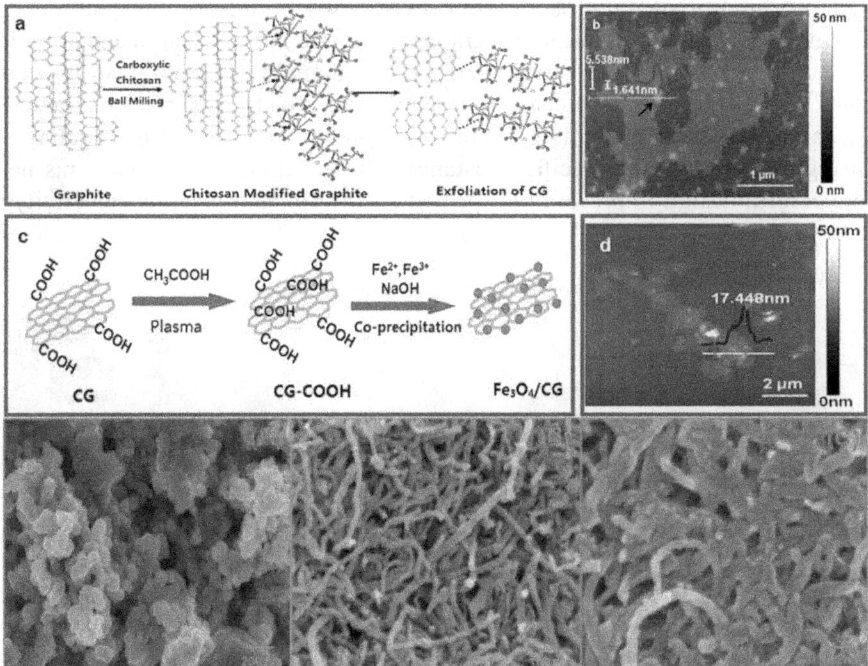

FIGURE 3.9 Formation of Fe_3O_4/CG nanocomposites. (a) Schematic synthesis of the CG. (b) A typical AFM image of CG nanosheets. Arrow indicates presence of chitosan. (c) Schematic preparation of Fe_3O_4/CG nanocomposites. (d) A typical AFM image of Fe_3O_4/CG nanocomposites (above). FE-SEM images of Mo nanoparticles (A), *f*-MWCNTs (B), and Mo NPs@*f*-MWCNT core-shell nanocomposite (C) (below) [60–63].

suggesting that a single/minority layer of graphene nanosheet may be manufactured through chitosan functionalization using ball milling technology, as reported previously. The presence of chitosan in chitosan-graphite is indicated by an arrow in Figure 3.9(b). The total thickness of chitosan-graphite was confirmed to be about 5.538 nm. It was confirmed that the dispersed acid of chitosan-graphite was well maintained even if stored in air for more than 15 days. Figure 3.9(d) shows the good distribution of Fe_3O_4 nanoparticles in chitosan-graphite nanosheets. It was confirmed that the thickness of the Fe_3O_4/chitosan-graphite nanocomposite was about 17.448 nm, and the average diameter of the Fe_3O_4 nanoparticles was about 12 nm. The average diameter of Fe_3O_4 nanoparticles on the nanosheet was about 12 nm, which was found to be well consistent with the AFM results presented previously (Figure 3.9).

3.3.4 Methodology of Sensing Performance

Various methods may be used to test and determine the behavior of sensors made of active material. Many researchers have evaluated CV tests, selectivity, and photocurrent characteristics or proposed other procedures to verify the effectiveness and activity of the sensors in these samples. Graphene is an electrode material that is human friendly and excellent in sensor application in the medical field to address environmental problems. However, new methods for the synthesis and accurate processing of high-quality graphene require more research and attention and should be studied from various perspectives in many basic applications in the future. Graphene is not responsive to biomolecules due to its surface characteristics, so the necessity and strategy of reforming them are effective in targeting the sensing and analysis of specific substances. In addition, sensor platforms need more careful design to prevent unexpected adsorption and reaction selectivity of

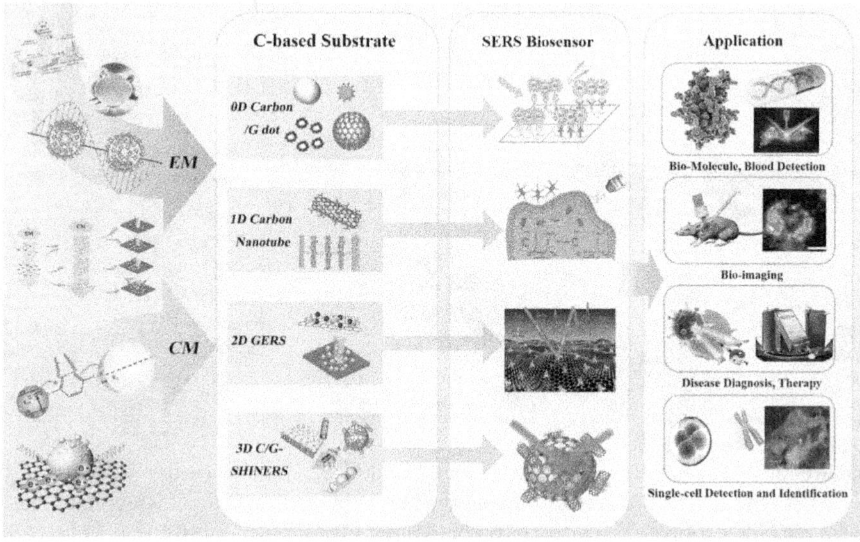

SCHEME 3.2 Antibiotic sensing mechanism [51].

molecules on the graphene surface and to improve factors to increase biomolecular reactivity on the graphene platform. Therefore, if graphene's original properties are controlled through surface modification, physical modification, and chemical properties, interaction with biomolecules will play an important role in terms of better interfacial reactions and physicochemical binding in graphene-based nanobiosensors. Detailed antibiotic detection mechanisms are presented in Scheme 3.2.

3.3.5 Cyclic Voltammetry Test

In a cyclic voltametric test using one of the electrochemical test methods, it was confirmed that the anode surface was successfully modified by INGS and ZAGS materials. Figure 3.10 illustrates the normal cycle voltage diagram of PBS in a test with active materials using BGSV and BGSVGS over the entire range of albumin, which is a material to be detected at a scan rate of 10 mV s^{-1} (pH 7.0). As illustrated in Figure 3.10, as a result of contacting the virus with INGS and ZAGS as the anodes, the cyclic voltage response decreased. These shifts were able to predict significantly improved electron transfer barriers. It is widely accepted that the development of a compound layer at the interface between the active material and the material to be sensed hinders electron transfer in the electrolyte. This is one indicator that can predict the formation of a resistance layer and the performance of the sensor. This is because these oxidative-reduced species penetrate through the layer and do not reach the conductive anode surface. As can be seen in Figure 3.10, this can be found from the coincidence and correlation of the oxidation peaks on the anode surfaces of INGS and ZAGS. Along this line, the oxidation peak was greatly reduced from 1.0×10^{-3} mA cm^{-2} for INGS anode to 3.0×10^{-3} mA cm^{-2} for albumin bonding to the cathode surface. As a base, it acts as a switching barrier for dormant electrons in the process of immobilizing and neutralizing the material to be bound and prevents the electrolyte from spreading to the cathode surface. This final result suggests that albumin was detected on the anode surface. For example, azithromycin concentrations at a constant current scan rate of 10 mV/s in the presence of 0.1 MPBS (pH = 7) in the reaction with the INGS electrode are shown in the CV results in Figure 3.10(b). This indicates that a significant decrease in current is directly proportional to the amount of the analyte [Figure 3.10(c)]. The cyclic test of the INGS and ZAGS electrodes measured in the absence of 0.1M PBS (phosphate buffer solution), pH 7.0, and E as other experimental conditions showed that the cathode wave was remarkably positive in the first scan at pH 7.0. The reduction peak was obtained when the scan was inverted. This phenomenon showed similar results in the electrode reaction of *E. coli* O157:H7 in another example. In the CV (current density) responses of INGS without *E. coli* (5.0×10^{-5} $mAcm^{-2}$), INGS with *E. coli* (1.5×10^{-4} $mAcm^{-2}$), and ZAGS (2.0×10^{-4} $mAcm^{-2}$), O157:H7 showed reversible results. CV reactions in INGS and ZAGS electrodes were measured after treatment and infection with *E. coli* O157:H7. In addition, this reaction shows that the peak current density of E is significantly reduced. This is believed to be because O157:H7 destroys the electron transfer capability compared to INGS and ZAGS, and it can be seen that the INGS current peak was greatly reduced due to the combination of bacteria after pathogens formed a film

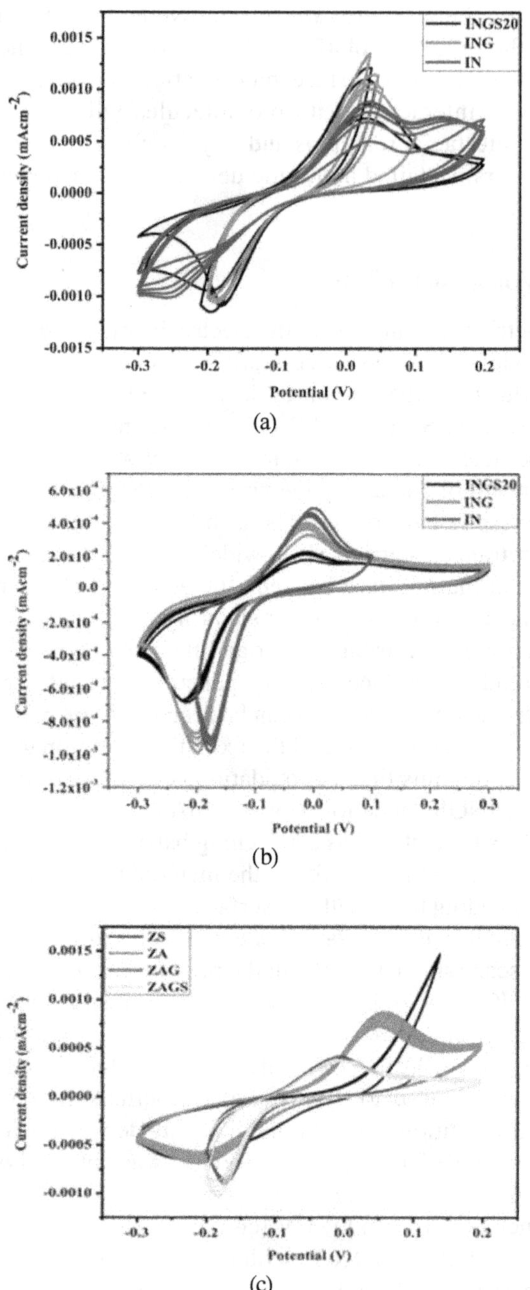

FIGURE 3.10 (a) Cyclic voltammogram measurement with In_2O_3 (IN), In_2O_3–G (ING), and In_2O_3-G-SiO_2 20% (INGS20) samples without *E. coli*. (b) Cyclic voltammogram measurement in the presence of *E. coli* for In_2O_3 (IN), In_2O_3–G (ING), and In_2O_3-G-SiO_2 20% (INGS20) and (c) synthesized ZrO_2 (ZS), ZrO_2–Ag (ZA), ZrO_2-Ag-G (ZAG), and ZrO_2-Ag-G-SiO_2 (ZAGS) [48–50].

on the active material surface and were detected (Scheme 3.1). Therefore, the hybrid composite–based INGS sensor provides an effective design for detecting *E. coli* O157:H7. This design is easy to operate, economical, and excellent in performance. Therefore, the reported results show electrochemical reduction of *E. coli* O157:H7, and the behavior of O157:H7 on the surface of INGS and ZAGS controlled by electrochemical redox blocking channels can be understood [58].

We have explained the mechanism of increasing current conductivity of GO and conductive polymer PPy nanocomposite. The electron transfer effect occurs as a result of a fine interaction between the interface between the PPy monomolecular layer and the GO monomolecular layer, leading to an increase in electron density. The long conjugate length of the NC leads to an increase in the electron transfer effect due to the delocalization of electrons and the increase in electron mobility. Moreover, the uniform distribution of GO within the conductive polymer PPy produces an extreme kinetic state of the localized free carrier that enables electron hopping, which leads to an increase in NC conductivity and contributes to the conductivity of the composite. The conductivity of the polymer PPy-GO NCs by such free electron induction is improved. A pyrrole monomer with a significant amount of positive charge around it interacts electrostatically with GO during surface polymerization and removes two protons to generate relatively free electrons from the monomer. Another study reported that the GO of the coil structure has more pronounced electrical conductivity than the GO uniformly arranged in PPy. In addition, the bandgap of NC (GO-PPY) of the composite is reduced, which improves the conductivity of the composite itself by forming a conductive network. Also, polymerization with PANI in the GO layer of NCs leads to increased conductivity. It was observed that the π-π^* transitions of PANI increased the conductivity of NC using graphene as an enhancer. The NCs of this study induce clear intermolecular interaction, distribution of GO dispersed on the polymer matrix, and polymerization of PANI and PPy on the GO sheet and are closely related to the presence of the π-π^* transitions in both polymers [61, 71].

In this work, we present a novel analytical method for POC electrochemical detection of metal proteins such as blood (Hb) dissolved in liquid. We propose a reaction of a novel receptor based on the aza heterocyclic compounds pyridine and imidazole to facilitate direct electron transfer between the central atom iron and the working electrode within Hb. Figure 3.11 shows the detection system of the entire sensor using pyridine-based liquid chemical reactions and imidazole-based meteorological chemical reactions. Figure 3.11(c,f) shows a cyclic voltammogram (CV) of a carbon electrode using a screen printing method obtained using hemolytic blood samples and alkaline hematin, respectively, using the same blood sample. It can be clearly seen from Figure 3.11(c) that dissolved blood does not show electrochemical activity in the carbon-printed electrode, but alkaline hematin shows irreversible CV, as illustrated in Figure 3.11(f). The reduction current of alkaline hematin appears to be a potential candidate for Hb detection, but it is not. It was confirmed that the reduction current peak of alkali hematin was saturated in a very low Hb concentration range, which was much lower than the clinically relevant value. The reason for this nonlinear reaction is that hematine forms a dimer in an aqueous solution and is reactive to it. Dimerization of hematin inhibits

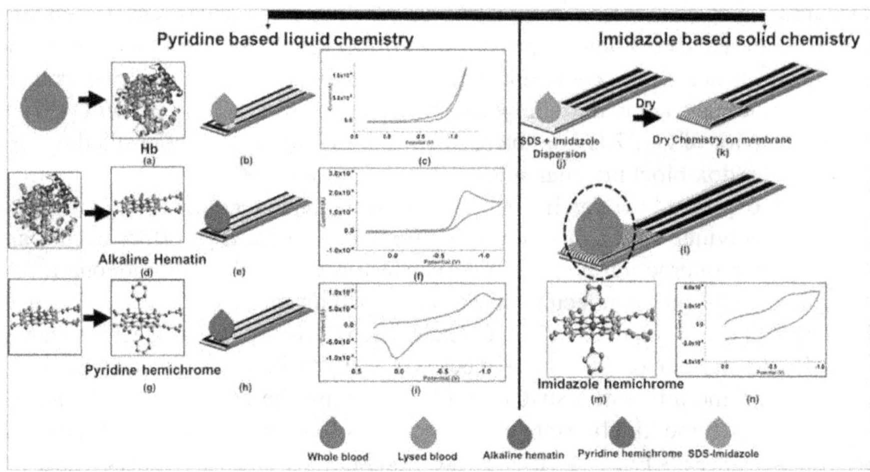

FIGURE 3.11 Pyridine and imidazole-based concept for electrochemical detection of Hb. (a) Hb after lysing. (b) 300 uL lysed blood on carbon printed electrode. (c) CV of lysed blood. (d) Structure of alkaline hematin. (e) 300 uL of alkaline hematin on carbon printed electrode. (f) CV of alkaline hematin. (g) Pyridine hemichrome. (h) 300 uL of pyridine hemichrome on carbon printed electrode. (i) CV of pyridine hemichrome. (j) Dispersion of imidazole-based chemistry on membrane-laminated electrode. (k) Ready-to-use electrode with dried chemistry. (l) Imidazole-metHb hemichrome. (m) CV of imidazole hemichrome [61].

diffusion-controlled electrochemical reactions, which work very well for biological detection [59–63].

The coordination compound of iron-containing porphyrins has attracted considerable attention due to the widespread formation of hemi proteins. Research on hemichrome began in the late 60s, and academic classification began. Hemichrome with relatively high solubility in aqueous solution is a spin metal compound in vivo. It is formed when the fifth and sixth coordination sites of iron covalently bond with the ligand. Reversible hemichrome may be formed in the sixth position of the metal iron and in the coordination of an internal ligand such as histidine. It can be confirmed that reversible hemichromes can be converted back into normal Hb biomolecules. It was confirmed that another type of hemichrome called irreversible hemichrome could not be converted back to normal Hb. It is represented as an irreversible hemichrome characterized by nitrogen bonds at the fifth and sixth positions of iron (III), and these hemichromes are characterized by electrochemical activity through good diffusion control without complex surface deformation. These unique properties of hemichrome can be used as a useful principle for developing Hb biosensors. Here it is proposed to convert Hb to hemichrome by adjusting nitrogen ligands such as pyridine and imidazole. Thereafter, a reversible CV that works well as the irreversible CV of alkali hematin can be obtained [Figure 3.11(i,n)]. In addition, nitrogen ligands enable free electron transfer to electrodes.

3.3.6 Sensing for Anti-Interference Ability

Induced protein engineering techniques through structural modification can induce intensive mutagenesis for amino acid residues around the pocket where ATP binding of the Perceval sensor was originally made. During this reaction, it was confirmed that the fluorescence reaction increased after several cycles of screening, and Perceval HR, an improved variant that detects a wider range of ATP, was obtained. Here it was confirmed that the ADP ratio and Perceval HR showed a significant improvement compared to the maximum signal change of 8 times or more for the optimal wavelength and the maximum signal change of 2 times for Perceval. The results of the measurement by applying this to the sensor confirmed two distinct peaks that responded differently to nucleotide bonds in excitation spectra of ~420 and ~500 nm (Figure 3.12). A conventional ATP binding showed fluorescence to

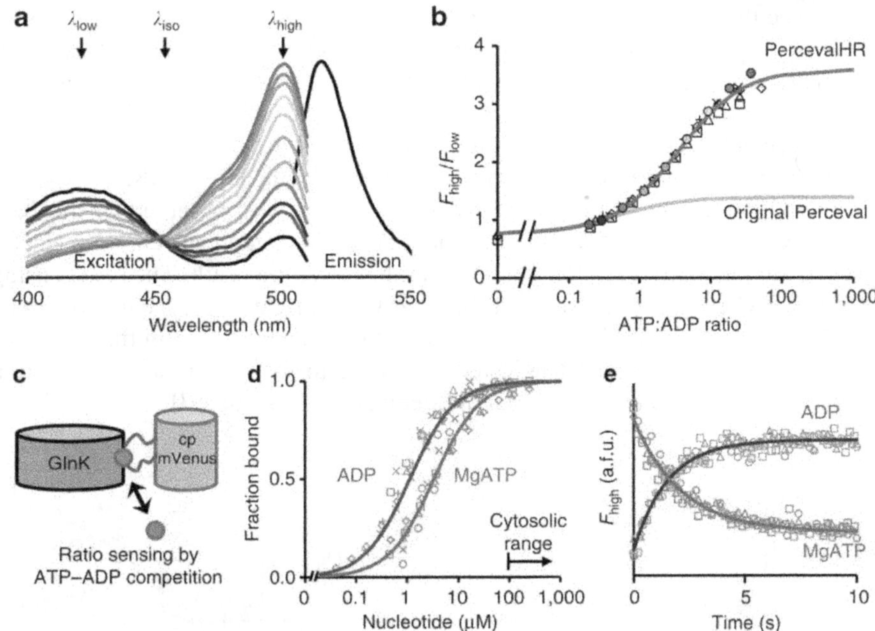

FIGURE 3.12 Perceval HR sensor characterization. (a) Fluorescence excitation and emission spectra. Excitation spectra from blue to red indicate increasing ATP:ADP and correspond to the colored data points in (b). (b) ATP:ADP dose-response of Perceval HR with Hill fit (green): $K_R = 3.5 \pm 0.2$; $n_H = 0.97 \pm 0.04$ ($N = 6$). Scaled original Perceval response for comparison (grey). Dose-response curves and kinetics were obtained in constant 0.5 mM free Mg^{2+} obtained by adjusting the total $MgCl_2$ concentration for the different nucleotide concentrations. (c) Diagram of Perceval HR ATP:ADP sensing. (d) Binding curves for MgATP (red: $K_{app} = 3.4 \pm 0.2\,\mu M$; $n_H = 1.2 \pm 0.2$; $N = 6$; [MgATP] calculated) and ADP in the presence of EDTA (blue: $K_{app} = 1.1 \pm 0.2\,\mu M$; $n_H = 1.0 \pm 0.1$; $N = 6$). Expected range of cytosolic [ATP] and [ADP] is indicated. (e) Apparent off rates for MgATP (red: $\tau = 2.1 \pm 0.2$ s; $N = 3$) and ADP (blue: $\tau = 1.5 \pm 0.1$ s; $N = 3$). Data are mean ± s.e.m. [52].

excitation at 500 nm, while ADP showed half of the increasing fluorescence properties at 420 nm excitation. It was confirmed that ~455 nm has a fixed isometric point. This reaction can be estimated to occur mainly due to changes in the absorbance of the sensor. By measuring the ratio of fluorescence at two different excitation wavelengths (F_{high}/F_{low} or F_{high}/F_{iso}), the occupancy of persistent HR can be determined regardless of the amount of sensor protein.

The sensor is represented by a very high affinity [1–3 μM, Figures 3.12(c, d)] and can be represented by ATP or ADP by physiological concentration variation from biological sensors to nucleotide bonds. Therefore, when ATP is occupied within the cell, the sensor is expressed as a ratio of ADP:ATP, not an absolute nucleotide concentration that determines the fluorescent signal, and the ratio of ADP:HR is 0.5 mM, indicating that the signal change (KR) is ~3.5 times greater than KR~0.5 of the original Perceval [Figure 3.12(b)]. In practice, Perceval HR clearly detects ATP. It can be seen that ADP cells are useful in mammalian cells within the reaction range. The time constants of ATP unloading and ADP unloading react at 25°C at ~2.1 and ~1.5 seconds, respectively. Since the kinetics of ligand bonds are generally faster at higher temperatures, it can be expected that sensors will be able to respond to changes in ATP. As originally discussed in the report on Perceval, ATP binding depends on the concentration of Mg^{2+}, and free ATP does not induce an increase in signal. Perceval HR, like other biosensors, exhibits pH-dependent detection, but, as described, it can roughly eliminate pH bias from actual ratio changes (ATP:ADP) when imaging living cells.

Although no clear oxidation and reduction current was observed in the bare electrode CV curve of Figure 3.13, the oxidation and reduction current were clearly shown in Ab (antibody)/AgNP in the PCT antibody culture result. The results using the SLG@ITO electrode showed that the clear CV (current density) curve obtained from the probe electrode (PCT/Ab/AgNP/SLG@ITO) generated from the PCT protein culture result was clearly indicated according to the oxidation and reduction current of the obtained antibody–protein interaction [12, 43, 61–63].

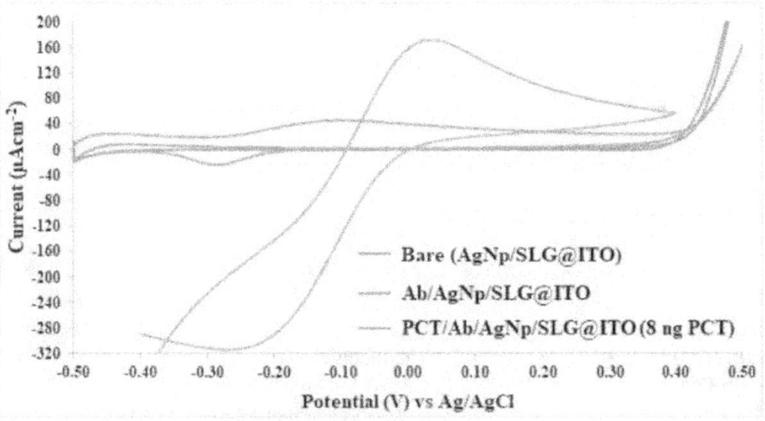

FIGURE 3.13 CV (current density) curves of bare, Ab/AgNp/SLG@ITO, and PCT/Ab/SLG/AgNp/@ITO electrodes (8 ng/mL PCT, scan rate: 20 mVs^{-1}) [62].

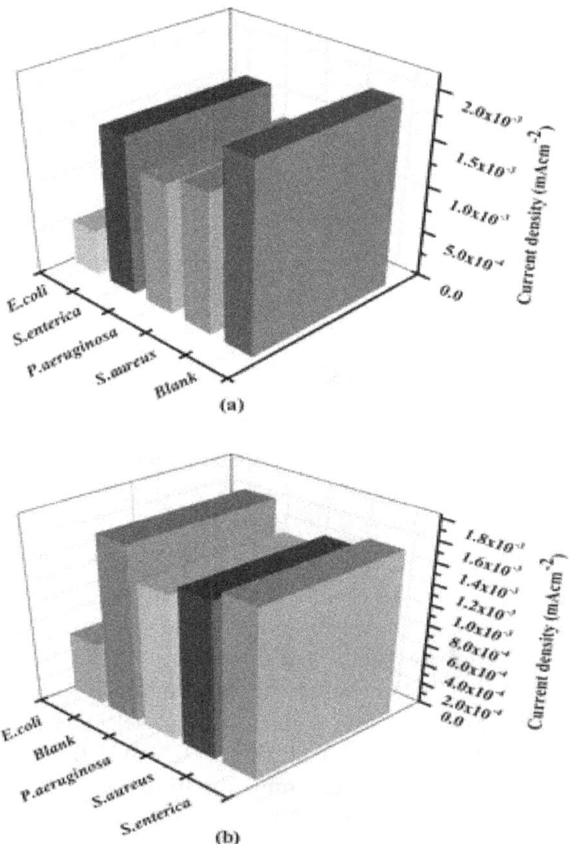

FIGURE 3.14 Selectivity test of *E. coli* with ZrO_2-Ag-G-SiO_2 (ZAGS) sensor (a) and In_2O_3-G-SiO_2 (IGS) sensor (b) [48].

3.3.7 Anti-Interference Capability of MACGS Sensors

To determine the selectivity of MACGS sensors, evaluation of the characteristics of azithromycin can proceed with the interaction of glucose or vitamin C with azithromycin and has been tested in the presence of several potentially interfering compounds. When 1 L of azithromycin was added, a decrease in current density of 1×10^{-5} $mAcm^{-2}$ was observed, which is important in the azithromycin sensor. In the presence of interference, the observed current density value increases, and the increased value was 2×10^{-4} $mAcm^{-2}$, showing that MACGS is still sensitive to azithromycin. The results are shown in Figure 3.14.

3.3.8 Anti-Biomedia Ability of the CMAGS Sensor

The CMAGS (hybrid composite) sensor was also evaluated in solutions including *E. coli O157*:H7 and is shown in Figure 3.15. CMAGS were soaked in a solution

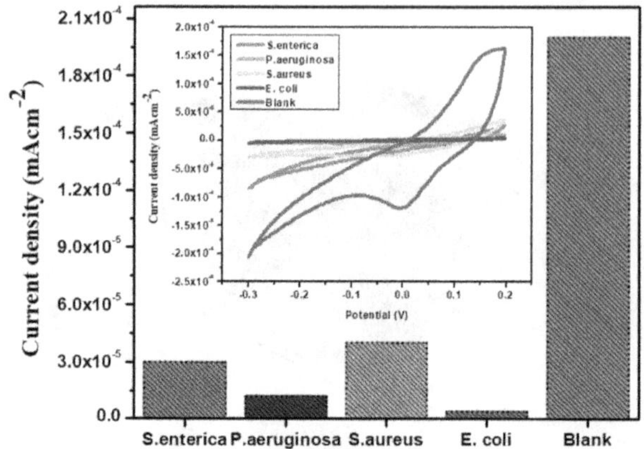

FIGURE 3.15 Selectivity results of CMAGS (hybrid nanocomposite) based sensor with azithromycin at RT. The relative response for various bacteria [48, 50].

including *E. coli O157* at a concentration of 101 CFU. After adding 101 CFU of *E. coli* O157, a current density of about 1.2×10^{-5} mA cm^{-2} was observed from the CV results. The increases of current density of about 2.1×10^{-4} mAcm^{-2} observed with the addition of the perturbation indicated that CMAGS was still E-limited. These results suggest that CMAGS can respond to detection of E. These obtained results demonstrate the suitability of CMAGS electrodes for selective detection of *E. coli* O157:H7 in a real sample. The results confirm that CMAGS (hybrid nanocomposite) electrodes are selective against biomass immunity to *E. coli* O157H7 [58, 59].

Figure 3.16(a) shows the detection results for *E. coli* O157:H7. O157:H7 detection was experimentally tested with *E. coli* O157:H7 at various pH conditions between 5.0 and 8.0, which was tested by the CMAGS active material electrode in the range of 10 mV/s in the presence of O157:H7. It is expressed as a peak for the current variation according to the change in current value. The figure shows that detection of *E. coli* O157:H7 showed a maximum value at pH 7 of PBS. The increases of peak current

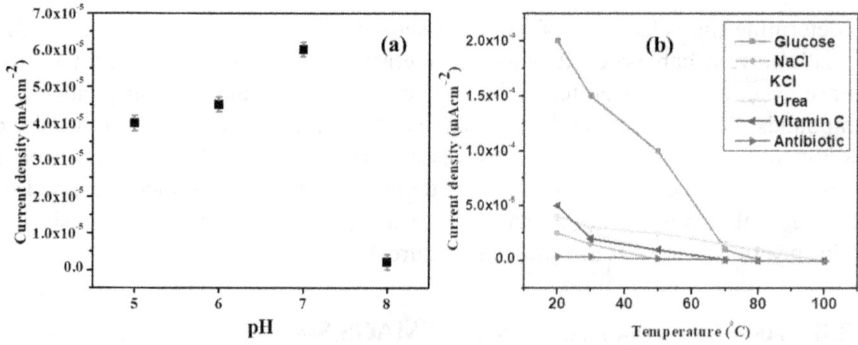

FIGURE 3.16 (a) *E. coli* O157:H7 sensing, with different pH conditions; (b) the temperature dependency for the test results of the CMAGS-based sensor [62, 63].

with pH 7 were reached before decreasing to pH 7 or below, which is presumed to be related to the protonation process on the electrode surface. After that, it was shown that the reduction current gradually decreased. *E. coli* O157:H7 showed anode peak current responses at various pH values when the CMAGS electrode active material was used. Therefore, the result means that detection of azithromycin attacking through H^+ ion concentration is better performed at pH 7. Therefore, all CV tests were performed at PBS of 0.1 M at optimal pH 7, and these test conditions were used for further CV tests [58, 61, 63]. Figure 3.16(b) shows the results of the temperature dependence test according to the temperature variation of the CMAGS-based electrode for biomedia (glucose, vitamin C, NaCl, KCl, urea, and azithromycin). The temperature range used in the sensor was measured at 20–100°C. During the measurement process, the best result was obtained in the 20–30°C range. This is a temperature condition that can mutate the solution. In particular, an increase in temperature leads to activation of the solution channel. The CMAGS-based electrode showed good results at 30°C, confirming that the RT detection ability was excellent. [61–63].

3.3.9 Impedimetric Biosensors

The surface of the hybrid nanocomposite based (PCT/Ab/AgNP/SLG@ITO) sensor used as the electrode can measure the impedance that measures the change in charge conductivity and capacitance for proteins. Proteins are formed by selectively binding to the biosensor AgNP/incubating antibody (AbPCT). As mentioned earlier, the CV technique proved its effectiveness with a simple electrochemical measurement method. In addition, the electrochemical impedance spectroscopy (EIS) technique is another method of evaluating characteristics, which can be used to evaluate electrochemical properties for PCT detection (Figure 3.17). Analysis of various concentrations measured the impedance sensor effect on PCT. Similar to the CV scheme, an electrolyte of 4 mM solution of $Fe(CN)_6^{3-}$ can be used as an redox electrode for monitoring electron transfer between the ITO electrode surfaces. Like the CV method, the EIS method is used as a very sensitive, simple, and practical technology to measure impedance changes using an active material local index of a biosensor manufactured using a relatively simple method as a variable [12, 46, 61–63]. In addition to the interaction

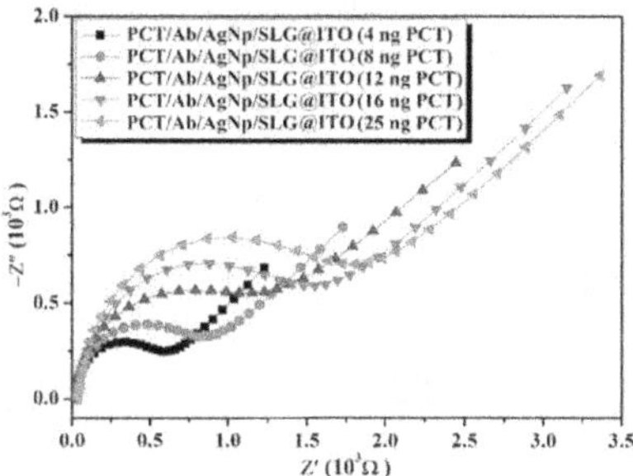

FIGURE 3.17 EIS plots of a biosensor incubated by controlling PCT concentrations (4–25 ng) [61, 62].

between the electrolyte and the electrode surface, the EIS technique can evaluate the behavior of these molecules, which can be explained as their interaction with the sensor active material surface as a result of Ab and PCT culture. The semicircle formed here represents the relationship between the electrolyte and the electronic behavior of the manufactured PCT electrode [46, 47, 61–63]. One of example for the EIS plots of a biosensor incubated by controlling PCT concentrations is shown in Figure 3.17.

3.3.10 Electrochemical Sensing Test for Biodetection

Cyclic voltammetry was obtained using bilirubin and cholesterol as sensing substances under different concentration conditions are shown in Figure 3.18. We described the callout in this part. these values measured using the maximum peak current values for the calibration curve obtained from different concentration variables of bilirubin and cholesterol showed a high level of linearity. The estimated linear regression values (R^2) of bilirubin and cholesterol, which were the substances to be detected, were 0.9787 and 0.98374, respectively. It can be seen that the oxidation of bilirubin and cholesterol attached to the surface of the electrode active material deviates from the linearity above the saturation level. The sensitivity of biosensors for the detected substances, cholesterol and bilirubin, was measured at 0.92 and 0.2

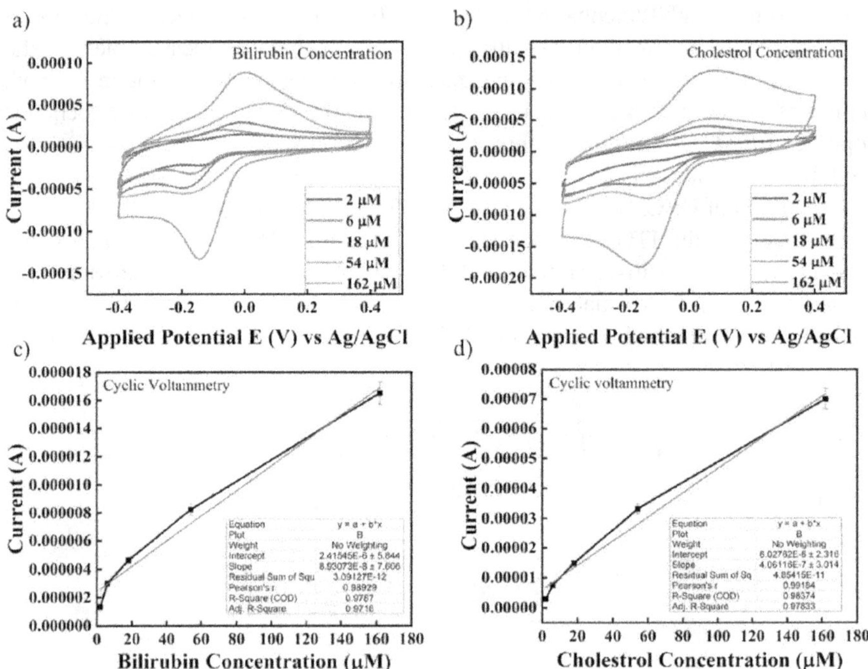

FIGURE 3.18 CV test data with the calibration for (a) GCE/NC in bilirubin and (b) GCE/NC in cholesterol at different concentrations; (c) calibration curve of bilirubin, and (d) calibration curve of cholesterol [62, 63].

μA μM^{-1} cm^{-2}, respectively. Electrochemical sensors showed excellent sensitivity when using electrodes of nanocomposite materials modified for biochemical substances such as cholesterol oxidase and bilirubin oxidase. The amount of electric current obtained by transporting free electrons generated during the reaction depends on the presence of bilirubin and cholesterol in the reaction system [57, 62, 63]. A key property of the proposed sensor is the potential use of cholesterol and bilirubin as biomarkers to provide important information for diagnosis of critical diseases.

3.4 CONCLUSION

In summary, sensors for diagnostic purposes require new sensor technologies such as miniaturization of compact biosensors and the economics of fabrication. Nanocomposite sensors are expected to be reliable and reproducible target materials with high sensitivity, accuracy, and specificity. Some of these sensors need to reduce costs and be relatively selective in order to improve availability in remote areas in case of an emergency. Biosensors have various applications in health, medicine, and medical fields in terms of disease diagnosis, environmental analysis, food inspection, drug discovery, biomedical study, forensics, and so on. These sensor devices require interaction in various fields and rely on special functions such as biomolecule recognition, interaction of their analysis and recognition components, device manufacture and useful design, chip-type electronics, sampling technology, and microfluidics. Integrating bionanoparticles into sensors can set up a new generation of sensing technologies.

REFERENCES

[1] Dungchai, W., Chailapakul, O. and Henry, C.S., 2009. Electrochemical detection for paper-based microfluidics. *Analytical Chemistry*, *81*(14), pp. 5821–5826.

[2] Tu, Y., Yu, Y., Zhou, Z., Xie, S., Yao, B., Guan, S., Situ, B., Liu, Y., Kwok, R.T., Lam, J.W. and Chen, S., 2019. Specific and quantitative detection of albumin in biological fluids by tetrazolate-functionalized water-soluble AIEgens. *ACS Applied Materials & Interfaces*, *11*(33), pp. 29619–29629.

[3] Jun, M., Perkovic, V. and Cass, A., 2011. Intensive glycemic control and renal outcome. *Diabetes and the Kidney*, *170*, pp. 196–208.

[4] Aoyagi, S., Iwata, T., Miyasaka, T. and Sakai, K., 2001. Determination of human serum albumin by chemiluminescence immunoassay with luminol using a platinum-immobilized flow-cell. *Analytica Chimica Acta*, *436*(1), pp. 103–108.

[5] Choi, S., Choi, E.Y., Kim, H.S. and Oh, S.W., 2004. On-site quantification of human urinary albumin by a fluorescence immunoassay. *Clinical Chemistry*, *50*(6), pp. 1052–1055.

[6] Comper, W.D., Jerums, G. and Osicka, T.M., 2004. Differences in urinary albumin detected by four immunoassays and high-performance liquid chromatography. *Clinical Biochemistry*, *37*(2), pp. 105–111.

[7] Marre, M., Claudel, J.P., Ciret, P., Luis, N., Suarez, L. and Passa, P., 1987. Laser immunonephelometry for routine quantification of urinary albumin excretion. *Clinical Chemistry*, *33*(2), pp. 209–213.

[8] Thakkar, H., Newman, D.J., Holownia, P., Davey, C.L., Wang, C.C., Lloyd, J., Craig, A.R. and Price, C.P., 1997. Development and validation of a particle-enhanced turbidimetric inhibition assay for urine albumin on the Dade aca® analyzer. *Clinical Chemistry*, *43*(1), pp. 109–113.

[9] Watts, G.F., Bennett, J.E., Rowe, D.J., Morris, R.W., Gatling, W., Shaw, K.M. and Polak, A., 1986. Assessment of immunochemical methods for determining low concentrations of albumin in urine. *Clinical Chemistry*, *32*(8), pp. 1544–1548.
[10] Goyal, R.N., Gupta, V.K. and Bachheti, N., 2007. Fullerene-C60-modified electrode as a sensitive voltammetric sensor for detection of nandrolone—an anabolic steroid used in doping. *Analytica Chimica Acta*, *597*(1), pp. 82–89.
[11] Goyal, R.N., Gupta, V.K. and Chatterjee, S., 2008. Simultaneous determination of adenosine and inosine using single-wall carbon nanotubes modified pyrolytic graphite electrode. *Talanta*, *76*(3), pp. 662–668.
[12] Goyal, R.N., Gupta, V.K. and Chatterjee, S., 2010. Voltammetric biosensors for the determination of paracetamol at carbon nanotube modified pyrolytic graphite electrode. *Sensors and Actuators B: Chemical*, *149*(1), pp. 252–258.
[13] Gupta, V.K., Jain, A.K., Maheshwari, G., Lang, H. and Ishtaiwi, Z., 2006. Copper (II)-selective potentiometric sensors based on porphyrins in PVC matrix. *Sensors and Actuators B: Chemical*, *117*(1), pp. 99–106.
[14] Jain, A.K., Gupta, V.K., Singh, L.P. and Raisoni, J.R., 2006. A comparative study of Pb2+ selective sensors based on derivatized tetrapyrazole and calix arene receptors. *Electrochimica Acta*, *51*(12), pp. 2547–2553.
[15] Omidfar, K., Dehdast, A., Zarei, H., Sourkohi, B.K. and Larijani, B., 2011. Development of urinary albumin immunosensor based on colloidal AuNP and PVA. *Biosensors and Bioelectronics*, *26*(10), pp. 4177–4183.
[16] Fatoni, A., Numnuam, A., Kanatharana, P., Limbut, W. and Thavarungkul, P., 2014. A novel molecularly imprinted chitosan—acrylamide, graphene, ferrocene composite cryogel biosensor used to detect microalbumin. *Analyst*, *139*(23), pp. 6160–6167.
[17] Liu, J., Ma, R.T. and Shi, Y.P., 2020. "Recent advances on support materials for lipase immobilization and applicability as biocatalysts in inhibitors screening methods": A review. *Analytica Chimica Acta*, *1101*, pp. 9–22.
[18] Lai, T., Hou, Q., Yang, H., Luo, X. and Xi, M., 2010. Clinical application of a novel sliver nanoparticles biosensor based on localized surface plasmon resonance for detecting the microalbuminuria. *Acta Biochimica et Biophysica Sinica*, *42*(11), pp. 787–792.
[19] Niu, X., Li, Y., Tang, J., Hu, Y., Zhao, H. and Lan, M., 2014. Electrochemical sensing interfaces with tunable porosity for nonenzymatic glucose detection: A Cu foam case. *Biosensors and Bioelectronics*, *51*, pp. 22–28.
[20] Yan, F., Zhang, M. and Li, J., 2014. Solution - gated graphene transistors for chemical and biological sensors. *Advanced Healthcare Materials*, *3*(3), pp. 313–331.
[21] Huang, X., Yin, Z., Wu, S., Qi, X., He, Q., Zhang, Q., Yan, Q., Boey, F. and Zhang, H., 2011. Graphene - based materials: Synthesis, characterization, properties, and applications. *Small*, *7*(14), pp. 1876–1902.
[22] do Amaral, C.E.F. and Wolf, B., 2008. Current development in non-invasive glucose monitoring. *Medical Engineering & Physics*, 30(5), pp. 541–549.
[23] Fu, S., Fan, G., Yang, L. and Li, F., 2015. Non-enzymatic glucose sensor based on Au nanoparticles decorated ternary Ni-Al layered double hydroxide/single-walled carbon nanotubes/graphene nanocomposite. *Electrochimica Acta*, *152*, pp. 146–154.
[24] Hoa, L.T., Sun, K.G. and Hur, S.H., 2015. Highly sensitive non-enzymatic glucose sensor based on Pt nanoparticle decorated graphene oxide hydrogel. *Sensors and Actuators B: Chemical*, *210*, pp. 618–623.
[25] Wu, G.-H., Song, X.-H., Wu, Y.-F., Chen, X.-M., Luo, F. and Chen, X., 2013. Non-enzymatic electrochemical glucose sensor based on platinum nanoflowers supported on graphene oxide. *Talanta*, *105*, pp. 379–385.
[26] Hou, L., Zhao, H., Bi, S., Xu, Y. and Lu, Y., 2017. Ultrasensitive and highly selective sand paper supported copper framework for non-enzymatic glucose sensor. *Electrochimica Acta*, *248*, pp. 281–291.

[28] Liu, X., Yang, W., Chen, L. and Jia, J., 2017. Three-dimensional copper foam supported CuO nanowire arrays: An efficient non-enzymatic glucose sensor. *Electrochimica Acta*, *235*, pp. 519–526.
[29] Zhong, Y., Shi, T., Liu, Z., Cheng, S., Huang, Y., Tao, X., Liao, G. and Tang, Z., 2016. Ultrasensitive non-enzymatic glucose sensors based on different copper oxide nanostructures by in-situ growth. *Sensors and Actuators B: Chemical*, *236*, pp. 326–333.
[30] Wang, Q., Wang, Q., Li, M., Szunerits, S. and Boukherroub, R., 2015. Preparation of reduced graphene oxide/Cu nanoparticle composites through electrophoretic deposition: Application for nonenzymatic glucose sensing. *RSC Advances*, *5*, pp. 15861–15869.
[31] Hai, B. and Zou, Y., 2015. Carbon cloth supported NiAl-layered double hydroxides for flexible application and highly sensitive electrochemical sensors. *Sensors and Actuators B: Chemical*, *208*, pp. 14–150.
[32] Li, X., Liu, J., Ji, X., Jiang, J., Ding, R., Hu, Y., Hu, A. and Huang, X., 2010. Ni/Al layered double hydroxide nanosheet film grown directly on Ti substrate and its application for a nonenzymatic glucose sensor. *Sensors and Actuators B: Chemical*, *147*, pp. 241–247.
[33] Shu, Y., Yan, Y., Chen, J., Xu, Q., Pang, H. and Hu, X., 2017. Ni and NiO nanoparticles decorated metal—organic framework nanosheets: Facile synthesis and high-performance nonenzymatic glucose detection in human serum. *ACS Applied Materials & Interfaces*, *9*, pp. 22342–22349.
[34] Zubair, N.F., Jamil, S., Bhatti, H.N. and Shahid, M., 2019. A comprehensive thermodynamic and kinetic study of synthesized rGO-ZrO_2 composite as a photocatalyst and its use as fuel additive. *Journal of Molecular Structure*, *1198*, p. 126869.
[35] Rahsepar, M., Foroughi, F. and Kim, H., 2019. A new enzyme-free biosensor based on nitrogen-doped graphene with high sensing performance for electrochemical detection of glucose at biological pH value. *Sensors and Actuators B: Chemical*, *282*, pp. 322–330.
[35] Venkatathri, N., 2007. Synthesis of silica nanosphere from homogeneous and heterogeneous systems. *Bulletin of Materials Science*, *30*(6), pp. 615–617.
[36] Wang, L., Zhao, J., Liu, H. and Huang, J., 2018. Design modification and application of semiconductor photocatalysts. *Journal of the Taiwan Institute of Chemical Engineers*, pp. 590–602.
[37] Song, Y., Gong, C., Su, D., Shen, Y., Song, Y. and Wang, L., 2016. A novel ascorbic acid electrochemical sensor based on spherical MOF-5 arrayed on a three-dimensional porous carbon electrode. *Analytical Methods*, *8*(10), pp. 2290–2296.
[38] Attallah, N., Osman-Malik, Y., Frinak, S. and Besarab, A., 2006. Effect of intravenous ascorbic acid in hemodialysis patients with EPO-hyporesponsive anemia and hyperferritinemia. *American Journal of Kidney Diseases*, *47*(4), pp. 644–654.
[39] Fatema, K.N., Sagadevan, S., Liu, Y., Cho, K.Y., Jung, C.H. and Oh, W.C., 2020. New design of mesoporous SiO_2 combined In_2O_3-graphene semiconductor nanocomposite for highly effective and selective gas detection. *Journal of Materials Science*, *55*(27), pp. 13085–13101.
[40] Fatema, K.N., Biswas, M.R.U.D., Bang, S.H., Cho, K.Y. and Oh, W.C., 2020. Electroanalytical characteristic of a novel biosensor designed with graphene–polymer-based quaternary and mesoporous nanomaterials. *Bulletin of Materials Science*, *43*(1).
[41] Janyasupab, M., Liu, C.W., Chanlek, N., Chio-Srichan, S., Promptmas, C. and Surareungchai, W., 2019. A comparative study of non-enzymatic glucose detection in artificial human urine and human urine specimens by using mesoporous bimetallic cobalt-iron supported N-doped graphene biosensor based on differential pulse voltammetry. *Sensors and Actuators B: Chemical*, *286*, pp. 550–563.
[42] Song, Y., Gong, C., Su, D., Shen, Y., Song, Y. and Wang, L., 2016. A novel ascorbic acid electrochemical sensor based on spherical MOF-5 arrayed on a three-dimensional porous carbon electrode. *Analytical Methods*, *8*(10), pp. 2290–2296.

[43] Yan, F., Zhang, M. and Li, J., 2014. Solution-gated graphene transistors for chemical and biological sensors. *Advanced Healthcare Materials*, *3*, pp. 313–331.

[44] Ohno, Y., Maehashi, K., Yamashiro, Y. and Matsumoto, K., 2009. Electrolyte-gated graphene field-effect transistors for detecting pH protein adsorption. *Nano Letters*, *9*, pp. 3318–3322.

[45] Sudibya, H.G., He, Q., Zhang, H. and Chen, P., 2011. Electrical detection of metal ions using field-effect transistors based on micropatterned reduced graphene oxide films. *ACS Nano*, *5*, pp. 1990–1994.

[46] He, R.X., Lin, P., Liu, Z.K., Zhu, H.W., Zhao, X.Z., Chan, H.L. and Yan, F., 2012. Solution-gated graphene field effect transistors integrated in microfluidic systems and used for flow velocity detection. *Nano Letters*, *12*(3), pp. 1404–1409.

[47] Huang, Y., Dong, X., Shi, Y., Li, C.M., Li, L.J. and Chen, P., 2010. Nanoelectronic biosensors based on CVD grown graphene. *Nanoscale*, *2*(8), pp. 1485–1488.

[48] Fatema, K.N., Liu, Y., Cho, K.Y. and Oh, W.-C., 2020. Comparative study of electrochemical biosensors based on highly efficient mesoporous ZrO_2-Ag-G-SiO_2 and In_2O_3-G-SiO_2 for rapid recognition of *E. coliO157:H7*. *ACS Omega*, *5*(36), pp. 22719–22730.

[49] Ezhil Vilian, A.T., Veeramani, V., Chen, S.-M., Madhu, R., Kwak, C.H., Huh, Y.S. and Han, Y.-K., 2015. Immobilization of myoglobin on Au nanoparticle-decorated carbon nanotube/polytyramine composite as a mediator-free H_2O_2 and nitrite biosensor. *Scientific Reports*, *5*, p. 18390.

[50] Abrar, M.A., Dong, Y., Lee, P.K. and Kim, W.S., 2016. Bendable electro-chemical lactate sensor printed with silver nano-particles. *Scientific Reports*, *6*, p. 30565.

[51] Liang, X., Li, N., Zhang, R., Yin, P., Zhang, C., Yang, N., Liang, K. and Kong, B., 2021. Carbon-based SERS biosensor: From substrate design to sensing and bioapplication. *NPG Asia Materials*, *13*, p. 8.

[52] Tantama, M., Martínez-François, J.R., Mongeon, R. and Yellen, G., 2013. Imaging energy status in live cells with a fluorescent biosensor of the intracellular ATP-to-ADP ratio. *Nature Communications*, *4*, p. 2550.

[53] Zou, R., Zhang, Z., Yuen, M.F., Sun, M., Hu, J., Lee, C.S. and Zhang, W., 2015. Three-dimensional-networked $NiCo_2S_4$ nanosheet array/carbon cloth anodes for high-performance lithium-ion batteries. *NPG Asia Materials*, *7*(6), pp. e195–e195.

[54] Wang, Q., Jiao, L., Han, Y., Du, H., Peng, W., Huan, Q., Song, D., Si, Y., Wang, Y. and Yuan, H., 2011. CoS_2 hollow spheres: Fabrication and their application in lithium-ion batteries. *The Journal of Physical Chemistry C*, *115*(16), pp. 8300–8304.

[55] Brinker, C.J., Lu, Y., Sellinger, A. and Fan, H., 1999. Evaporation - induced self - assembly: Nanostructures made easy. *Advanced Materials*, *11*(7), pp. 579–585.

[56] Li, G., Huo, H. and Xu, C., 2015. $Ni_{0.31}Co_{0.69}S_2$ nanoparticles uniformly anchored on a porous reduced graphene oxide framework for a high-performance non-enzymatic glucose sensor. *Journal of Materials Chemistry A*, *3*(9), pp. 4922–4930.

[57] Zhou, Y., Zheng, H., Chen, X., Zhang, L., Wang, K., Guo, J., Huang, Z., Zhang, B., Huang, W., Jin, K. and Tonghai, D., 2009. The Schistosoma japonicum genome reveals features of host-parasite interplay. *Nature*, *460*(7253), p. 345. Yang, J. *Journal of Materials Chemistry A*, 2014, 2, 3031–3037.

[58] Zhan, B., Liu, C., Chen, H., Shi, H., Wang, L., Chen, P., Huang, W. and Dong, X., 2014. Free-standing electrochemical electrode based on Ni (OH) 2/3D graphene foam for nonenzymatic glucose detection. *Nanoscale*, *6*(13), pp. 7424–7429.

[59] Fatema, K.N., Lim, C.S. and Oh, W.C., 2021. High surface area mesoporous BiZnSbV-G-SiO_2-based electrochemical biosensor for quantitative and rapid detection of microalbuminuria. *Journal of Applied Electrochemistry*, *51*(9), pp. 1345–1360.

[60] Zhang, W., Li, X., Zou, R., Wu, X., Shi, H., Yu, Y. and Liu, Y., 2015. Multifunctional glucose biosensors from Fe_3O_4 nanoparticles modified chitosan/graphene nanocomposites. *Scientific Reports*, *5*, p. 11129.

[61] Kumar, V., Nikhila Kashyap, D.M., Hebbar, S., Swetha, R., Prasad, S., Kamala, T., Srikanta, S.S., Krishnaswamy, P.R. and Bhat, N., 2017. Aza-heterocyclic receptors for direct electron transfer hemoglobin biosensor. *Scientific Reports*, *7*, p. 42031.

[62] Fatema, K.N. and Oh, W.C., 2021. A comparative electrochemical study of non-enzymatic glucose, ascorbic acid, and albumin detection by using a ternary mesoporous metal oxide (ZrO_2, SiO_2 and In_2O_3) modified graphene composite based biosensor. *RSC Advances*, *11*(7), pp. 4256–4269.

[63] Keerthi, M., Boopathy, G., Chen, S.-M., Chen, T.-W. and Lou, B.-S., 2019. A core-shell molybdenum nanoparticles entrapped *f*-MWCNT*s* hybrid nanostructured material based non-enzymatic biosensor for electrochemical detection of dopamine neurotransmitter in biological samples. *Scientific Reports*, *9*, p. 13075.

[64] Fatema, K.N., Sagadevan, S., Liu, Y., Cho, K.Y., Jung, C.H. and Oh, W.C., 2020. New design of mesoporous SiO_2 combined In_2O_3-graphene semiconductor nanocomposite for highly effective and selective gas detection. *Journal of Materials Science*, *55*(27), pp. 13085–13101.

[65] Riazimehr, S., Kataria, S., Bornemann, R., Haring Bolívar, P., Ruiz, F.J.G., Engström, O., Godoy, A. and Lemme, M.C., 2017. High photocurrent in gated graphene–silicon hybrid photodiodes. *ACS Photonics*, *4*(6), pp. 1506–1514.

[66] Liu, S., Kang, S., Wang, H., Wang, G., Zhao, H. and Cai, W., 2016. Nanosheets-built flowerlike micro/nanostructured Bi_2O_2. 33 and its highly efficient iodine removal performances. *Chemical Engineering Journal*, *289*, pp. 219–230.

[67] Yngman, S., McKibbin, S.R., Knutsson, J.V., Troian, A., Yang, F., Magnusson, M.H., Samuelson, L., Timm, R. and Mikkelsen, A., 2019. Surface smoothing and native oxide suppression on Zn doped aerotaxy GaAs nanowires. *Journal of Applied Physics*, *125*(2), p. 025303.

[68] Martini, I., Chevallay, E., Heßler, C., Nistor, V., Neupert, H., Fedosseev, V. and Taborelli, M., 2015. *Surface characterization at CERN of photocathodes for photoinjector applications*, Proceedings of IPAC 2015, Richmond, VA (No. CERN-ACC-2015-290, p. TUPJE040).

[69] Tomanin, P.P., Cherepanov, P.V., Besford, Q.A., Christofferson, A.J., Amodio, A., McConville, C.F., Yarovsky, I., Caruso, F. and Cavalieri, F., 2018. Cobalt phosphate nanostructures for non-enzymatic glucose sensing at physiological pH. *ACS Applied Materials & Interfaces*, *10*(49), pp. 42786–42795.

[70] Kumar, A., Gupta, G.H., Singh, G., More, N., Keerthana, M., Sharma, A., Jawade, D., Balu, A. and Kapusetti, G., 2023. Ultrahigh sensitive graphene oxide/conducting polymer composite based biosensor for cholesterol and bilirubin detection. *Biosensors and Bioelectronics: X*, *13*, p. 100290.

[71] Pothipor, C., Wiriyakun, N., Putnin, T., Ngamaroonchote, A., Jakmunee, J., Ounnunkad, K., Laocharoensuk, R. and Aroonyadet, N., 2019. Highly sensitive biosensor based on graphene–poly (3-aminobenzoic acid) modified electrodes and porous-hollowed-silver-gold nanoparticle labelling for prostate cancer detection. *Sensors and Actuators B: Chemical*, *296*, p. 126657.

Section B

Drug Delivery

4 Bionanomaterials for Ocular Drug Delivery Applications

Venkateshwaran Krishnaswami[1] *and Saravanakumar Arthanari*[2]

[1]Department of Pharmaceutics, SA Raja Pharmacy College, Vadakangulam, Tirunelveli, Tamil Nadu, India.

[2]Department of Pharmaceutics including Pharmaceutical Biotechnology, Vellalar College of Pharmacy, Thindal, Erode, Tamil Nadu, India.

4.1 INTRODUCTION

The complex organ eye holds an unique anatomy and physiology which is partitioned into anterior and posterior poles. The anterior segment is composed of about one-third of the portion of the eye, and the posterior segment consists of the remaining portion. The cornea, conjunctiva, aqueous humor, iris, ciliary body, and lens constitute the anterior segment. The sclera, choroid, retinal pigment epithelium, neural retina, optic nerve, and vitreous humor occupy the posterior segment of the eye [1]. Diseases associated with eye include age-related macular degeneration, glaucoma, diabetic retinopathy, allergic conjunctivitis, anterior uveitis, and cataracts. The sclera, which refracts and transmits the light to the lens. Cornea, the outermost layer, provides protection to the eye [2]. Infection protection to the eye and structural damage to the deeper parts is afforded by the retina. The connective tissue coat sclera provides protection to the eye from internal and external forces; further, it is covered by a transparent mucous membrane, the conjunctiva. The middle part of the eye is composed of the iris, ciliary body, and choroid. The size of the pupil is controlled by the iris. The ciliary body controls the power and shape of the lens and supplies the aqueous humor, oxygen and nutrients to the outer retinal layers. The retina occupies the inner layer of the eye, is complex, and captures light [3,4].

The diverse drug absorption obstacles connected to the various routes unique to the eye make it difficult to deliver drugs to the eye. Topical delivery is a non-invasive method of medicine administration for conditions affecting the anterior segment. Due to their ease of administration and high patient compliance, eye drops make up nearly 90% of the commercially available ophthalmic formulations. Because of its extremely limited absorption, the topical route suffers. Drug absorption through the cornea is hampered by anatomical and physiological factors such as tear turnover, nasolacrimal drainage, reflex

DOI: 10.1201/9781003425427-6

blinking, and ocular static and dynamic barriers. Less than 5% of the dose administered topically overall penetrates the cornea and reaches the interior structures.

Ways to deliver the drugs to the posterior portion of the eye include intravitreal, periocular, and systemic injections. The main barrier to drug transport to the posterior pole of the eye is the blood retinal barriers. Even though many ocular illnesses are treated using intravitreal injections. Endophthalmitis, hemorrhage, retinal detachment, and low patient tolerance are problems with this method. Researchers use a variety of techniques, such as gels and polymeric networks, to prolong the period that drugs remain in the precorneal layer, thereby improving their bioavailability. The three-dimensional, hydrophilic, polymeric networks found in hydrogels, which may ingest water or biological fluids due to their hydrophilic groups, also play a crucial function.

4.2 OCULAR BARRIERS

Ocular anatomy and physiology are occupied with specific barriers that are inherent and unique. Based on the route of administration, such as topical, systemic, and injectable, the barriers are varied. These anatomical and physiological barriers protect the eye from toxins.

Topical formulation bioavailability is significantly impacted by precorneal variables including solution drainage, blinking, tear film, tear turnover, lacrimation, and anatomical obstacles. The cornea serves as the front surface of the eye and functions as a mechanical barrier to prevent foreign objects from entering the eye and to safeguard the ocular tissues. The endothelium, Bowman's membrane, Descemet's membrane stroma, and epithelium of the cornea are lipoidal in nature and have different polarities, making them rate-limiting structures for drug permeability. Since the

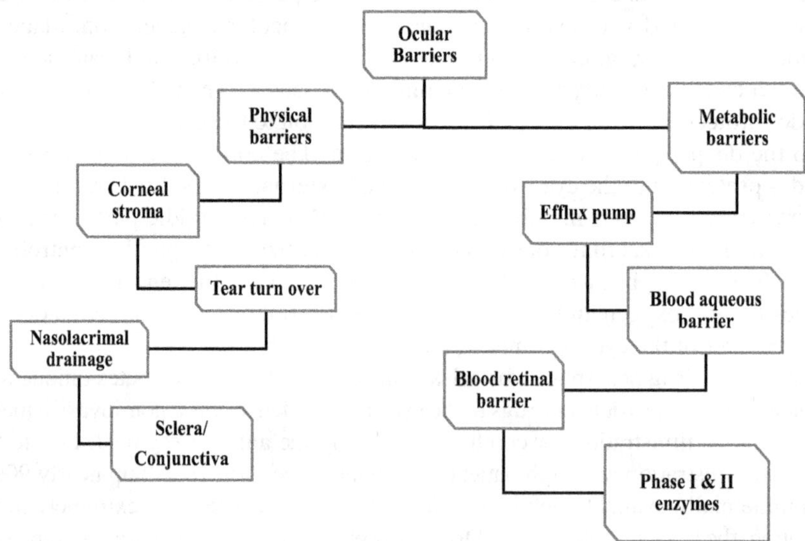

FIGURE 4.1 Ocular drug delivery barriers restricting the delivery of drugs to the interior of the eye.

cornea is lipoidal in nature, hydrophilic medications applied topically do not penetrate it. Desmosomes connected to the zona occludens connect the corneal epithelial cells. Tight junctional complexes slow down drug penetration into paracellular membranes [5]. Topically administered ocular formulations enter through either or both corneal and non-corneal routes. Ocular drug delivery barriers are shown in Figure 4.1.

The cornea's epithelium serves as a barrier. The five to six layers of columnar epithelial cells connected by extremely tight junctions have been linked to greater paracellular resistance of 12–16 kcm. Collagen fibers enable the stroma to support the passage of hydrophilic medications across the cornea, which also has aqueous pores or channels. To achieve the best bioavailability, a balance between lipophilic and hydrophilic molecules is necessary. The non-corneal pathway, which avoids the cornea, is supported by the conjunctiva and the sclera (larger molecules, hydrophilic molecules peptides, proteins, and siRNA). Because of the extremely low expression of tight junction proteins, hydrophilic molecules are more permeable [6]. Moreover, the endothelium divides the stroma and aqueous humor, and because of its secretory nature and selective carrier-mediated transport, it maintains aqueous humor and corneal transparency. The endothelium allows macromolecules to pass through. As a result, the epithelium and stroma of the cornea, in particular, are regarded as significant obstacles to the administration of ocular drugs. Drug permeability is supported by the collagen fibers and proteoglycans embedded in the extracellular matrix of the sclera; in this case, drug penetration is inversely proportional to molecular radius.

A drug is washed off by lacrimation when applied topically, which reduces drug concentration and absorption and, eventually, reduces the amount of time the drug molecules are in contact with the tissue. Nasolacrimal drainage and injection-related formulation leaking will both cause clearing. The precise amount of time required for drug absorption will be cut short by lower bioavailability. The blood-aqueous barrier and blood-retinal barrier are the two main impediments to the transport of ocular medications to both the anterior and posterior segments. The iris/ciliary blood vessel endothelium and the nonpigmented ciliary epithelium in the anterior part of the eye are the two separate cell layers that make up the blood-aqueous barrier. Drug entrance from blood into the posterior segment is restricted by the retinal pigment epithelium (RPE) and retinal capillary endothelial cells in the blood-retinal system. The RPE facilitates the biochemical processes by selectively transporting chemicals between photoreceptors and choriocapillaris.

The physicochemical properties of the drug molecule are crucial for ocular drug delivery because of the many anatomical and physiological restraints. The solubility, lipophilicity, level of ionization, molecular weight, and ocular tissue shape all affect how quickly medicines are absorbed. The inner layer of the cornea (stroma) supports hydrophilic medications for maximal penetration, while the epithelial layers of the cornea support lipophilic pharmaceuticals. Log P should fall between 2 and 4 for optimum corneal permeability. Two to three layers of wing cells and one or two outer layers of squamous cells can be found in the stratified corneal epithelium. Intercellular tight junctions (zonula occludens) occupy the superficial cells. The junction proteins ZO-1, cingulin, ZO-2, and occluden of anastomotic strands that are present at tight junctions serve as a barrier. The extracellular and intracellular amounts of calcium support the permeability of medicines. Although ethylene

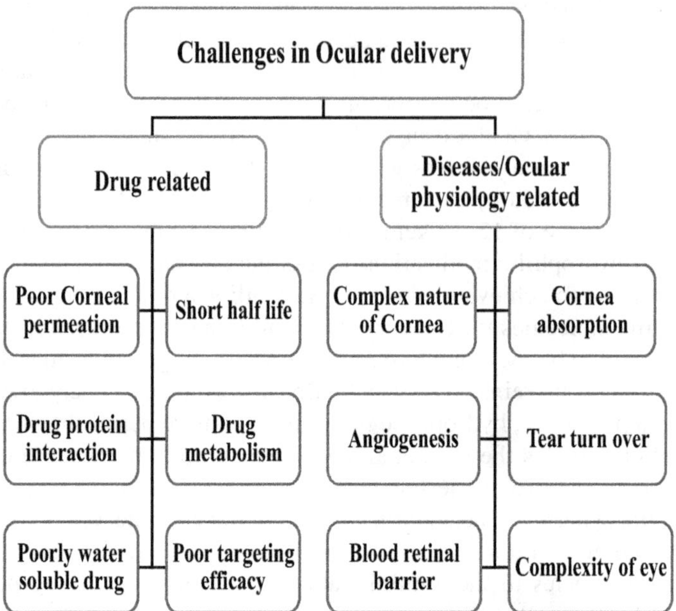

FIGURE 4.2 Challenges encountered in the ocular drug delivery of drugs.

diamine tetra acetic acid (EDTA) can be used to remove extracellular proteins, doing so compromises the integrity of the tight junction membrane, increasing drug penetration. Figure 4.2 depicts the difficulties in drug distribution to the eyes.

4.3 DISEASES AFFECTING THE EYE

Diseases affecting the eye are shown in Figure 4.3. The major ocular diseases affecting the eye are discussed in the following.

4.3.1 Age-Related Macular Degeneration

Age-related macular degeneration (AMD) is an eye condition that harms the macula and impairs central vision. AMD is the main cause of vision loss in adults over 60. Wet and dry variants of AMD both exist. Wet AMD is caused by angiogenesis, which is the formation of aberrant blood vessels under the macula that leak blood and fluid [7–9] and produces a loss of central vision. In extreme cases, dry AMD can cause significant blindness and vision loss. With dry AMD, the macula thins, which over time causes central vision to become blurry. The dry type of AMD is more prevalent than the wet form in 70% to 90% of cases. Dark patches in the macula and blurring central vision are signs of AMD. Therapies might stop the disease's spread and stop catastrophic visual loss. Anti-VEGF intraocular injections may slow the progression of AMD. The photoreceptor, retinal pigment epithelium, Bruch's membrane, and choroidal complex are all affected negatively by AMD [10]. AMD can result in neovascularization or atrophy [11,12]. The risk factors for AMD include environmental variables like smoking and sun exposure and nutritional factors such micronutrients, dietary fish intake, and alcohol consumption.

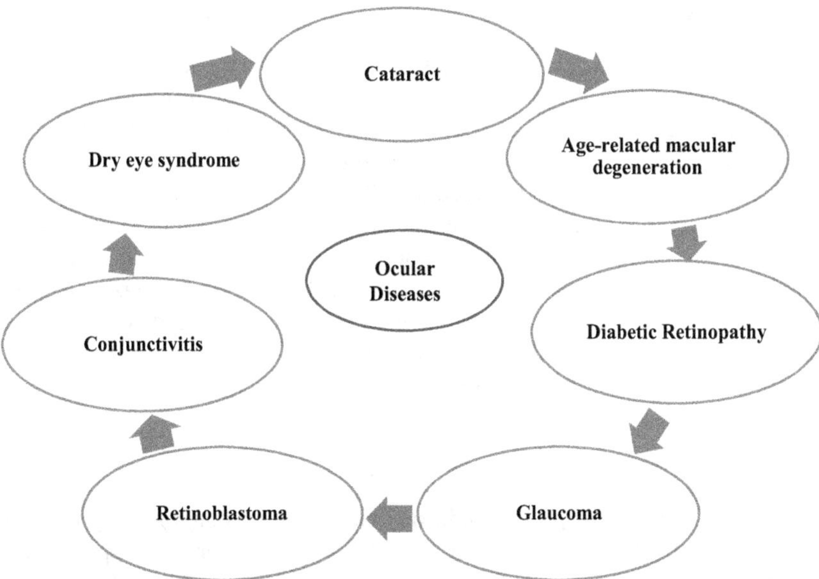

FIGURE 4.3 Diseases affecting the ocular globe.

Sociodemographic, genetic, economical, and systemic variables are additional causes of AMD [13–15].

4.3.2 Diabetic Retinopathy

Diabetes complications lead to diabetic retinopathy (DR), where too much glucose narrows the blood vessels leading to the eyes, causing edema and fluid leakage. These issues lead to moderate to severe visual impairment. The retina may not receive enough oxygen, which can lead to DR. Proliferative DR (PDR) and non-proliferative DR (NPDR) are the two kinds of DR. Treatment for DR involves the use of anti-VEGF medications. Fundus pictures are utilized to assess the severity of DR. The characteristics that are frequently examined for DR include microaneurysms, hemorrhagic lesions, exudates, and cotton wool spots [16].

4.3.3 Glaucoma

Glaucoma is one of the main cause of irreversible blindness globally. It is characterized by optic neuropathies and leads to gradual degeneration of retinal ganglion cells, which are neurons in the central nervous system. Multifactorial illness will result in the progressive deterioration of the optic nerve, associated loss of vision field, and finally blindness. Primary, secondary, juvenile, and congenital glaucoma are the several types of glaucoma that can occur. The clinical phenotypes of primary glaucoma are angle-closure glaucoma and open-angle glaucoma. The primary risk factor for the onset of glaucoma is an increase in intraocular pressure (IOP). In the United States, open-angle glaucoma is documented in more than 80% of cases, but angle-closure glaucoma is linked to significant visual loss. Trauma, drugs (corticosteroids), inflammation, tumors, and other conditions such as pigment dispersion or pseudo-exfoliation may cause secondary glaucoma.

4.3.4 Cataracts

Reversible blindness known as a cataract is brought on by the lens becoming opaque as we age as a result of cardiovascular, metabolic, and environmental conditions. Children can develop cataracts as well [17]. Secondary cataracts are brought on by ocular surgery and trauma due to hyperglycemia, which can result in oxidative stress and eventually induce cataract, intraocular inflammation, infection, or posterior capsule opacification. Cataract symptoms include difficulty recognizing colors, hazy or double vision, sensitivity to strong lights, and difficulty seeing at night. UV ray exposure and smoking are risk factors for cataracts [18]. Surgery is a possible form of cataract treatment. The side effects of cataract surgery can include endophthalmitis and retinal detachment. Patients suffer a sharp decline in vision here. This will result in faults in one visual field and blurred vision,

4.3.5 Conjunctivitis

Conjunctivitis is an inflammatory condition that affects the conjunctiva portion of the eye and is typically brought on by allergens from the environment. Specifically, an immunological hypersensitivity disease is the source of the allergy. The prevalence of allergic conjunctivitis has been estimated to be between 3 and 40% of the population, and it is projected that this number will rise steadily over time. Photophobia, dryness, swollen eyelids, conjunctival hyperemia, mucous or watery discharge, conjunctival swelling (chemosis), and tarsal conjunctival papillary response are some of the clinical symptoms of conjunctivitis.

4.4 NANOTECHNOLOGICAL APPROACH FOR OCULAR DRUG DELIVERY SYSTEMS

4.4.1 Nanoparticles

In addition to providing better platforms for diagnosis, imaging, and treatment, nanotechnology will also ensure a deeper understanding of disease molecular pathophysiology [19]. These developments may boost the therapeutic effectiveness of medication and improve patient outcomes. The use of nanoparticle drug delivery in glaucoma patients has enormous potential. Due to the numerous built-in defenses, it can be difficult to get enough therapeutic chemicals to the target tissues of the eye.

A nanoparticle is made up of three layers: the surface layer, the support layer, and the functionalization layer. An object can be functionalized using small molecules, metal ions, surfactants, and polymers. The shell and the core make up the other two layers. The center of the nanoparticle is where the core is found [20]. Di Huang et al. (2017) integrated peptide into hyaluronic acid (HA)–coated human serum albumin (HSA) nanoparticles and found the sustained release of peptide to boost the bioactivity of Cx43 MP with targeted delivery to the retina. They discovered that HA-coated HSA NPs exhibited better ex vivo retinal penetration and in vitro cellular absorption as compared to uncoated particles. In interactions mediated by the HA-CD44 receptor, they noticed this impact. Their findings revealed bioactivity that lasted longer and was protected from deterioration. HA-coated HSA NPs with positive potential to treat diverse retinal inflammatory diseases showed sustained and targeted delivery impact [21].

In order to increase ocular drug bioavailability by enhancing the stability of peptides, Vasconcelos et al. developed poly(lactic-co-glycolic acid) (PGLA)-polyethylene glycol (PEG) nanoparticles (NPs) loaded with connexin43 mimetic peptide in 2015. This method may prevent retinal ischemic and inflammatory disorders. They started by making PLGA NPs; then they added PEG and peptide conjugation. Proton nuclear magnetic resonance spectroscopy and a colorimetric test were used to validate the PEG and peptide conjugation. They used flurbiprofen as a control to examine the in vitro properties of the created bionanoparticles for their morphology, in vitro release, cell survival, and ocular tolerance. The anti-inflammatory effectiveness of sodium arachidonate was tested in rabbit eyes following topical application. The discovered peptide-loaded nanoparticles have a positive charge that makes it easier for medications to penetrate the cornea. The hen's egg testCAM assay [22] showed that the produced nanoparticles are not harmful to the eyes.

In a 2019 study by Kim et al., they compared the antioxidant effects of curcumin and rosmarinic orursolic acid in retinal epithelial cells. They found that rosmarinic or ursolic acid significantly increased drug solubility while reducing reactive oxygen species (ROS) formation. They came to the conclusion that nanoformulations made with albumin could improve the bioavailability and anti-oxidative effects of rosmarinic or ursolic acid in the retina [23].

Tetrandrine (TET) was created in 2022 by El-Sayed Radwan et al. in bovine serum albumin with chitosan-loaded nanoparticles for the topical administration of glaucoma. Investigations on pH and cross-linking agent concentrations continued. They measured the particle size as 237.9 nm, the zeta potential 24 mV, and the sustained release pattern 95%. Enhanced transcorneal permeability was demonstrated in this chitosan-based formulation. For chitosan-based formulations on corneal stromal fibroblasts, improved biocompatibility with improved cellular absorption and improved antioxidant and anti-proliferative action were reported. As compared to free drugs, CS-TET-BSA-NPs were shown to be twice as prevalent in rabbit aqueous humor. The optimal model for treating glaucoma was suggested by the observation of a prolonged decrease in intraocular pressure in the rabbit glaucoma model [24].

4.4.2 Nanomicelles

Nanomicelles are amphiphilic molecules that self-assemble in aqueous media to produce arranged supramolecular structures. The size of nanomicelles can range from 10 to 1000 nm, and they can have a star-shaped, spherical, or cylindrical morphology. At a particular concentration, these nanomicelles may produce critical micelle concentrations (CMCs). The force promoting self-assembly and maintaining supramolecular assembly is crucial for the development of nanomicelles in conventional micellar structures. A corona-forming block has succeeded in achieving the solubility of micelles in water. The ability of nanomicelles to solubilize and stable hydrophobic substances allows for longer retention of these particles [25]. Core cross-linked polymeric micelle technology has been established to stop premature disintegration and release of therapeutic cargo. The cross-linking of the core and corona involves the covalent bond, hydrogen bond, or stacking.

To increase the solubility and transcorneal penetration of quercetin into the eye, Moghimipour et al. (2022) developed quercetin microemulsions. They used the phase diagram method to generate the formulation. The formulation was created using a variety of excipients, including co-surfactant (Tween 20), surfactant (oleic acid-Transcutol P), and surfactants (Tween 80) (propylene glycol). As a result of flow (58.8 g/cm^2/h) and diffusivity coefficients (0.009 cm^2/h), their findings showed that 98.06% of the quercetin was released after 24 h. They noticed 16.11% of the medication passing through the cornea. By improving transcorneal permeability, the formulation made with Transcutol P lowered the amounts of phospholipids across epithelial cells [26]. Table 4.1 displays the nanoformulation strategies reported for the management of glaucoma.

Since drugs cannot cross the blood-retinal-barrier efficiently, various types of nanoparticles have been developed for improving DR treatment. Gold NPs of 20 nm could pass through the blood-retinal-barrier, but 100-nm AuNPs were not observed in any of the retinal layers when they were administered intravenously. These findings suggested that nanoparticles with a small enough diameter might overcome the limited penetration of therapeutic agents with larger sizes across the inner blood-retinal barrier. Similarly, the periocular route is also a promising path for therapeutic nanoparticles, because the sclera has a large surface area and relatively high permeability. Nanoparticles could penetrate into the outer blood-retinal-barrier. Current therapeutic attempts to use various nanoparticles in the treatment of angiogenesis, retinal degeneration, and uvetis are summarized in Tables 4.2, 4.3, and 4.4. PLGA

TABLE 4.1
Novel Nanoformulations for Treatment of Glaucoma

Particle Type	Formulation Approaches	Ocular Applications	Reference
Nanoparticles	Dorzolamide-loaded PLGA/vitamin E TPGS nanoparticles	To enhance the encapsulation efficiency with two-emulsifier polyvinyl alcohol and vitamin E TPGS	27
Nanoparticles	Methazolamide-loaded solid lipid chitosan nanoparticles	Lowers the IOP (38%)	28
DNA-nanoparticles	Non-viral surfactant-phospholipid nanoparticles for intravitreal and topical administration	Thermodynamically stable and non-aggregated particles	29
Chitosan nanoparticles	Dorzolamide-loaded chitosan nanoparticles	Enhances the bioavailability and efficacy of dorzolamide for glaucoma treatment	30
Superparamagnetic nanoparticles	Local magnetic hyperthermia using engineered superparamagnetic nanoparticles	Diffuses NPs through the vitreous body to the retina in rat eye for neuroprotection modality in glaucoma treatment	31

TABLE 4.2
Novel Nanoformulations for Treatment of Angiogenesis

Material	Type	Size (nm)	Administration	Application	References
Gold	Nanoparticle	20	IVT	Retinal neovascularization inhibition	32
Gold	Nanoparticle	50	–	Block VEGF induced retinal neovascularization	33
Nanoceria	Nanoparticle	3–5	IVT	Prevents downstream effects of oxidative stress in vivo	34
PLA/ PLA-PEO	Nanocapsule	302	IVT	Potential for intravitreal therapy	35
PLGA	Nanoconjugate	270–420	IV	Targeted gene delivery	36
PLGA-Chitosan	Nanocapsule	260	IVT	Receptor and signaling pathway mediating anti-inflammatory activity	37
PLGA-Chitosan	Nanocapsule	260	IVT	Nanoparticle-mediated gene delivery	38
Silicate	NPs	57	IVT	antiangiogenic effect	39
Silver	NPs	50	–	Inhibit cell survival via PI3K/ Akt-dependent pathway	40, 41

Note: IVT: Intravitreal.

TABLE 4.3
Novel Nanoformulations for Treatment of Retinal Degeneration

Material	Type	Size (nm)	Administration	Application	References
Gelatin	Nanoconjugate	585	IVT	Potential short-term treatment for photoreceptor degeneration in humans	42
Lipid	Nanocapsule	40–200	Topical	Potential to slow the progression of retinitis pigmentosa in a mammalian model	43
Nanoceria	Nanoparticle	–	IVT	Direct therapy for multiple diseases	44
Nanoceria	Nanoparticle	–	Intracardial	Therapeutic treatment of inherited retinal degeneration	45
PEG	Nanoconjugate	175.9	Subretinal	Nonviral gene therapy for retinal diseases	46
PEG	Nanocapsule	–	Subretinal	Gene replacement therapy	47

Note: IVT: Intravitreal.

TABLE 4.4
Novel Nanoformulations for Treatment of Uvetis

Material	Type	Size (nm)	Administration	Application	References
PEG	Nanocapsule	95–112	IVT	Tamoxifen-loaded NPs may represent a new option for the experimental treatment of uveitis	48, 49
PLA	Nanocapsule	100–200	IV	Systemic administration of betamethasone phosphate PLA NPs may lead to a new therapeutic strategy in controlling intraocular inflammation	50

Note: IVT: Intravitreal.

nanocapsules encapsulating drugs also ameliorate retinal neovascularization in an oxygen-induced retinopathy mice model and vascular leakage in a streptozocin-induced diabetes model. These studies showed that nanocapsules of size around 300 nm effectively deliver antiangiogenic molecules to the retina.

Few nanomedicines have received US Food and Drug Administration (FDA) approval after being tested in clinical studies. The development of formulations for use in ocular medication delivery applications has favorable effects on clinical studies. Cyclosporin is the first commercially available ocular nanoformulation (Restasis) used for the treatment of dry eye disorders. It is legal to use Durezol, a difluprednate-loaded nanoemulsion, to treat ocular inflammation. FDA approval for intravenous injection in the treatment of choroidal neovascularization was given to the liposomal formulation of Visudyne, an ocular nanomedicine formulation known as verteporfin commercialized by Novartis Pharma for age-related macular generation. The FDA approved the intravitreal injection of polyethylene glycol anti-VEGF aptamer, also known as Macugen, for the treatment of wet age-related macular degeneration in 2004. Phase II clinical trials using latanoprost-loaded liposome for the treatment of ocular primary open-angle glaucoma are now underway.

4.5 CONCLUSION

There is great potential for using nanotechnology in ocular research. Nanomedicine has the potential to increase the bioavailability of several pharmaceuticals. The many forms of nanomedicines used include liposomes, dendrimers, nanomicelles, nanodispersions, nanoemulsions, and others. Ocular nanomedicine has the potential to improve bioavailability and pharmacokinetic properties. This chapter examined the physiology of the eye, the associated ocular barriers that pose a challenge for medication administration, ocular disorders, and novel bionanoformulations described for the treatment of serious ocular diseases. The simple scale-up and reproducibility of ocular nanomedicine necessitate technological innovation and cost-effective commercialization tactics.

REFERENCES

[1] Patel A, Cholkar K, Agrahari V, Mitra AK. Ocular drug delivery systems: an overview. *World J Pharmacol*. 2013;2(2):47–64.

[2] Gaudana R, Ananthula HK, Parenky A, Mitra AK. Ocular drug delivery. *AAPS J*. 2010;12(3):348–360.

[3] Willoughby CE, Ponzin D, Ferrari S, Lobo A, Landau K, Omidi Y. Anatomy and physiology of the human eye: effects of mucopolysaccharidoses disease on structure and function—a review. *Clin Experiment Ophthalmol*. 2010;38(1):2–11.

[4] Gaudana R, Ananthula HK, Parenky A, Mitra AK. Ocular drug delivery. *AAPS J*. 2010;12(3):348–360.

[5] Kwatra D, Mitra AK. Drug delivery in ocular diseases: barriers and strategies. *World J Pharmacol* 2013;2(4):78–83. https://doi.org/10.5497/wjp.v2.i4.78.

[6] Bachu RD, Chowdhury P, Al-Saedi ZHF, Karla PK, Boddu SHS. Ocular drug delivery barriers-role of nanocarriers in the treatment of anterior segment ocular diseases. *Pharmaceutics*. 2018;10(1):28.

[7] Gheorghe A, Mahdi L, Musat O. Age-related macular degeneration. *Rom J Ophthalmol*. 2015;59(2):74–77.

[8] Mullins RF, Russell SR, Anderson DH, et al. Drusen associated with aging and age-related macular degeneration contain proteins common to extracellular deposits associated with atherosclerosis, elastosis, amyloidosis, and dense-deposit disease. *Faseb J*. 2000;14:835–846.

[9] Johnson LV, Leitner WP, Staples MK, et al. Complement activation and inflammatory processes in Drusen formation and age related macular degeneration. *Exp Eye Res*. 2001;73:887–896.

[10] Green WR, McDonnell PJ, Yeo JH. Pathologic features of senile macular degeneration. *Ophthalmology*. 1985;92:612–627.

[11] Pauleikhoff D, Harper CA, Marshall J, et al. Aging changes in Bruch's membrane. A histochemical and morphologic study. *Ophthalmology*. 1990;97:171–178.

[12] Martin D, Maguire M, Fine S, et al. Ranibizumab and bevacizumab for neovascular age-related macular degeneration (AMD). *N Engl J Med*. 2011;364:1897–1908.

[13] Klein R, Klein BE, Knudtson MD, et al. Fifteen-year cumulative incidence of age-related macular degeneration: the Beaver Dam Eye Study. *Ophthalmology*. 2007;114:253–262.

[14] Cohen SY, Dubois L, Tadayoni R, et al. Prevalence of reticular pseudodrusen in age-related macular degeneration with newly diagnosed choroidal neovascularisation. *Br J Ophthalmol*. 2007;91:354–359.

[15] Schmitz-Valckenberg S, Steinberg JS, Fleckenstein M, et al. Combined confocal scanning laser ophthalmoscopy and spectral-domain optical coherence tomography imaging of reticular drusen associated with age-related macular degeneration. *Ophthalmology*. 2010;117:1169–1176.

[16] Bidwai P, Gite S, Pahuja K, Kotecha K. A systematic literature review on diabetic retinopathy using an artificial intelligence approach. *Big Data Cogn Comput*. 2022;6:152.

[17] Weinreb RN, Aung T, Medeiros FA. The pathophysiology and treatment of glaucoma: a review. *JAMA*. 2014;311(18):1901–1911.

[18] Ioanna M, Tsinopoulos I. A critical appraisal of new developments in intraocular lens modifications and drug delivery systems for the prevention of cataract surgery complications. *Pharmaceuticals*. 2020;13(12):448.

[19] Yan, T., Ma, Z., Liu, J., et al. Thermoresponsive Genistein NLC-dexamethasone-moxifloxacin multi drug delivery system in lens capsule bag to prevent complications after cataract surgery. *Sci Rep*. 2021;11:181.

[20] Khan I, Saeed K, Khan I. Nanoparticles: properties, applications and toxicities. *Arab J Chem*. 2019;12(7):908–931.

[21] Huang D, Chen Y-S, Rupenthal ID. Hyaluronic acid coated albumin nanoparticles for targeted peptide delivery to the retina. *Mol Pharmaceutics*. 2017;14(2):533–545.
[22] Vasconcelos A, Vega E, Pérez Y, Gómara MJ, García ML, Haro I. Conjugation of cell-penetrating peptides with poly(lactic-co-glycolic acid)-polyethylene glycol nanoparticles improves ocular drug delivery. *Int J Nanomedicine*. 2015;10(1):609–631.
[23] Kim D, Maharjan P, Jin M, Park T, Maharjan A, Amatya R, Yang JW, Ah Min K, Cheol Shin M. Potential albumin-based antioxidant nanoformulations for ocular protection against oxidative stress. *Pharmaceutics*. 2019;11:297.
[24] El-Sayed Radwan S, El-Moslemany RM, Mehanna RA, Thabet EH, Abdelfattah E-ZA, El-Kamel A. Chitosan-coated bovine serum albumin nanoparticles for topical tetrandrine delivery in glaucoma: in vitro and in vivo assessment. *Drug Deliv*. 2022;29(1): 1150–1163.
[25] Vadlapudi AD, Mitra AK. Nanomicelles: an emerging platform for drug delivery to the eye. *Ther Deliv*. 2013;4(1):1–3.
[26] Moghimipour E, Farsimadan N, Salimi A. Ocular delivery of quercetin using microemulsion system: design, characterization, and ex-vivo transcorneal permeation. *Iran J Pharm Res*. 2022;21(1):e127486.
[27] Mengtan C, Kun Z, Yongbin Q, Xinrong L, Yuanwei C, Xianglin L. pH and redox-responsive mixed micelles for enhanced intracellular drug release. *Colloids Surf B Biointerfaces*. 2014;116:424–431.
[28] Wang F, Chen L, Zhang D, Jiang S, Shi K, Huang Y, Li R, Xu Q. Methazolamide-loaded solid lipid nanoparticles modified with low-molecular weight chitosan for the treatment of glaucoma: vitro and vivo study. *J. Drug Target*. 2014;22:849–858.
[29] Alqawlaq S, Sivak JM, Huzil JT, Ivanova MV, Flanagan JG, Beazely MA, Foldvari M. Vehicles, delivery systems, pharmacokinetics, formulation. *Nanomedicine*. 2014; 10:1637–1647.
[30] Katiyar S, Pandit J, Mondal RS, Mishra AK, Chuttani K, Aqil M, Ali A, Sultana Y. In situ gelling dorzolamide loaded chitosan nanoparticles for the treatment of glaucoma. *Carbohydr Polym*. 2014;15:117–124.
[31] Jeun M, Jeoung JW, Moon S, Kim YJ, Lee S, Paek SH, Chung KW, Park KH, Bae S. In vitro application of Mn-ferrite nanoparticles as novel magnetic hyperthermia agents. *Biomaterials*. 2011;32:387–394.
[32] Kim JH, Kim MH, Jo DH, Yu YS, Lee TG, Kim JH. The inhibition of retinal neovascularization by gold nanoparticles via suppression of VEGFR-2 activation. *Biomaterials*. 2011;32:1865–1871.
[33] Kalishwaralal K, Sheikpranbabu S, BarathManiKanth S, Haribalaganesh R, Ramkumarpandian S, Gurunathan S. Gold nanoparticles inhibits vascular endothelial growth factor- induced angiogenesis and vascular permeability via src dependent pathway in retinal endothelial cells. *Angiogenesis*. 2011;14:29–45.
[34] Zhou X, Wong LL, Karakoti AS, Seal S, McGinnis JF. Nanoceria inhibit the development and promote the regression of pathologic retinal neovascularization in the Vldlr knockout mouse. *PLoS One*. 2011;6:e16733.
[35] Kim H, Csaky KG. Nanoparticle-integrin antagonist C[16]. peptide treatment of choroidal neovascularization in rats. *J Control Release*. 2010;142:286–293.
[36] Singh SR, Grossniklaus HE, Kang SJ, Edelhauser HF, Ambati BK, Kompella UB. Intravenous transferrin, RGD peptide and dual-targeted nanoparticles enhance anti-VEGF intraceptor gene delivery to laser-induced CNV. *Gene Ther*. 2009;16:645–659.
[37] Jin J, Zhou KK, Park K, Hu Y, Xu X, Zheng Z, Tyagi P, Kompella UB, Ma JX. Anti-inflammatory and antiangiogenic effects of nanoparticle-mediated delivery of a natural angiogenic inhibitor. *Invest Ophthalmol Vis Sci*. 2011;52:6230–6237.

[38] Park K, Chen Y, Hu Y, Mayo AS, Kompella UB, Longeras R, Ma JX. Nanoparticle-mediated expression of an angiogenic inhibitor ameliorates ischemia-induced retinal neovascularization and diabetes-induced retinal vascular leakage. *Diabetes*. 2009;58:1902–1913.

[39] Jo DH, Kim JH, Yu YS, Lee TG, Kim JH. Antiangiogenic effect of silicate nanoparticle on retinal neovascularization induced by vascular endothelial growth factor. *Nanomedicine*. 2012;8:784–791.

[40] Kalishwaralal K, Banumathi E, Ram Kumar Pandian S, Deepak V, Muniyandi J, Eom SH, Gurunathan S. Silver nanoparticles inhibit VEGF induced cell proliferation and migration in bovine retinal endothelial cells. *Colloids Surf B*. 2009;73:51–57

[41] Kalishwaralal K, Barathmanikanth S, Pandian SR, Deepak V, Gurunathan S. Silver nano – a trove for retinal therapies. *J Control Release*. 2010;145:76–90.

[42] Sakai T, Kuno N, Takamatsu F, Kimura E, Kohno H, Okano K, Kitahara K. Prolonged protective effect of basic fibroblast growth factor–impregnated nanoparticles in royal college of surgeons rats. *Invest Ophthalmol Vis Sci*. 2007;48:3381–3387.

[43] Strettoi E, Gargini C, Novelli E, Sala G, Piano I, Gasco P, Ghidoni R. Inhibition of ceramide biosynthesis preserves photoreceptor structure and function in a mouse model of retinitis pigmentosa. *Proc Natl Acad Sci USA*. 2010;107:18706–18711.

[44] Chen J, Patil S, Seal S, McGinnis JF. Rare earth nanoparticles prevent retinal degeneration induced by intracellular peroxides. *Nat Nanotechnol*. 2006;1:142–150.

[45] Kong L, Cai X, Zhou X, Wong LL, Karakoti AS, Seal S, McGinnis JF. Nanoceria extend photoreceptor cell lifespan in tubby mice by modulation of apoptosis/survival signaling pathways. *Neurobiol Dis*. 2011;42:514–523.

[46] Read SP, Cashman SM, Kumar-Singh R. POD nanoparticles expressing GDNF provide structural and functional rescue of light-induced retinal degeneration in an adult mouse. *Mol Ther*. 2010;18:1917–1926.

[47] Cai X, Conley SM, Nash Z, Fliesler SJ, Cooper MJ, Naash MI. Gene delivery to mitotic and postmitotic photoreceptors via compacted DNA nanoparticles results in improved phenotype in a mouse model of retinitis pigmentosa. *FASEB J*. 2010;24:1178–1191.

[48] Cai X, Nash Z, Conley SM, Fliesler SJ, Cooper MJ, Naash MI. A partial structural and functional rescue of a retinitis pigmentosa model with compacted DNA nanoparticles. *PLoS ONE*. 2009;4:e5290.

[49] de Kozak Y, Andrieux K, Villarroya H, Klein C, Thillaye-Goldenberg B, Naud MC, Garcia E, Couvreur P. Intraocular injection of tamoxifen-loaded nanoparticles: a new treatment of experimental autoimmune uveoretinitis. *Eur J Immunol*. 2004;34:3702–3712.

[50] Sakai T, Kohno H, Ishihara T, Higaki M, Saito S, Matsushima M, Mizushima Y, Kitahara K. Treatment of experimental autoimmune uveoretinitis with poly(lactic acid) nanoparticles encapsulating betamethasone phosphate. *Exp Eye Res*. 2006;82:657–663.

5 Oxide-Based Nanomaterials for Drug Preparation and Delivery

Karna Wijaya, Wahyu Dita Saputri, Amalia Kurnia Amin, Latifah Hauli, Hilda Ismail, Budhijanto Budhijanto, Aldino Javier Saviola, and Won-Chun Oh

5.1 INTRODUCTION

In the world of materials science, the development of nanotechnology continues to increase, and it significantly impacts human life. Materials at the nano scale provide physical and chemical properties that are unique and different from materials in bulk size. The application of oxide-based nanomaterials is increasingly expanding into various fields of life, such as drug preparation and drug delivery into the human body. For drug preparation, using nanocatalysts to accelerate organic reactions offers multiple advantages that bring the pharmaceutical industry closer to the green industry. The catalytic properties of materials at the nano scale are determined by the size, shape, distribution, and support of the nanomaterials and the reaction conditions (Challener, 2016). As materials with a size of 1 to 100 nm, nanomaterials have many atoms on their surface and form weak bonds with lattice atoms. Therefore, these atoms have high surface energy and are not physically stable, so they are chemically active against chemical reactions (Zhang, 2016; Khan *et al.*, 2021). Due to their much smaller size than their bulk form, nanocatalysts have a larger area, and as a consequence, they have a greater reactivity to many organic reactions.

Research on developing nanomaterials as catalysts in drug preparation reactions has been widely carried out. One example of oxide-based nanomaterials developed as nanocatalysts in drug preparation includes sulfated silica (SO_4/SiO_2). Paracetamol, a drug that has long been known as an analgesic-antipyretic, can be synthesized through several synthesis pathways, one of which uses nitrobenzene compounds. Up to now, the production of nitrobenzene in the industry uses concentrated sulfuric acid as a catalyst that will help the formation of active electrophile sites from concentrated nitric acid, resulting in electrophilic substitution reactions to benzene. However, the use of concentrated sulfuric acid in the industry causes various serious problems caused by its high toxicity and regeneration difficulty at the end of the reaction (Koskin *et al.*, 2016). Solid acid catalyst materials such as sulfated silica can be an alternative

 DOI: 10.1201/9781003425427-7

solution to replace the role of sulfuric acid catalysts in the nitrobenzene industry. SO_4/SiO_2 materials have also been reported to have good function in catalyzing the formation of α-amino phosphonates, compounds that play an essential role as drug preparation intermediates (Gawande *et al.*, 2014).

Another oxide-based nanomaterial developed in drug preparation is a magnetic material. The use of magnetic nanocatalysts has been widely reported in the development of anticancer drugs because it offers advantages such as easy separation from the product and high reusability, high selectivity to prevent the formation of by-products, and lower production costs in the industry (Bardajee *et al.*, 2016; Kaur *et al.*, 2019). For example, Ni-Fe_2O_4 material synthesized by the hydrothermal method was successfully applied as a nanocatalyst in synthesizing acetylferoccene chalcones, an anti-colon cancer candidate compound with good activity and selectivity that can be used up to six times running (Abu-Dief *et al.*, 2016). Due to the recoverable character of the reaction medium using magnetic utilization, magnetic nanocatalysts may be introduced as promising catalysts for the pharmaceutical industry.

Oxide-based nanomaterials also play an essential role as biomaterials in drug delivery. Over the past few years, the field of nanomedicine, which applies materials science to the medical field and the treatment of diseases, has continued to proliferate (Patra *et al.*, 2018). Figure 5.1 schematically illustrates targeted drug delivery by oxide-based nanoparticles. Regarding drug delivery, computational chemistry methods are necessary to improve the efficiency of drug performance in target tissues in the body while minimizing accumulation in non-target tissues (Garofalo *et al.*, 2020). Nanoscale materials can move more freely in the human body than bulk materials because they have unique physicochemical properties that can optimize the therapeutic effect on the target tissue. The interaction between nanomaterials and biomolecules can occur both on the surface and inside the cell. Nanomaterials have more substantial penetration power and have been successfully applied in diagnosing and treating cancer cells (Hong *et al.*, 2011). Metal oxide nanomaterials are reported with a large area/volume ratio, making them very useful for cancer treatments. Metal oxide nanomaterials can encapsulate therapeutic agents, thereby solving the problem of drug stability and solubility if used alone (Sharma *et al.*, 2016). This chapter will provide a review of the research journey on oxide-based nanomaterials and their applications in drug preparation and delivery in the human body.

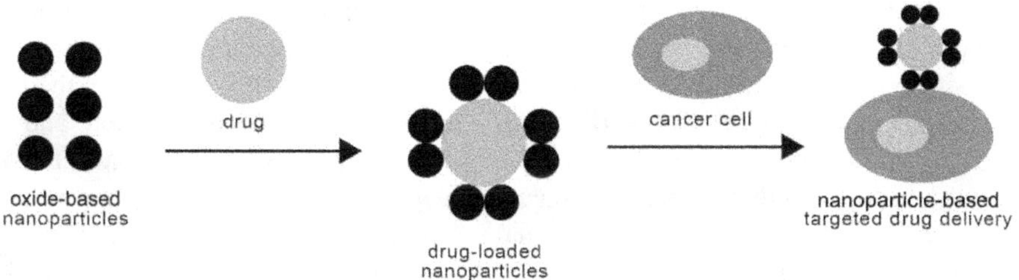

FIGURE 5.1 Schematic illustration of nanomaterials for targeted drug delivery.

5.2 OXIDE-BASED NANOMATERIALS

Nanotechnology is defined as the design, manufacture, characterization, and application of a structure or system by controlling the shape and size at the nanometer scale. Many experts have tried to define the limits of nanomaterial grouping. However, at present, the terminology of nanomaterials, generally, is materials with dimensions in the range of 1–100 nm. The investigation and development of nanomaterials do not rely on just one scientific discipline but are interdisciplinary and multidisciplinary (Wijaya, Heraldy *et al.*, 2021). As in nanomedicine, the scope of preparation includes designing, synthesizing, and delivering drugs in the human body, combining scientific fields such as materials chemistry, computational chemistry, pharmacy, biology, and chemical engineering. After being modified to nanometer dimensions, the improvement in material properties is why nanomaterial fabrication for drug preparation and delivery has increased in recent years.

The use of nanomaterials in manufacturing nanocatalysts is very concerned with pore size and surface area. Therefore, research on porous nanomaterials is increasing in line with its needs, such as in the industrial field. Oxide-based nanomaterials are generally porous materials with micropore and mesopore dimensions. Setting the pore dimensions is also necessary to improve the catalytic performance of the nanocatalyst. This chapter will provide some examples of oxide-based nanomaterials that have been studied by us.

5.2.1 Silica

Silica (SiO_2) is one of the most promising oxide-based nanomaterials for pharmaceutical applications. Mesoporous silica is a kind of silica that may be prepared, and it is considered particularly effective in delivering drugs (Budiman and Aulifah, 2021). Many methods have been used to synthesize mesoporous silica. Mirzaei *et al.* (2020) highlighted several conditions that can impact the mesoporous silica synthesis process, including silica source used, pH, temperature, solvents, surfactants, and co-surfactants. Several sources of silica that are often used for the preparation of mesoporous structure are tetraethyl orthosilicate (TEOS), sodium silicate, sodium metasilicate, tetramethyl ortho-silicate (TMOS), tetrapropyl ortho-silicate (TPOS), trimethoxysilane (TMS), and tetrabutoxysilane (TBOS) (Liu *et al.*, 2007). Organosilanes like tetraethyl ortho-silicates, sodium silicates, and colloidal solutions are all examples of precursors that could be used to create ordered mesoporous materials (Blin *et al.*, 2001). pH conditions are one of the important factors for synthesizing mesoporous materials. Park *et al.* (2012) reported that the charge density at the interface between organic and inorganic molecules most likely had a role in how the hexagonal mesophase formed. Thus, the water content, guest ion concentration, and pH were crucial factors in the synthesis of mesoporous materials since they all affected charge density. By reducing pH to a range between 10.4 and 10.6, which improved the condensation of silicate, the best hexagonal mesoporous material was created.

Borówka and Skrzypiec (2021) also investigated mesoporous silica materials that were templated using ionic surfactant and how the structure was affected by the temperature at which they were created. The result revealed that at the step of forming the template framework, a too-low synthesis temperature causes local structural disorder (in the form of crystallites or lamellar phase), which causes the production of more pores after calcination. When silica material was precipitated at a temperature just a little bit higher than the Krafft point of the surfactant employed as a porosity template, the best structure with a smooth surface, homogenous pores, and long-range ordering developed. Surfactants and co-surfactants certainly have a significant role in the synthesis of mesoporous silica structures. Surfactants of many types, including ionic surfactants, cationic surfactants, non-ionic surfactants, and amphoteric surfactants, are applied to create mesoporous silica. Alcohols, one of the types of co-surfactants, not only have a major impact on the ductility and size of the pores but also, as their concentrations rise, reduce the proportion of spherical forms in the corresponding mesoporous materials (Mirzaei *et al.*, 2020). In biomedical applications, mesoporous silica has been developed as a nano-drug delivery system (nano-DDS). It has become an alternative method for improving the solubility and bioavailability of drugs, and their absorption by the body through the digestive system (Dening *et al.*, 2019; Mu *et al.*, 2020). Mesoporous silica nanomaterials may be used to transport bioactive molecules and preserve the drug molecules from deterioration while allowing for targeted release, longer blood circulation, enhanced disease targeting, and minimized adverse effects (Casasús *et al.*, 2004).

Silica nanoparticles can be synthesized through two approaches, bottom-up and top-down. For the bottom-up approach, the sol-gel method synthesized silica nanoparticles using tetraethyl orthosilicate (TEOS) precursor. In the sol-gel method, there are two reaction stages, hydrolysis and condensation. Figure 5.2 explains the mechanism for the formation of silica nanoparticles from TEOS precursors using the sol-gel method which consists of hydrolysis and condensation stages (Colleoni et al., 2016). This reaction involves using solvents for TEOS, such as ethanol, to reduce the hydration of water. In the hydrolysis stage, silanol groups (Si-O-H) formation occurs. The silanol groups will undergo a condensation reaction as a gel as siloxane groups (Si-O-Si).

The concentrated sulfuric acid solution is charged on SiO_2 nanoparticles in ethanol to obtain SO_4/SiO_2 solids. In the process of adding sulfuric acid, SO_4^{2-} ions as ligands will donate their free electron pairs through O atoms so that they will form coordination bonds with Si^{4+} as the central atom (Wijaya, Lammaduma Malau *et al.*, 2021). Figure 5.3 illustrates the structure of sulfated silica (Zarei et al., 2017). Modifying silica material by adding sulfate ions will increase its acid strength and reactivity as a catalyst (Aneu *et al.*, 2021). The presence of SO_4^{2-} ions on the catalyst surface from the sulfation process will provide Brønsted acid sites that increase the reaction product (Christiansen *et al.*, 2013). Sulfated silica has the potential to be used as a catalyst material in the manufacture of nitrobenzene where nitrobenzene is a raw material for the synthesis of paracetamol.

5.2.2 Zirconia

Zirconia (ZrO_2) is a material that has been utilized extensively in the field of dentistry. This material is extremely promising for use as a drug delivery system in the biomedical field (Priyadarsini *et al.*, 2018; Leonetti *et al.*, 2021). Sponchia *et al.*

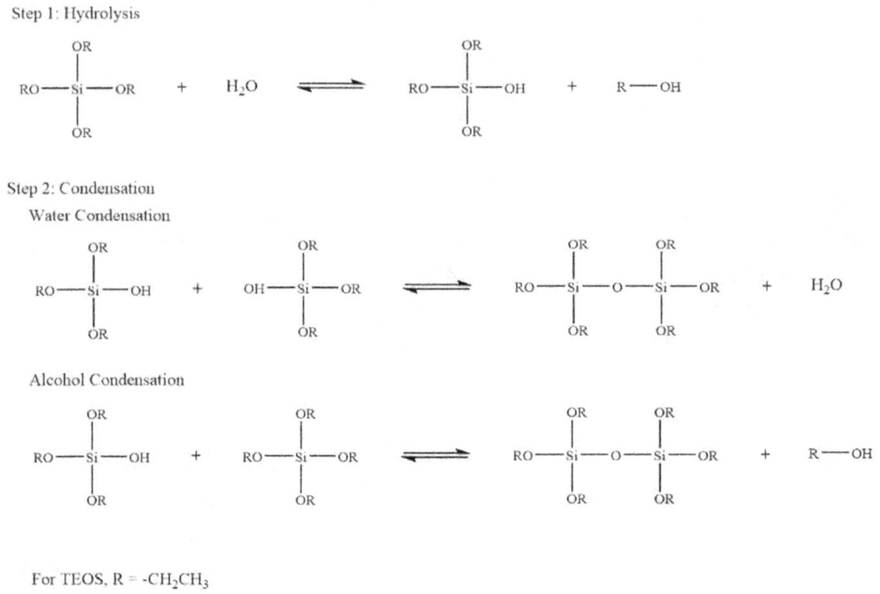

FIGURE 5.2 Reaction mechanism for the synthesis of silica nanoparticles from TEOS precursor.

Source: Reprinted from Colleoni *et al.* (2016). Copyright 2016 Royal Society of Chemistry.

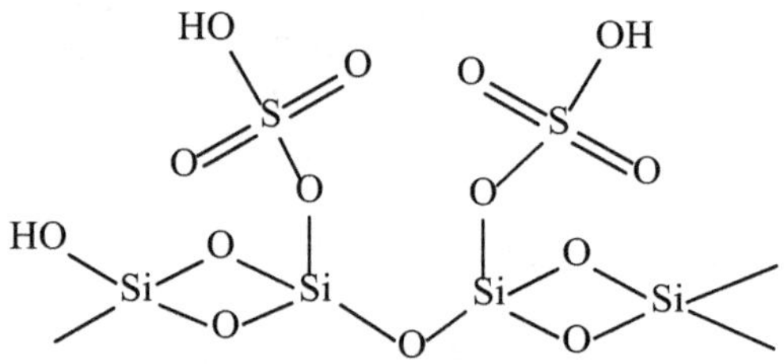

FIGURE 5.3 Structure of sulfated silica.

Source: Reprinted from Zarei *et al.* (2017). Copyright 2017 Elsevier.

(2015) reported on the controlled synthesis of novel mesoporous zirconia nanoparticles (MZNs) and showed that they were biocompatible, cell permeable, and degradable, making them an excellent choice for theranostic applications. Mabrouk *et al.* (2022) synthesized mesoporous zirconia nanoparticles and titania (MTNs) for bone regeneration and drug delivery system applications. These mesoporous nanocarriers demonstrated excellent biocompatibility and proliferation characteristics, according to the findings. Another study found that sulfate-modified zirconia promoted by iron-manganese metals could be used to treat cancer (Al-fahdawi *et al.*, 2015). Sulfated zirconia has also a high potential for use in this biomedical application as well.

Sulfated zirconia can be prepared by modifying zirconia with sulfate compound agents. The sulfate agent, the precursor employed, the process used, and the calcination temperature are some few factors that may determine how well the sulfated zirconia performs. All of these factors play a crucial part in enhancing the effectiveness of the material. Some of the precursors used for the preparation of sulfated zirconia are $ZrCl_4$, $ZrOCl_2{\cdot}8H_2O$ (Saravanan *et al.*, 2014), sulfated $Zr(OH)_4$ (Drago and Kob, 1997), $ZrO(NO_3)_2$ (Katada *et al.*, 2000), zirconium propoxide (Raissi *et al.*, 2015), and zirconium oxychloride (Patel *et al.*, 2013). Some researchers even employ ZrO_2 as a raw material to create sulfated zirconia without any intermediate steps. The difference in precursors certainly has an influence on the performance of the material. This is evident from the experiments carried out to synthesize ZrO_2 from different precursors, namely $ZrOCl_2{\cdot}8H_2O$ and $Zr(SO_4)_2{\cdot}H_2O$; the results showed that there were differences in morphology and particle size from the use of the two precursors (Siddiqui *et al.*, 2012). Sulfate agents that can be used in various studies on the preparation of sulfated zirconia material are H_2SO_4, $(NH_4)_2SO_4$, SO_3, SO_2, and so on. However, the most commonly used are sulfuric acid and ammonium sulfate. Observations of Hammet's acidity of $H_0 \leq -14.50$ were obtained on sulfated zirconia material with sulfur compounds which are in the highest oxidation state (SO_4^{2-}) (Sohn and Kim, 1989).

The wet impregnation, sol-gel, and hydrothermal methods are some of the commonly employed procedures for the production of sulfated zirconia. The wet impregnation method is one of the most frequently used methods for the production of sulfated zirconia material. This process is very simple to carry out, especially when using ZrO_2 as a direct precursor. ZrO_2 precursors can be directly reacted with a sulfate agent, then stirred and calcined at a certain temperature. The sulfate concentration and calcination temperature used will determine the performance of the resulting material (Hauli *et al.*, 2018; Utami *et al.*, 2017). When using other precursors to produce ZrO_2, this method is typically carried out in two steps, first preparing $Zr(OH)_4$ by reacting the zirconia precursor with precipitation agent and then a wet impregnation process to produce sulfated zirconia material by adding sulfate agent to $Zr(OH)_4$. Another well-established technique for synthesizing sulfated zirconia is the sol-gel method. This method has the advantage of allowing control over the homogeneity and physical characteristics of the material and can obtain more pure materials. In general, the sol-gel method involves the formation of sol by means of hydrolysis and condensation of precursors, especially metal alkoxide precursors. Condensation of sol particles into a three-dimensional network can produce zirconium hydroxide gel, and drying of the gel will produce zirconia solid products (Mishra *et al.*, 2003). This method can be used to prepare sulfated zirconia in either one or two steps. In the two-step stage, the first step involves hydrolysis and condensation of the zirconium precursor, followed by a sulfation process on the zirconia material as the second step. In a one-step process, sulfuric acid can be added to the zirconium precursor in propanol, then dried and calcined (Mishra *et al.*, 2004). The hydrothermal method is also widely used for the synthesis of sulfated zirconia material. The hydrothermal method also has the advantage of being able to produce catalysts with high purity through the use of low temperatures and a simple preparation process. Using this hydrothermal method, sulfated zirconia synthesis is accomplished in two steps. In the hydrothermal process, extracting $Zr(OH)_4$ from the precursor is often the initial step. Then, sulfated process is performed with a straightforward impregnating technique (La Ore *et al.*, 2020). Sulfated zirconia has also the potential to be used as a catalyst material in the manufacture of nitrobenzene where nitrobenzene is a raw material for the synthesis of paracetamol.

5.2.3 Titania

Titania (TiO_2) is a metal oxide material that has been widely used for photocatalytic activity. In addition, the utilization of this material can be developed for various applications, including in the biomedical field. The development of functionalized-titania (F-TiO_2) has been widely studied as a cancer treatment. López *et al.* (2014) created a functionalized-nanostructured titania for the treatment of cancer. Sulfates, amines, and phosphates are used to change the surface of nanostructured titania. The goal of this alteration is to produce a good biocompatible material that can interact with the charges of the hydrophobic heads on the lipid bilayer that compose the cell membrane. As a result of the finding that F-TiO_2 material had no discernible adverse effects on the cancer cell, it can be concluded that the nanomaterial is a strong candidate to be used as a drug delivery system. Modification of titania with sulfate is also widely used to synthesize the organic compounds. Nivetha *et al.* (2022) investigated the chalcone preparation using sulfated titania. Chalcone and its derivatives can be used as pharmaceuticals and agricultural chemicals. Because of the diverse range of biological activities of chalcones, their production has attracted a significant amount of attention from medicinal and organic chemists.

Sulfated titania is a material that can be synthesized from titanium with sulphate precursors. Sulfated titania is known as an acid catalyst that has good catalytic activity in various reactions such as esterification, isomerization, alkylation, acylation, and etherification. The increase in acidity of sulfated titania is due to the presence of Brønsted acid sites and Lewis acid sites (Hu *et al.*, 2015). Experimental factors influencing the formation of Brønsted and Lewis acid sites are preparation method, sulfate content, and calcination temperature (Noda *et al.*, 2005). Several methods have been used for the synthesis of sulfated titania, including impregnation, sol-gel, hydrothermal, ultrasonic, and seeding. Calcination temperature, acid concentration, and precursor used are important factors for good catalyst performance. The advantages of these methods include increased surface area, high purity of the catalyst, uniform particle size, enhanced thermal stability, well-defined distribution over the support, their accessibility, and the possibility of obtaining better microstructural control of the supporting materials (Nivetha *et al.*, 2022). Krishnakumar and Swaminathan (2011) reported the synthesis of sulfated titania by the sol-gel method to produce quinoxaline and dipyridophenazine derivatives. The results showed that sulfated titania as a catalyst produces excellent yields with a short reaction time and is solvent free. Additionally, this catalyst is affordable, environmentally friendly, reusable, and particularly effective at producing derivatives of the quinoxaline and dipyridophenazine. As well as sulfated silica and sulfated zirconia, sulfated titania has also the potential to be used as a catalyst material in the manufacture of nitrobenzene where nitrobenzene is a raw material for the synthesis of paracetamol.

5.2.4 Ceria

Cerium dioxide (CeO_2) is an important oxide-based nanomaterial that finds applications in biomedical engineering; for instance, chitosan/cerium oxide nanoparticles were synthesized to improve the bioavailability of the drug based on the assembly of inorganic nanoparticles at the polymer–nanoparticle interfaces.

There are several methods available for the preparation of CeO_2 nanomaterial, each with its own advantages and disadvantages. The choice of the specific method depends on the desired properties of the CeO_2 nanomaterial and the intended application. Some commonly used methods for the preparation of CeO_2 nanomaterial are conventional sol-gel methods, chemical precipitation, and hydrothermal synthesis. In the sol-gel method, precursors such as cerium nitrate hexahydrate undergo hydrolysis, resulting in the production of a cerium hydroxide. The metal hydroxide then undergoes condensation, thus forming a gel. The chemical precipitation method involves the precipitation of CeO_2 nanoparticles from a solution of cerium salts, such as cerium nitrate or cerium chloride, using a precipitating agent, such as ammonia or sodium hydroxide. The resulting precipitate is washed and dried, followed by calcination at high temperatures to obtain CeO_2 nanoparticles (Nyoka *et al.*, 2020). In the hydrothermal synthesis method, a solution of cerium salt is mixed with a hydrothermal reaction medium, typically alkaline solution, in a sealed autoclave, which is heated to high temperatures and pressures. The resulting product is washed and dried, followed by calcination to obtain CeO_2 nanoparticles (Kovacevic *et al.*, 2016).

5.3 DRUG DELIVERY NANOMATERIALS

Designing site-specific stimuli-responsive controlled drug delivery systems (DDSs) is essential for the growth of the biomedical and pharmaceutical and personal healthcare industries. Researchers from all around the world are very interested in DDSs that can control biological properties like pharmacokinetics, pharmacodynamics, therapeutic index, biodistribution, and tissue uptake of therapeutic drugs. The development and application testing of nanomaterial-based drug delivery systems, in particular oxide-based nanomaterials such as iron oxides, titanium oxides, zinc oxides, zirconium oxides, silicon dioxides, and cerium oxides, has received considerable attention in the field of biomedical materials research because it is considered to have high effectiveness, sensitivity, efficiency, and efficacy.

Regarding intracellular drug delivery carriers, the cell membrane is the main barrier to be overcome. Nanocarrier DDSs, due to their small size, can increase permeability, allowing cell membranes and other bio-barriers to be penetrated more easily (Rizvi and Saleh, 2018; Xu *et al.*, 2021). The high ratio of surface area to volume makes nanomaterials superior in drug transportation because they allow more drug attachment to the surface of the material than its bulk form (Vangijzegem *et al.*, 2019; Singh *et al.*, 2020). Moreover, the surface of nanomaterials is easily modifiable with such surface functional modification, allowing drug biomolecules to adhere to and be delivered more precisely targeted on localized tissue with appropriate dosage and drug release controllability as a function of environmental parameters (Gupta and Gupta, 2005; Ebadi *et al.*, 2021).

Generally, metal oxide nanomaterials have very low toxicity and are biocompatible and biodegradable, making them relatively safe for the human body with few side effects (Vivero-Escoto *et al.*, 2010; Chen *et al.*, 2011; Liu *et al.*, 2016). Surface modification enhances their biocompatibility and stability in human aqueous systems, which electrophoretic mobility determinations as a function of pH were used to manage all stages of the surface modification (Rudzka *et al.*, 2013). Furthermore, DDS nanomaterial improved drug stability in the face of self-degradation or macrophage clearance during transport by encapsulating, conjugating, and so on (Zhang *et al.*, 2023). The use of DDS nanomaterial needs accurate design of nanocarriers and fabrication strategies in terms of numerous characteristics that must be evaluated in tandem to achieve maximum therapeutic effectiveness (Vangijzegem *et al.*, 2019).

Iron oxide nanoparticles (IONPs) are proficient in drug delivery systems, as they have outstanding and unique physicochemical properties, especially their superparamagnetic properties and surface function, as well as promising biocompatibility, biodegradability, high stability, nontoxicity, ease of fabrication, and low cost of fabrication. Maghemite (γ-Fe_2O_3), hematite (α-Fe_2O_3), and magnetite (Fe_3O_4) surrounded by surface modification like an organic and/or inorganic coating are the main forms of IONPs used, as they have a stable crystal structure and are magnetically active (Ansar *et al.*, 2015; Vangijzegem *et al.*, 2019; Ebadi *et al.*, 2021). It is well known that decreasing IONPs size to less than 100 nm creates superparamagnetic characteristics (Ansar *et al.*, 2015). Iron oxide particles (IOPs) were initially widely utilized as contrast agents in magnetic resonance imaging (MRI), and they later appeared for a variety of DDS therapeutic purposes.

Rudzka *et al.* (2013) reported surface functionalization of maghemite IOPs into magnetic core/silica/gold shell nanospheres that have efficient release of doxorobucin (DOX) drug in liver and colon cancer cells, accumulated in cell nuclei and cytoplasm. BSA-magnetite nanotorpedo (BMNT), which is a composite of magnetite nanoparticles and doxorubicin molecules encapsulated in a protein polymer nanocage of six sub-units of bovine serum albumin (BSA), was successfully fabricated and showed the ability of chemotherapeutic drug delivery, in which this nanomaterial can effectively overcome the problem of leakage of hydrophobic antitumor drug molecules because the interaction of DOX drugs with nanocarrier materials is stable in the pH range of blood or extracellular lysosomal environment of healthy tissues (7.5–5.5) and extends the half-life of drugs in the blood circulation so that drugs can be more targeted for release at localized sites or diseased tissue that tends to have a low pH (intracellular lysosomal environment). In addition to showing remarkable therapeutic effects, DOX loaded in BMNT DDSs can also be tracked by T1-weighted MRI and has been clinically tested to be highly biocompatible (Zhao *et al.*, 2023). Conjugated doxorubicin antitumor drug on iron core coated magnetite and polyethylenglycol (PEG) DDSs also shows reasonable realizing rate as a function of initial DOX concentration and pH condition (Gómez-Sotomayor *et al.*, 2015). Another anticancer or antitumor drug, sorafenib, has been successfully developed on iron oxide DDSs. Sorafenib coated on magnetite core-shell nanoparticles loaded into Zn/Al hydroxide double-layered and coated with polyvinyl alcohol (MPVASO-Zn/Al-LDH) fabricated by coprecipitation technique has physicochemical properties that support the enhancement of the drug transportation rate to the target tissue, allowing drug dosage to be reduced while still achieving an appropriate and easily controllable drug release. This nanocarrier exhibits zero cytotoxicity against healthy 3T3 fibroblast cells (Ebadi *et al.*, 2021).

The use of mesoporous silica nanoparticles (MSNs) as a base material for drug delivery is usually based on metal nanoparticles functionalized on its mesoporous surface, such as AuNPs, AgNPs, CeO_2NPs, and capped IOPs, as well as capped organic polymers like dendrimers, proteins, and so on. This modification is very useful for controlling the location and timing of drug release based on stimuli-responsive linkers. Xu *et al.* (2013), for instance, has successfully developed DSS based β-cyclodextrin@CeO_2 nanoparticle-capped ferrocene-MSN in which CeO_2 NPs triggered synergetic drug release via host–guest interactions and switchable enzymatic activity and cellular effects of CeO_2. Ferrocenyl moieties are converted to ferrocenium ions by CeO_2 lids after being internalized into lung carcinoma cells via a lysosomal pathway, which may lead to the uncapping of the CeO_2 and the release of the drug. The CeO_2 in this scenario functions as a multifunctional substance that not only serves as a lid but also has a synergistic antitumor impact on cancer cells because of the pH-dependent toxicity. In the meantime, it has been shown that CeO_2 nanoparticles on their own have a protective impact on cells, ensuring that the dissolved CeO_2 nanoparticles will not harm healthy cells. Figure 5.4 illustrates how the pH stimuli-responsive CeO_2-capped MSN drug release mechanism was designed and built.

5.4 COMPUTATIONAL FRAMEWORKS OF OXIDE-BASED NANOMATERIALS IN DRUG DESIGN

Biological systems generally have very large chemical molecular components that can change over time and are sensitive to in vivo conditions, such as temperature and environmental conditions. Several limitations are faced in experimental studies in the design of drug compounds that are then used further in biological systems, i.e., costly, and, in particular studies, we need a sophisticated laboratory instrument. On the other hand, these limitations need to be resolved (Hindré *et al.*, 2012).

Discovering effective drugs and vaccines is still in various research or study processes, so currently, the public is still required to take preventive actions to control the spread of viruses or bacteria. This reality underlies various scientists' need to conduct studies or evaluate more efficiently in discovering drugs or vaccines before viruses or bacteria spread. In a study on drug design, apart from using experimental pathways, such as synthesis, computational and bioinformatics methods are also required for faster and more efficient drug discovery frameworks (Hindré *et al.*, 2012).

Computational chemistry or biology can bridge theory and experiment by investigating the static or dynamic response of a molecular system based on its initial condition. In the initial state, the molecular system model is created based on the relevant molecular theory and solving mathematical equations calculated using a certain algorithm (Allen and Tildesley, 2017). One of the computational methods that is usually used in research or studies in drug discovery is molecular dynamics (MD)

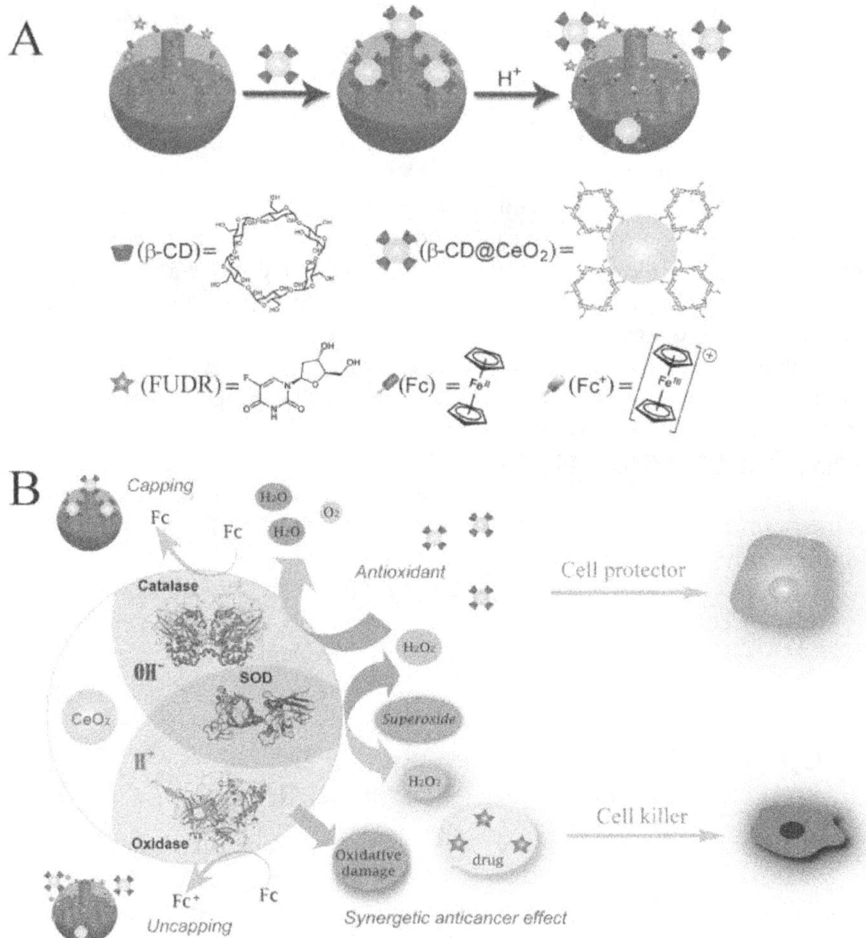

FIGURE 5.4 (a) Schematic illustration for pH-triggered release of the anticancer drug from β-CD@CeO_2 capped Fc-MSN to the cytosol. The controlled-release mechanism of the system is based on oxidation of the ferrocenyl moieties by CeO_2 in the acidic lysosomes of cancer cells, which decreases the binding affinity of ferrocene to β-CD and triggers the release of the CeO_2 nanoparticle caps. (b) The different enzymatic activity of CeO_2 at different pH values. At neutral or high pH conditions, the CeO_2 behaves as SOD and mimics catalase, which makes it display antioxidant activity and act as a cell protector. At low pH values, it behaves as an SOD and mimics oxidase, which makes it a cell killer.

Source: Reprinted from (Xu *et al.*, 2013). Copyright 2013 Wiley.

simulation. This method is applied for studying the properties of a system from the structure and microscopic interactions of the molecules within it. MD simulation also combines empirical data and theory to model the behavior of molecules so that it can be used to study a system composed of these molecules. MD simulation can be carried out based on the theory of classical mechanics and/or quantum mechanics.

The motion of molecules in molecular simulations is determined using the equations of motion. The position of the molecule determines its potential energy, and its movement determines its kinetic energy (Leach, 2001)

In a study conducted by Grasso *et al.* (2019), MD simulations were carried out to study the adsorption mechanism of several cell-penetrating peptides (CPPs) on the silica surface to study the effect of the surface ionization state on the adsorption mechanism. CPPs are used to coat silica to increase the colloidal stability and binding affinity of magnetic nanoparticles with various types of small organic molecules and peptides. Through this method, the nanoparticles can be modified by combining physicochemical properties with cell-penetrating abilities. Functionalization of silica-coated CPPs is an efficient step to increase the absorption of magnetic nanoparticles due to the intrinsic ability of CPPs to cross cell membranes without causing significant damage (Sanders *et al.*, 2011).

The starting point of the MD simulation is prepared by creating the coordinates or the topology of the simulation cell. In the work of Grasso *et al.* (2019), they prepared the silica surface topology with various degrees of ionization and then set up the force field. The force field is important because it contains the functional mathematics and parameter sets employed to calculate the potential energy of the simulation systems (Leach, 2001). After getting the best force field, the computational calculation of the peptide-silica interaction can be achieved accurately. MD simulation was conducted in various conditions, that is, the different types of CPP (each CPP brings a different peptide charge) and the percentage of silica atoms and water molecules. TIP3P was employed to describe the water model.

These simulation systems must simultaneously be minimized and equilibrated in NVT in the NPT ensemble. Each letter in NVT and NPT means the simulation system has the same values in the thermodynamic properties of the number of atoms (N), volume (V), temperature (T), and pressure (P). The simultaneous NVT and NPT ensemble is a condition in the isothermal-isobaric simulation system that needs to be applied for position restraint on CPP and surface silica. After the minimization and equilibration process, the MD simulation was performed for 100 ns. The isothermal-isobaric ensemble was applied during the simulation by employing the Nosè-Hoover temperature coupling and Parrinello-Rahman pressure coupling (Grasso *et al.*, 2019).

The MD simulation produces data in terms of the interaction between the nanomaterials and the biology component, like peptides. Figure 5.5 reveals that the adsorption of the peptides onto silica surfaces is performed by ammonium of the N-terminal residues (Grasso *et al.*, 2019). These data can help in selecting the best forms of oxide-based nanomaterial and the type of ligand or peptides that interact with the nanomaterial. It can show efficient ways to design a drug with that specific ligand in the future.

5.5 CONCLUSIONS AND FUTURE PERSPECTIVES

These studies clearly demonstrate that oxide-based nanomaterials have shown great potential in drug preparation and delivery applications due to their unique physicochemical properties, biocompatibility, and ease of synthesis. Oxide-based

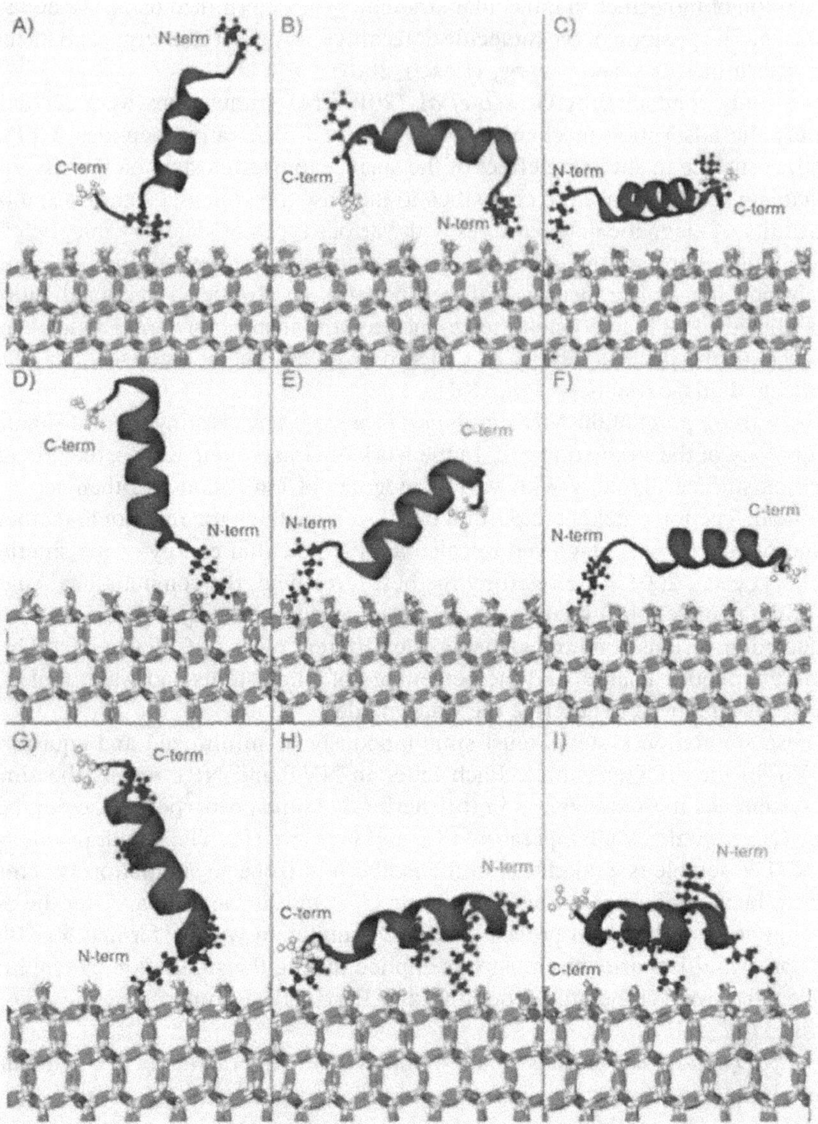

FIGURE 5.5 Picture of the contacts between the MAP peptide and the SiOH (a–c), 9SiO (H,Na) (d–f), and 18SiO (H,Na) (g–i) surfaces: (a, d, g) initial contact, (b, e, h) during contact, (c, f, i) final configuration. Nonpolar residues are highlighted in white (ALA); basic residues are in black (LYS).

Source: Reprinted from (Grasso *et al.*, 2019). Copyright 2019 ACS Publications.

nanomaterials can be used as carriers for drugs, allowing for controlled release of the drug over an extended period. This approach can improve the efficacy of the drug and reduce the likelihood of side effects. Oxide-based nanomaterials can improve the bioavailability of drugs that have poor solubility or low permeability. The small size of nanoparticles allows them to penetrate cell membranes more easily, leading to improved drug uptake and distribution. Many oxide-based nanomaterials are biocompatible and can be used in vivo without causing adverse effects. However, further studies are needed to assess the long-term safety of these materials. The synthesis and functionalization of oxide-based nanomaterials are relatively simple and cost-effective, making them an attractive option for drug delivery applications.

In general, oxide-based nanomaterials show great potential in drug preparation and delivery. However, more research is needed to fully understand their safety, efficacy, and potential clinical applications.

5.6 ACKNOWLEDGMENTS

This research was funded by PTUPT Grant, Universitas Gadjah Mada (Contract Number: 1679/UN1/DITLIT/DIT-LIT/PT.01.03/2022).

REFERENCES

Abu-Dief, A.M., Nassar, I.F., and Elsayed, W.H., 2016, Magnetic $NiFe_2O_4$ nanoparticles: Efficient, heterogeneous and reusable catalyst for synthesis of acetylferrocene chalcones and their anti-tumour activity, *Appl. Organomet. Chem.*, 30, 917–923.

Al-Fahdawi, M.Q., Rasedee, A., Al-Qubaisi, M.S., Alhassan, F.H., Rosli, R., El Zowalaty, M.E., Naadja, S., Webster, T.J., and Taufiq-Yap, Y.H., 2015, Cytotoxicity and physicochemical characterization of iron–manganese-doped sulfated zirconia nanoparticles, *Int. J. Nanomedicine.*, 10, 5739–5750.

Allen, M.P., and Tildesley, D.J., 2017, *Computer Simulation of Liquids*. Oxford: Oxford University Press.

Aneu, A., Wijaya, K., and Syoufian, A., 2021, Silica-based solid acid catalyst with different concentration of H_2SO_4 and calcination temperature: Preparation and characterization, *Silicon*, 13, 2265–2270.

Ansar, M.Z., Atiq, S., Riaz, S., and Naseem, S., 2015, Magnetite nano-crystallites for anticancer drug delivery, *Mat. Today: Proc.*, 2, 5410–5414.

Bardajee, G.R., Mohammadi, M., and Kakavand, N., 2016, Copper(II)-diaminosarcophagine-functionalized SBA-15: A heterogeneous nanocatalyst for the synthesis of benzimidazole, benzoxazole and benzothiazole derivatives under solvent-free conditions, *Appl. Organomet. Chem.*, 30, 51–58.

Blin, J.L., Léonard, A., and Su, B.L., 2001, Synthesis of large pore disordered MSU-type mesoporous silicas through the assembly of C16(EO)10 surfactant and TMOS silica source: Effect of the hydrothermal treatment and thermal stability of materials. *J. Phys. Chem.*, 105, 6070–6079.

Borówka, A., and Skrzypiec, K., 2021, Effects of temperature on the structure of mesoporous silica materials templated with cationic surfactants in a nonhydrothermal short-term synthesis route, *J. Solid State Chem.*, 299, 1–7.

Budiman, A., and Aulifah, D.L., 2021, Encapsulation of drug into mesoporous silica by solvent evaporation: A comparative study of drug characterization in mesoporous silica with various molecular weights, *Heliyon*, 7, 1–7.

Casasús, R., Marcos, M.D., Martínez-Máñez, R., Ros-Lis, J.V., Soto, J., Villaescusa, L.A., Amorós, P., Beltrán, D., Guillem, C., and Latorre, J., 2004, Toward the development of ionically controlled nanoscopic molecular gates. *J. Am. Chem. Soc.*, 126, 8612–8613.

Challener, C.A., 2016, Nanotechnology shows promise for API synthesis and delivery, *Pharm. Technol.*, 40, 48–51.

Chen, L., Xie, J., Aatre, K.R., Yancey, J., Chetan, S., Srivatsan, M., and Varadan, V.K., 2011, Synthesis of hematite and maghemite nanotubes and study of their applications in neuroscience and drug delivery, *Nanosensors, Biosensors, Info-Tech Sensors Syst. 2011*, 7980, 798008.

Christiansen, M.A., Mpourmpakis, G., and Vlachos, D.G., 2013, Density functional theory-computed mechanisms of ethylene and diethyl ether formation from ethanol on γ-Al_2O_3 (100), *ACS Catal.*, 3(9), 1965–1975.

Colleoni, C., Esposito, S., Grasso, R., Gulino, M., Musumeci, F., Romeli, D., Rosace, G., Salesi, G., and Scordino, A., 2016, Delayed luminescence induced by complex domains in water and in TEOS aqueous solutions, *Phys. Chem. Chem. Phys.*, 18, 772–780.

Dening, T., Zemlyanov, D., and Taylor, L.S., 2019, Application of an adsorption isotherm to explain incomplete drug release from ordered mesoporous silica materials under supersaturating conditions, *J. Contr. Release*, 307, 186–199.

Drago, R.S., and Kob, N., 1997, Acidity and reactivity of sulfated zirconia and metal-doped sulfated zirconia, *J. Phys. Chem. B*, 101, 3360–3364.

Ebadi, M., Buskaran, K., Bullo, S., Hussein, M.Z., Fakurazi, S., and Pastorin, G., 2021, Drug delivery system based on magnetic iron oxide nanoparticles coated with (polyvinyl alcohol-zinc/aluminium-layered double hydroxide-sorafenib), *Alexandria Eng. J.*, 60, 733–747.

Garofalo, M., Grazioso, G., Cavalli, A., and Sgrignani, J., 2020, How computational chemistry and drug delivery techniques can support the development of new anticancer drugs, *Molecules*, 25, 1–22.

Gawande, M., Hosseinpour, R., and Luque, R., 2014, Silica sulfuric acid and related solid-supported catalysts as versatile materials for greener organic synthesis, *Curr. Org. Synth.*, 11, 526–544.

Gómez-Sotomayor, R., Ahualli, S., Viota, J.L., Rudzka, K., and Delgado, Á.V., 2015, Iron/magnetite nanoparticles as magnetic delivery systems for antitumor drugs, *J. Nanosci. Nanotechnol.*, 15, 3507–3514.

Grasso, G., Mercuri, S., Danani, A., and Deriu, M.A., 2019, Biofunctionalization of silica nanoparticles with cell-penetrating pepties: Asorption mechanism and binding energy estimation, *J. Phys. Chem. B*, 123, 10622–10630.

Gupta, A.K., and Gupta, M., 2005, Synthesis and surface engineering of iron oxide nanoparticles for biomedical applications, *Biomaterials*, 26, 3995–4021.

Hauli, L., Wijaya, K., and Armunanto, R., 2018, Preparation and characterization of sulfated zirconia from a commercial zirconia nanopowder, *Orient. J. Chem.*, 34, 1559–1564.

Hindré, T., Knibbe, C., Beslon, G., and Schneider, D., 2012, New insights into bacterial adaptation through in vivo and in silico experimental evolution, *Nat. Rev. Microbiol.*, 10, 352–365.

Hong, H., Shi, J., Yang, Y., Zhang, Y., Engle, J.W., Nickles, R.J., Wang, X., and Cai, W., 2011, Cancer-targeted optical imaging with fluorescent zinc oxide nanowires, *Nano Lett.*, 11, 3744–3750.

Hu, Y., Guo, B., Fu, Y., Ren, Y., Tang, G., and Chen, X., 2015, Facet-dependent acidic and catalytic properties of sulfated titania solid superacids, *Chem. Commun.*, 1, 1–4.

Katada, N., Endo, J.I., Notsu, K.I., Yasunobu, N., Naito, N., and Niwa, M., 2000, Superacidity and catalytic activity of sulfated zirconia, *J. Phys. Chem. B*, 104, 10321–10328.

Kaur, G., Devi, P., Thakur, S., Kumar, A., Chandel, R., and Banerjee, B., 2019, Magnetically separable transition metal ferrites: Versatile heterogeneous nano-catalysts for the synthesis of diverse bioactive heterocycles, *ChemistrySelect*, 4, 2181–2199.

Khan, S., Sharifi, M., Hasan, A., Attar, F., Edis, Z., Bai, Q., Derakhshankhah, H., and Falahati, M., 2021, Magnetic nanocatalysts as multifunctional platforms in cancer therapy through the synthesis of anticancer drugs and facilitated Fenton reaction, *J. Adv. Res.*, 30, 171–184.

Koskin, A.P., Mishakov, I.V., and Vedyagin, A.A., 2016, In search of efficient catalysts and appropriate reaction conditions for gas phase nitration of benzene, *Resour. Technol.*, 2, 118–125.

Kovacevic, M., Mojet, B.L., Van Ommen, J.G., and Lefferts, L., 2016, Effects of morphology of cerium oxide catalysts for reverse water gas shift reaction, *Catal. Lett.*, 146, 770–777.

Krishnakumar, B., and Swaminathan, M., 2011, A recyclable solid acid catalyst sulfated titania for easy synthesis of quinoxaline and dipyridophenazine derivatives under microwave irradiation, *Bull. Chem. Soc. Jpn.*, 84, 12611266.

La Ore, M.S., Wijaya, K., Trisunaryanti, W., Saputri, W.D., Heraldy, E., Yuwana, N.W., Hariani, P.L., Budiman, A., and Sudiono, S., 2020, The synthesis of SO_4/ZrO_2 and Zr/CaO catalysts via hydrothermal treatment and their application for conversion of low-grade coconut oil into biodiesel, *J. Environ. Chem. Eng.*, 8, 104205.

Leach, A.R., 2001, *Molecular Modelling Principles and Applications*. Upper Saddle River: Prentice Hall.

Leonetti, B., Perin, A., Ambrosi, E.K., Sponchia, G., Sgarbossa, P., Castellin, A., Riello, P., and Scarso, A., 2021, Mesoporous zirconia nanoparticles as drug delivery systems: Drug loading, stability and release, *J. Drug Deliv. Sci. Technol.*, 6, 1–9.

Liu, J., Ma, X., Jin, S., Xue, X., Zhang, C., Wei, T., Guo, W., and Liang, X.J., 2016, Zinc oxide nanoparticles as adjuvant to facilitate doxorubicin intracellular accumulation and visualize pH-responsive release for overcoming drug resistance, *Mol. Pharm.*, 13, 1723–1730.

Liu, J., Yang, Q., Zhao, X.S., and Zhang, L., 2007, Pore size control of mesoporous silicas from mixtures of sodium silicate and TEOS, *Micropor. Mesopor. Mater.*, 106, 62–67.

López, T., Ortiz, E., Guevara, P., Gómez, E., and Novaro, O., 2014, Physicochemical characterization of functionalized-nanostructured titania as a carrier of copper complexes for cancer treatment, *Mater. Chem. Phys.*, 146, 37–49.

Mabrouk, M., Moaness, M., and Beherei, H.H., 2022, Fabrication of mesoporous zirconia and titania nanomaterials for bone regeneration and drug delivery applications, *J. Drug Deliv. Sci. Technol.*, 78, 1–11.

Mirzaei, M., Zarch, M.B., Darroudi, M., Sayyadi, K., Keshavarz, S.T., Sayyadi, J., Fallah, A., and Maleki, H., 2020, Silica mesoporous structures: Effective nanocarriers in drug delivery and nanocatalysts, *Appl. Sci.*, 10, 1–36.

Mishra, M.K., Tyagi, B., and Jasra, R.V., 2003, Effect of synthetic parameters on structural, textural, and catalytic properties of nanocrystalline sulfated zirconia prepared by sol-gel technique, *Ind. Eng. Chem. Res.*, 42, 5727–5736.

Mishra, M.K., Tyagi, B., and Jasra, R.V., 2004, Synthesis and characterization of nano-crystalline sulfated zirconia by sol-gel method, *J. Mol. Catal. A Chem.*, 223, 61–65.

Mu, W., Chu, Q., Liu, Y., and Zhang, N., 2020, A review on nano-based drug delivery system for cancer chemoimmunotherapy. *Nano Micro Lett.* 12, 142.

Nivetha, N., Thangamani, A., and Velmathi, S., 2022, Sulfated titania (TiO_2-SO_4^{2-}) as an efficient catalyst for organic synthesis: Overarching review from 2000 to 2021, *ChemistrySelect*, 7, 1–43.

Noda, L.K., De Almeida, R.M., Probst, L.F.D., and Gonçalves, N.S., 2005, Characterization of sulfated TiO_2 prepared by the sol-gel method and its catalytic activity in the n-hexane isomerization reaction, *J. Mol. Catal. A Chem.*, 225, 39–46.

Nyoka, M., Choonara, Y.E., Kumar, P., Kondiah, P.P.D., and Pillay, V., 2020, Synthesis of cerium oxide nanoparticles using various methods: Implications for biomedical applications, *Nanomaterials*, 10, 1–21.

Park, J., Han, Y., and Kim, H., 2012, Formation of mesoporous materials from silica dissolved in various NaOH concentrations: Effect of pH and ionic strength, *J. Nanomater.*, 1–10.

Patel, A., Brahmkhatri, V., and Singh, N., 2013, Biodiesel production by esterification of free fatty acid over sulfated zirconia, *Renew. Energy*, 51, 227–233.

Patra, J.K., Das, G., Fraceto, L.F., Campos, E.V.R., Rodriguez-Torres, M.D.P., Acosta-Torres, L.S., Diaz-Torres, L.A., Grillo, R., Swamy, M.K., Sharma, S., Habtemariam, S., and Shin, H.S., 2018, Nano based drug delivery systems: Recent developments and future prospects, *J. Nanobiotechnology*, 16, 1–33.

Priyadarsini, S., Mukherjee, S., and Mishra, M., 2018, Nanoparticles used in dentistry: A review, *J. Oral Biol. Craniofac. Res.*, 8, 58–67.

Raissi, S., Kamoun, N., Younes, M.K., and Ghorbel, A., 2015, Effect of drying conditions on the textural, structural and catalytic properties of Cr/ZrO_2-SO_4: N-hexane conversion, *React. Kinet. Mech. Catal.*, 115, 499–512.

Rizvi, S.A.A., and Saleh, A.M., 2018, Applications of nanoparticle systems in drug delivery technology, *Saudi Pharm. J.*, 26, 64–70.

Rudzka, K., Viota, J.L., Muñoz-Gamez, J.A., Carazo, A., Ruiz-Extremera, A., and Delgado, Á.V., 2013, Nanoengineering of doxorubicin delivery systems with functionalized maghemite nanoparticles, *Colloids Surf. B Biointerfaces*, 111, 88–96.

Sanders, W.S., Johnston, C.I., Bridges, S.M., Burgess, S.C., and Willeford, K.O., 2011, Prediction of cell penetrating peptides by support vector machines, *PLoS Comput. Biol.*, 7.

Saravanan, K., Tyagi, B., and Bajaj, H.C., 2014, Catalytic activity of sulfated zirconia solid acid catalyst for esterification of myristic acid with methanol, *Indian J. Chem.—Sect. A Inorganic, Phys. Theor. Anal. Chem.*, 53, 799–805.

Sharma, H., Kumar, K., Choudhary, C., Mishra, P.K., and Vaidya, B., 2016, Development and characterization of metal oxide nanoparticles for the delivery of anticancer drug, *Artif. Cells, Nanomedicine Biotechnol.*, 44, 672–679.

Siddiqui, M.R.H., Al-Wassil, A.I., Al-Otaibi, A.M., and Mahfouz, R.M., 2012, Effects of precursor on the morphology and size of ZrO_2 nanoparticles, synthesized by sol-gel method in non-aqueous medium, *Mater. Res.*, 15, 986–989.

Singh, T.A., Das, J., and Sil, P.C., 2020, Zinc oxide nanoparticles: A comprehensive review on its synthesis, anticancer and drug delivery applications as well as health risks, *Adv. Colloid Interface Sci.*, 286, 102317.

Sohn, J.R., and Kim, H.W., 1989, Catalytic and surface properties of ZrO_2 modified with sulfur compounds, *J. Mol. Catal.*, 52, 361–374.

Sponchia, G., Ambrosi, E.K., Rizzolio, F., Hadla, M., Del Tedesco, A., Spena, C.R., Toffoli, G., Riello, P., and Benedetti, A., 2015, Biocompatible tailored zirconia mesoporous nanoparticles with high surface area for theranostic applications, *J. Mater. Chem.*, 3, 7300–7306.

Utami, M., Wijaya, K., and Trisunaryanti, W., 2017, Effect of sulfuric acid treatment and calcination on commercial zirconia nanopowder, *Key Eng. Mater.*, 757, 131–137.

Vangijzegem, T., Stanicki, D., and Laurent, S., 2019, Magnetic iron oxide nanoparticles for drug delivery: Applications and characteristics, *Expert Opin. Drug Deliv.*, 16, 69–78.

Vivero-Escoto, J.L., Slowing, I.I., Lin, V.S.Y., and Trewyn, B.G., 2010, Mesoporous silica nanoparticles for intracellular controlled drug delivery, *Small*, 6, 1952–1967.

Wijaya, K., Heraldy, E., Hakim, L., Suseno, A., Loekitowati Hariani, P., Utami, M., and Dita Saputri, W., 2021, Synthesis and application of nanolayered and nanoporous materials, *ICS Phys. Chem.*, 1, 1.

Wijaya, K., Lammaduma Malau, M.L., Utami, M., Mulijani, S., Patah, A., Wibowo, A.C., Chandrasekaran, M., and Rajabathar, J.R., 2021, Synthesis, characterizations and catalysis of sulfated silica and nickel modified silica catalysts for diethyl ether (DEE) production from ethanol towards renewable energy applications, *Catalysts*, 11(12), 1511.

Xu, C., Lin, Y., Wang, J., Wu, L., Wei, W., Ren, J., and Qu, X., 2013, Nanoceria-triggered synergetic drug release based on CeO_2-capped mesoporous silica host-guest interactions and switchable enzymatic activity and cellular effects of CeO_2, *Adv. Healthc. Mater.*, 2, 1591–1599.

Xu, Y., Zhao, C., Zhang, X., Xu, J., Yang, L., Zhang, Z., Gao, Z., and Song, Y.Y., 2021, Engineering tailorable TiO_2 nanotubes for NIR-controlled drug delivery, *Nano Res.*, 14, 4046–4055.

Zarei, A., Khazdooz, L., Aghaei, H., Gheisari, M.M., Alizadeh, S., and Golestanifar, L., 2017, Synthesis of phenols by using aryldiazonium silica sulfate nanocomposites, *Tetrahedron*, 73, 6954–6961.

Zhang, Z.J., 2016, Lattice mismatch induced curved configurations of hybrid boron nitride—carbon nanotubes, *Phys. E Low-Dimensional Syst. Nanostructures*, 84, 372–377.

Zhang, Z.J., Hou, Y.K., Chen, M.W., Yu, X.Z., Chen, S.Y., Yue, Y.R., Guo, X.T., Chen, J.X., and Zhou, Q., 2023, A pH-responsive metal-organic framework for the co-delivery of HIF-2α siRNA and curcumin for enhanced therapy of osteoarthritis, *J. Nanobiotechnology*, 21, 1–19.

Zhao, X., Xu, S., Jiang, Yuan, Wang, C., ur Rehman, S., Ji, S., Wang, Jiarong, Tao, T., Xu, H., Chen, R., Cai, Y., Jiang, Yanyi, Wang, H., Ma, K., and Wang, J., 2023, BSA-magnetite nanotorpedo for safe and efficient delivery of chemotherapy drugs, *Chem. Eng. J.*, 454, 140440.

6 Advancement in Bionanomaterials toward Drug Delivery and Sensor Applications

Kefayat Ulla, Salah Uddin, Amin Ur Rashid, and Won-Chun Oh

6.1 INTRODUCTION

Bionanomaterials are materials that have both biological and nanoscale properties, making them ideal for use in a variety of applications in medicine and biotechnology. Bionanomaterials are nano-sized materials made from biological products such as plants, peptides, bacteria viruses, and nucleic acid. In recent years, the field of bionanomaterials has seen significant advancements, particularly in the areas of drug delivery and sensor and imaging technology [1]. Bionanomaterials can be made from a variety of materials, including metals, polymers, and ceramics, and can be functionalized with biological moieties such as enzymes, antibodies, and RNA/DNA [2, 3].

Due to their small size and unique properties, bionanomaterials have gained attention for their potential applications in drugs delivery, imaging, and sensing.

In drug delivery, bionanomaterials can be used to target specific cells and tissues, increasing the efficiency of treatments while reducing side effects. For example, during the initial phase of cancer, conventional treatment methods like surgery, chemotherapy, and radiotherapy were the primary basis of the treatment strategy [4, 5]. As research advanced, it became evident that relying solely on traditional treatment methods like surgery, chemotherapy, and radiotherapy could result in several issues such as drug resistance in tumors, toxic side effects due to chemotherapy drugs affecting healthy cells, recurrence of tumors, and metastasis resulting from the presence of residual tumor foci after surgery [6]. Over the past few years, there has been a growing trend toward utilizing nano-materials to enhance the effectiveness of tumor immunotherapy. By employing inorganic nano-materials as tiny carriers, it is possible to transport a range of tumor antigens and immunomodulators, enabling targeted delivery to specific tumor sites and maximizing the potential benefits of immunotherapy [7–9].

In sensing, bionanomaterials can be used to detect a wide range of biological signals, including disease markers, diagnosis, drug detection, glucose monitoring,

 DOI: 10.1201/9781003425427-8

food safety, monitoring environmental pollutants, and defense and security [10]. The device used for this purpose is the biosensor; it is an analytical device comprising three basic components, a detector, transducer, and signal processing system. The first ever biosensor device was invented by Le and Clark in the year 1950, and the name biosensor was introduced in 1977 by Cammann [11–13]. During the last few decades, researchers from diverse fields have developed modifications both in the technology and the application of biosensors. Biosensors make it easier for humankind to explore biological processes at a molecular level to recognize a particular disease progression and, on the basis of this identification, develop next-generation therapeutics [14–16]. These biological processes allow sensing agents to first detect and then transduce output signals upon identifying a physiological target, with diverse applications ranging from detecting analytes such as viral particles or glucose to labeling cancer affected cells [17–20]. In order to minimize the detection limit and revamp the signal, these signal transducers are generally implanted on nanoparticles [21]. The most commonly used nanoparticles are bionanoparticles [22–25]; extracellular vesicles and protein nanoparticles are two naturally occurring bionanoparticles that are mostly utilized in biosensing [26, 27]. These nanoparticles have the capability to serve as scaffolds for cargo encapsulation [22] and are good candidates for molecular transport [24].

The size of protein nanoparticles varies from 10 to 100 nm, and they have a wide range of applications ranging from docking proteins for assembling enzyme complexes to delivering cargoes, such as ions, small molecules, and nucleic acid [23]. These bionanoparticles can be chemically modified to accommodate sensing agents and target moieties for a wide range of bioanalytical applications [28, 29]. Protein nanoparticles such as ferritin and virus-like particles are particularly attractive because of their monodispersity and physical stability and serve as nanocages [30, 31]. Extracellular vesicles nanoparticles range from 20 to 500 nm in size and are released by all living cells starting from microorganism bacteria to macro-organism mammalian cells [25]. The flexible lipid-bound structure of these nanoparticles makes them distinct from protein nanocages that can encapsulate cytoplasmic biomolecules, nucleic acids, and proteins to perform diverse functions, for example, acting as a mediator for intercellular communications in mammalian cells [32].

In imaging, bionanomaterials can be used to visualize biological structures and processes in real time. An incongruent agent is always needed to aid proper visualization of physiological structures inside a body. Recent advances and developments in bionanomaterials have become important in generating high-contrast and high-resolution images needed for precise and accurate diagnostics [33]. Molecular imaging is one of the important tools in the diagnostic branch. Molecules of living bodies undergo a number of changes during any disease process, and these changes can be observed and detected through molecular imaging by using nuclear medicine and radiology [34]. In order to enhance the functioning of molecular imaging, bionanomaterials are widely used because they provide selective binding with surface receptors, biocompatibility, and high surface area per unit volume [35]. The selective binding of bionanomaterials with biological molecules and high surface area to volume ratio produces specific signals, which results in the detection of changes at the molecular level. There are two main factors that play a vital role in molecular imaging: the payload, which basically means the smaller the nanoparticle size, the higher the payload, and the proper distribution of bionanomaterials in the affected living system and precise targeting by nanomaterials can be by both passive targeting and active means [35].

Despite the exciting potential of bionanomaterials, there are still challenges that need to be addressed, such as improving their stability and biocompatibility, as well as optimizing their targeting and delivery capabilities. Nevertheless, the field of bionanomaterials is rapidly growing, and there is much anticipation for future advancements and innovations in this area.

6.1.1 Drug Delivery Applications of Bionanomaterials

One of the main challenges in traditional drug delivery methods is the inefficient transport of therapeutic agents to targeted cells and tissues. Bionanomaterials have the potential to overcome this challenge by providing a more targeted and efficient means of delivering drugs to the body. Some of the most promising bionanomaterials for drug delivery include nanoparticles, liposomes, and dendrimers. Nanoparticles are spherical structures with a diameter ranging from 1 to 100 nanometers. They have a large surface area to volume ratio, which makes them ideal for encapsulating drugs and other therapeutic agents. Liposomes are spherical lipid bilayer structures that can be utilized to encapsulate medications and target precise cells or regions. Dendrimers are nanoscale polymers that are highly branched and can be functionalized with variety of medicinal agents and targeted moieties.

Nanopharmaceuticals are used in two categories in drug delivery processes: nanofabricated and nanocarriers. In the former category, the fabricated therapeutic molecules behave as known drugs, while in the latter, the molecules are attached to the nanoparticles, which act as carriers. In this contest, a single robot may sort numerous cargos repeatedly by taking advantage of this capability. In one test, different cargo-sorting activities can be carried out simultaneously to localized on DNA origami, and multiple robots can work together to complete the same task [36]. In an effort to offer fresh perspectives the developments in polymer-based antibiotics in drug delivery system for the treatment of bacterial infections on stressing the design principles and targeting capacity was carried out [37]. Owing to their special characteristics, such as their small size, high surface-to-volume ratio, and tunable surface functionality, dendrimers, which are branched nanoscale polymers, have drawn a lot of interest as drug delivery system [38]. Besides their several benefits, there are some disadvantages of the dendrimers, like controlled fabrication, shape affecting pharmakinstic properties, and interaction with the immune system that causes toxicity and hence limits their uses in various applications [39]. Drug delivery has also been carried out using inorganic nanocarriers, including gold, magnetic nanoparticles, and silica.

An external magnetic field can be used to guide a magnetic nanocarrier toward a target cell in the human body. Similarly, moieties can be used to functionalize silica nanoparticles in order to reduce the degradation during the therapeutic processes [40]. Due to their distinctive properties, gold nanoparticles can be used for the treatment of cancer cells during hyperthermal and imaging therapy [41].

It is well known that ineffective penetration of therapeutic nanoparticles into tumors and insufficient coordination of drug release at the affected site pose the biggest obstacles to the development of effective nanomedicine-based anticancer medications [42, 43].

The physiological barriers that contribute to this problem include microvessel permeability and intratumoral diffusion. The increased permeability and retention (EPR) effect, which is thought to be the major mechanism involved in this process, is primarily responsible for the accumulation of nanocarriers in tumors. Recent studies suggest that the EPR effect is clinically significant [44]. However, the level of nanoparticle congregation in different organs varies. According to a recent analysis of a broad body of data published over the last decade, across multiple formulations and tumor models, less than 1% of the injected nanoparticle dose reaches the malignant tissue on average [45].

6.1.2 Sensor Applications of Bionanomaterials

Bionanomaterials have also found a variety of applications in sensor technology. They can be used to detect a wide range of biological signals, including disease markers and environmental pollutants. For example, nanoscale materials such as gold nanoparticles, quantum dots, magnetic nanoparticles, graphene, carbon nanotubes, and photonic crystals can be used in biosensors to detect specific proteins or other biomolecules. Other bionanomaterials, such as nanoparticles and dendrimers, can be functionalized with recognition elements and used to detect specific targeted molecules in real time. A biosensor is a sensing device that is used for different purposes, such as detection of pathogens, measurement of glucose level in the blood, DNA analysis, detection of drug concentration in the body, and monitoring of food quality. The main function of a biosensor is to sense a specific biological material. These biological materials are immunological molecules, antibodies, enzymes, proteins, and so on. A biosensor is a three-component device, the bioreceptor, transducer, and detector. Figure 6.1 gives detailed information of the three components of biosensors. A bioreceptor is the component that serves as a template for the detected biomaterial. The second component is the transducer, which converts the interaction between the transducer and bioanalyte into an electrical signal. The name transducer is the combination of two words, "trans" meaning change and "ducer" meaning energy. From the name, it is clear that a transducer converts energy. The first type of energy produced by the interaction between bioanalyte and bioreceptor is biomechanical energy, while the second type is typically electrical energy. The detector is the final component of a biosensor; it accepts electrical signals from the output of the transducer and amplifies them to an appropriate range so that the associated response can be investigated precisely. Apart from these three components, there is another requirement for the biosensor to operate faster and smooth is the availability of immobilizer [46].

A group of bionanomaterials including gold, silicon, silver, copper nanoparticles, and carbon-based materials such as graphene, graphite, carbon nanospheres, carbon nanotubes, and quantum dots are used for developing biosensor immobilization

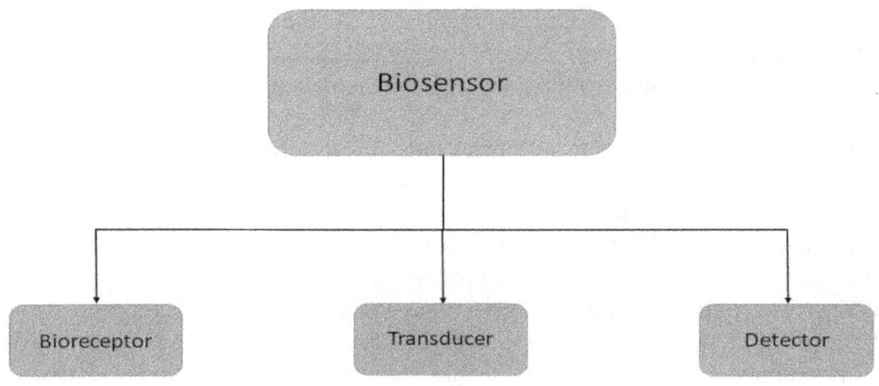

FIGURE 6.1 Block diagram of a biosensor.

[47–52]. Bionanomaterial-based materials also offer high sensitivity for developing electrochemical and other types of sensors.

Electrochemical biosensors based on metallic nanoparticles normally work to analyze biochemical reactions with the help of modified electrical means. The use of metallic nanoparticles helps in carrying out the biochemical reaction between the molecules easily and efficiently, which helps in obtaining immobilization of the reactant. This ability of metallic nanoparticles makes such a reaction very specific and removes any possibility of obtaining harmful and undesirable side products. Among metallic nanoparticles, gold-based particles are the most stable metallic nanomaterial with negligible toxicity [53], while other metallic nanoparticles such as silicon, silver, and copper oxidize and have the potential for toxicity, if used for internal purposes in medicine such as drug delivery [54]. Gold nanoparticles are mostly used in electrochemical biosensors because of their relatively simple production, biocompatibility, and electrical and optical properties [55, 56]. Gold nanoparticles have been utilized to increase DNA immobilization on gold electrodes, which can further increase the overall efficiency of the electrochemical biosensor and lower the detection limit [57]. For the detection and identification of hydrogen peroxide, glucose, and xanthine, a special kind of biosensor has been developed using enzyme-conjugated gold nanoparticles [58–60]. These biosensors showed a much lower detection limit and improved sensitivity as compared to those without gold nanoparticles.

The gold surface has a distinctive optical behavior; when the surface is exposed to light with a specific wavelength, it produces an optical effect. When the size of the particle is smaller than the wavelength of the incident light, the oscillated electrons in the conduction band cannot propagate along the surface, a phenomenon called surface plasmons resonance (SPR) setup. Due to this non-propagating behavior, the electrons now accumulate on one side of the nanoparticle (Figure 6.2), while the plasmons start oscillation with light frequency [61, 62]. This SPR phenomenon is strongly dependent on both the size and shape of the nanoparticles and also on the dielectric constant of the environment surrounding the nanoparticles [63]. The gold nanoparticles can interact with bio-analytics, and the recognition event undergoes a change

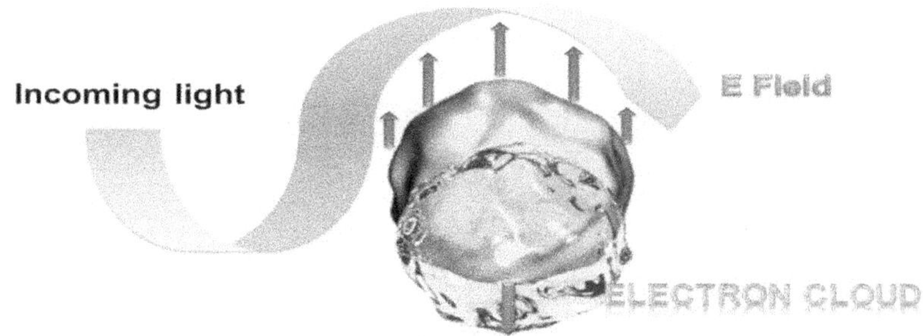

FIGURE 6.2 Schematic illustration of surface plasmon resonance.

Source: Reprinted from reference [21], under Created Commons (CC-BY) license.

in frequency and therefore a change in color of the gold nanoparticles is observable to the naked eye. Due to these unique optical properties of gold nanoparticles, a number of efficient colorimetric biosensors have been developed for oligonucleotide or DNA detection or immunosensors [64–67]. Apart from these outstanding and unique optical properties, gold nanoparticles also behave as electron carriers and transform electrons between biologically electroactive species and electrodes. This property of gold nanoparticles is used in redox enzyme biosensing, where the bioreceptor of the biosensor catalyzes the oxidation or reduction of the analyte [68].

Gold nanoparticles also play an important role in bioanalysis using SPR transduction. This analysis works on the principle of replacing the dielectric constant of the environment around the gold film, and the analyte can be detected in different ways, like changing the angle, phase, or intensity of reflected light [69]. An intense and clear SPR signal enhancement can be achieved by replacing a pure gold nanoparticle with gold film. In this process, gold nanoparticles (size ≈40 nm) and gold film (at a distance of ≈50 nm from the gold nanoparticle) are coupled in a sandwiched configuration, as shown in Figure 6.3 [70]. The gold nanoparticles behave as labels when connected to DNA strands or other secondary antibodies, and the surface plasmons provoke a perturbation of the evanescent field of gold film. In this process, the signal enhancement approaches several order of magnitudes, which is a convincing and strong argument for this approach to analyte detection.

Another prominent example of a bionanomaterial used in biosensors are magnetic nanoparticles (MNPs). MNPs have potential applications in imaging, gene therapy, target therapy, tissue engineering, and biosensors, as shown in Figure 6.4. Their size ranges from 1 to 100 nm, and they can be produced on large scale from iron, nickel, or cobalt or from their mixed oxides [71]. MNPs are made up of two parts: the core and the shell. MNP magnetic properties are related to both the core and the shell, which are active in catalysis, binding, and biomolecule recognition [72]. MNPs have small and random particle sizes, as well as a wide exposed surface area, and are simple to synthesize. The origin of magnetism is the motion of particles with mass and electric charge. In the absence of a magnetic field, the dipole moments of each domain (a region in the material where all atoms have their magnetic moments

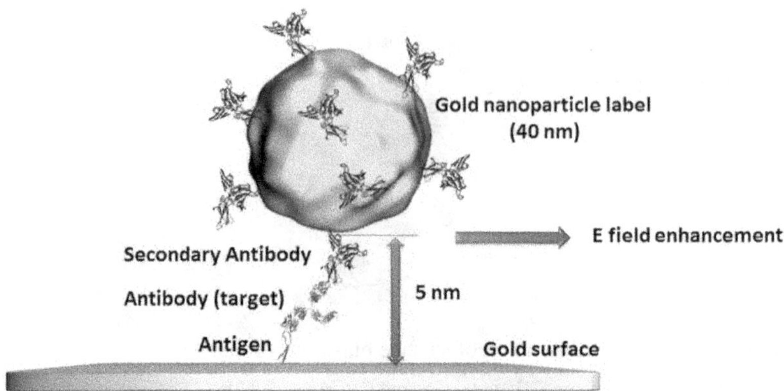

FIGURE 6.3 Schematic presentation of the propagating surface plasmon perturbation on gold films provoked by defined-sized gold nanoparticles causing a change in the evanescent field and producing enhanced signals.

Source: Reprinted with the permission from reference [70]. Copyright ©2013 Elsevier.

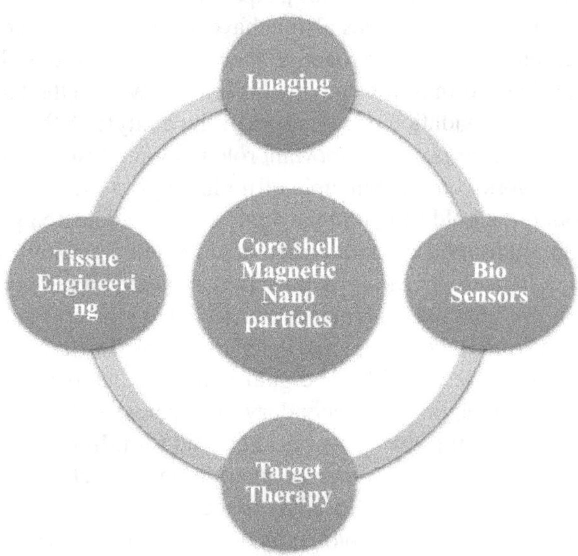

FIGURE 6.4 Biomedical application of magnetic nanoparticles.

parallel to each other such that the given region has a net magnetic dipole moment) are randomly oriented so that the net dipole moment becomes zero. When an external magnetic field is applied above a specific temperature, the dipole moment of each domain aligns itself in the direction of field, and the material has a net magnetic moment. However, when the field is removed, the magnetic moment becomes randomly oriented again [73].

MNPs show different behaviors as compared to bulk magnetic material due to the reduced number of domains, which leads to the formation of superparamagnetic behavior with high magnetic susceptibility [74]. This superparamagnetic behavior minimizes the force of attraction or repulsion between MNPs as long as no magnetic field is applied. Magnetic biosensor devices utilize some specially designed MNPs, which are prominent alternatives to fluorescent labels.

Among a wide range of MNPs, iron oxide MNPs are the most commonly used nanoparticles in biosensor applications [75]. The advantage of using MNPs in biosensors is the possibility of collecting the analyte before the event been detected. In this process, the MNPs are mixed with the solution of analyte and interact with a precise target. Under the influence of the magnetic field, the nanoparticle accumulates at a particular region and then can be detached from the analyte solution. Proper separation of DNA strands from the analyte solution can be obtained in an efficient and fast manner using gold coated core/shell MNPs [76–78].

Besides electrochemical [79] and optical [80] detection techniques in addition to other nanoscale labels, MNPs offer highly sensitive detection methods in the area of magnetic resonance. For this purpose, high-performance spin valve, large magnetoresistance, or magnetic tunnel junction biosensors were developed [81, 82]. Magnetic labels are much important for biosensing applications because biological entities are materials with no magnetic behavior, and hence no noise or interference is to be observed at the time of signal capturing [83].

Quantum dot (QD)–based technology can be utilized to study the delivery of nano-medicine and for the understanding of the tumor microenvironment [84]. The most commonly used QDs are based on cadmium chalcogenides, which have a broad excitation with a size0dependent (5–10 nm in diameter) tunable emission spectrum [85]. These unique properties are due to the variable energy gaps of the material for various nano-crystal sizes (the larger the particle size, the smaller the energy gap), which leads to the emission of distinct wavelength [86]. Because of these extraordinary properties QDs, are used in widespread fields such as biosensors, solar cells, and electronics [87–89]. The availability of different sizes and a broad range of wavelengths for QDs enables efficient multidimensional analysis using optical transduction [90]. The surface modification, broad excitation and narrow tunable emission band width of QDs have inspired the growth of multimodal probe-based biosensor devices through connecting with targeting ligands, nucleic acids, or peptides. QDs have a highly stable and sensitive fluorescence intensity; thus fluorescence transduction based on physical or chemical interaction occurs on the surface through quenching or direct photoluminescent [91]. QDs are widely used for sensing pH, biomolecules (protein, nucleic acids, and enzymes), organic compounds, and ions [92–94]. The successful achievements of biosensing research in recent years are developments in the application of QDs in tissue engineering to detect biomolecules or enzymes.

Graphene-based biosensors are a topic of great interest from the last decade and have attracted significant interest due to their unique characteristics, such as high electrical conductivity, large specific area, excellent electrochemical stability, good biocompatibility, and low production cost [95, 96]. Graphene is a two-dimensional material that is a favorable choice for different biosensor devices due to its surface availability for other chemical species [97]. Graphene quantum dots (GQDs) derived from carbon derivatives are photoluminescent materials having unique optical and quantum confinement properties [98, 99]. GQDs are prominent candidates for applications in electrochemical and photoluminescence sensor fabrication for a number of kinds of biological analysis [100]. The superiority of GQDs over other optical image agents such as cadmium-based QDs is because of their low toxicity, biocompatibility, and high photostability against photobleaching [101]. Due to these extraordinary properties, GQDs are used for different purposes in a number of biosensors, such as photoluminescent biosensors, electrochemical biosensors, and electronic biosensors [102].

Glucose biosensors are particularly used in tissue engineering for monitoring metabolic activities in the cell. To increase the efficiency of glucose biosensors, graphene oxide or graphene-based nanocomposites have been extensively used as a highly efficient enzyme electrode for the detection of glucose concentration [103, 104]. The detection limit and sensing efficiency of graphene-based glucose biosensors increase when a gold nanoparticle/silver nanoparticle hybrid is used to catalyze electrochemical reactions [105]. Cholesterol is present in the cell membranes of all humans, and its normal range is 1.0–2.2 mM. The excessive accumulation of cholesterol in the human body causes some fatal diseases. Gholivand and Khodadadian fabricated a cholesterol biosensor by modifying carbon electrodes using graphene/ionic liquid [106]. A DNA biosensor was prepared using a graphene/polyaniline nanocomposite. The inclusion of graphene and polyaniline and their synergistic effect enhance the electrode response due to faster travel of electrons at the surface of the electrode [107].

The beneficial properties of carbon nanotubes (CNTs) make them a extensively utilized bionanomaterial as an electronic or electrochemical transducer in biosensing devices [108]. CNTs have extraordinary properties, among which the most important are high electrical conductivity, high mechanical strength, and flexible geometrical structure. Due to these extraordinary properties, CNTs have become the most-renewed bionanomaterial right now in the fields of material science and optoelectronic applications. CNTs can be synthesized as both single-walled or multi-walled nanotubes commercially and have length in the *μm* range and diameter in the *nm* range, which can be used for designing biosensors for better performance [109]. The most important advances are developments in the design of glucose-monitoring biosensors in which CNTs have been utilized as immobilizing surfaces for glucose enzyme oxidase. The sensors using enzymes normally detect the glucose concentration from major body tissues; however, the introduction of CNTs as assemblies for immobilization has led to the prediction of glucose concentration even from scarce fluids such as saliva and tears [110]. Due to smoother electron flow characteristics, CNTs have been used to improve the electrical detection of biosensors. The chemoelectroluminiscent effect of biosensors has also been enhanced by coupling sensing molecules to CNTs through smooth conductance of electrons and controlling the flow characteristics of charge carriers [111]. The well-documented organic functionalization of CNTs [112] creates unique properties in nanostructured electrodes such as redox mediation of electrochemical reactions or particular

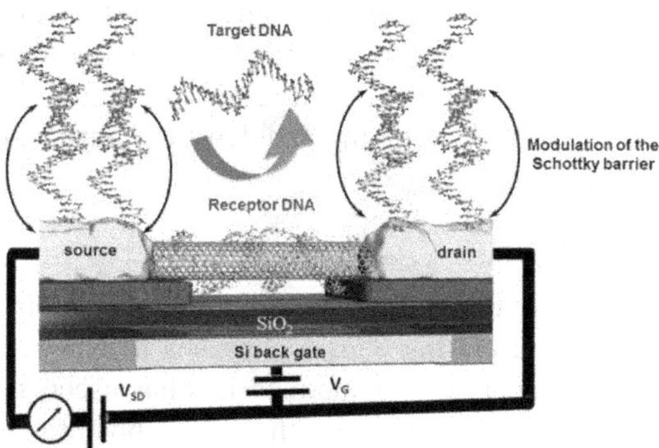

FIGURE 6.5 Schottky barrier modulation at the source and drain electrodes of CNT-FET due to hybridization of both receptor and target DNAs.

Source: Reprinted with permission from ref [117]. Copyright ©2006.

docking sites for molecules. Furthermore, CNT films provide a sensitive electroactive surface area because of the formation of a sensitive porous 3D network, which consequently leads to high sensitivities of biosensors [113]. The unique electric conductivity of CNTs was also utilized in field effect transistor biosensors in which variations in the conductivity of the CNT channel after the recognition event [114] led to low detection limits and high sensitivities [115]. The high affinity of CNTs and DNA strands led to the original proposal of CNT-FET–based DNA biosensors. The nucleic basis of DNA gets attached to CNTs through π-π stacking, which leads to the wrapping of DNA around CNTs [116] and minimizes the identification capacity of the DNA toward its counterpart. A group led by Tang observed a change in electrical conductance due to the alignment of modulated energy level between gold contacts and CNTs of a CNT-FET biosensor. The receptor unit of this biosensor was attached to the gold-coated electrode via thioethers where the DNA wrapped around the CNT channel. The recognition process is localized on source and drain, which modulates the conductance of the CNT channel, as shown in Figure 6.5 [117]. CNT-FET biosensors are generally used for enzyme and immunosensing application [118]. Raman scattering and NIR photoluminescence features are the spectroscopic characteristics of CNTs, also used in biosensing applications [56]. Thus CNTs and graphene are the most promising carbon bionanomaterials for biosensor devices, where each allotrope plays a particular role as a transducer element.

6.1.3 Recent Advances and Future Directions

Recently, there has been noteworthy progress in the development of bionanomaterials for drug delivery and sensor applications. New materials, such as upconverting nanoparticles and plasmonic nanostructures, have been developed with the potential to improve drug delivery system targeting and efficacy. In addition, new strategies for functionalizing bionanomaterials have been developed, including the use of RNA, DNA, and peptide-based recognition elements.

However, several obstacles must be overcome before the full potential of bionanomaterials in drug delivery and sensor applications can be realized. For example, further work is needed to improve the stability and biocompatibility of bionanomaterials, as well as to optimize their targeting and delivery capabilities.

6.2 CONCLUSION

Bionanomaterials have the potential to revolutionize medicine and biotechnology by providing novel and cutting-edge techniques to pharmaceutical industries and sensor applications. Recently bionanomaterials have seen significant advancement, and there is much excitement about the future potential of these materials. However, further research is needed to fully realize their potential and overcome the remaining challenges. Bionanomaterials can be functionalized with therapeutic agents and targeting moieties, allowing medications to be delivered directly to specific cells and tissues, improving treatment efficacy while decreasing negative effects. Bionanomaterials can be utilized as contrast agents in real-time imaging of biological structures and processes, providing important information for diagnosis and treatment. Bionanomaterials can be functionalized with recognition elements, allowing for the detection of a wide range of biological signals, including disease markers and environmental pollutants. These materials can be used as scaffolds to support the growth and function of cells, promoting tissue regeneration and repair. We highlighted a few advantages and uses of bionanomaterials in different medical applications. There is a lot of space and potential for bionanomaterials in future medicine and advanced imaging technologies.

REFERENCES

[1] Song, W., S.N. Musetti, and L. Huang, *Nanomaterials for cancer immunotherapy.* Biomaterials, 2017. **148**: p. 16–30.

[2] Jiang, Z., et al., *Nanomaterial-based drug delivery systems: a new weapon for cancer immunotherapy.* International Journal of Nanomedicine, 2022: p. 4677–4696.

[3] Zhou, L., et al., *Nano drug delivery system for tumor immunotherapy: next-generation therapeutics.* Frontiers in Oncology, 2022. **12**.

[4] Peng, T., T. Xu, and X. Liu, *Research progress of the engagement of inorganic nanomaterials in cancer immunotherapy.* Drug Delivery, 2022. **29**(1): p. 1914–1933.

[5] Sang, W., et al., *Recent advances in nanomaterial-based synergistic combination cancer immunotherapy.* Chemical Society Reviews, 2019. **48**(14): p. 3771–3810.

[6] Chen, Z., et al., *Nanomaterials: small particles show huge possibilities for cancer immunotherapy.* Journal of Nanobiotechnology, 2022. **20**(1): p. 1–38.

[7] Liu, Q., et al., *Nano-immunotherapy: unique mechanisms of nanomaterials in synergizing cancer immunotherapy.* Nano Today, 2021. **36**: p. 101023.

[8] Guo, R., et al., *Engineered nanomaterials for synergistic photo-immunotherapy.* Biomaterials, 2022. **282**: p. 121425.

[9] Bian, W., et al., *Review of functionalized nanomaterials for photothermal therapy of cancers.* ACS Applied Nano Materials, 2021. **4**(11): p. 11353–11385.

[10] Vigneshvar, S., et al., *Recent advances in biosensor technology for potential applications—an overview.* Frontiers in Bioengineering and Biotechnology, 2016. **4**: p. 11.

[11] Nurunnabi, M., et al., *Targeted near-IR QDs-loaded micelles for cancer therapy and imaging.* Biomaterials, 2010. **31**(20): p. 5436–5444.
[12] Krishnaperumal, I. and R. Lakshmanan, *An approximate analytical method for the evaluation of the concentrations and current for hybrid enzyme biosensor.* International Scholarly Research Notices, 2013. **2013**.
[13] Karunakaran, C., R. Rajkumar, and K. Bhargava, Introduction to biosensors, in *Biosensors and Bioelectronics.* 2015, Elsevier. p. 1–68.
[14] Pu, K., *Biosensors and Bioimaging.* 2019, Wiley Online Library. p. 420–421.
[15] Wolfbeis, O.S., *An overview of nanoparticles commonly used in fluorescent bioimaging.* Chemical Society Reviews, 2015. **44**(14): p. 4743–4768.
[16] Meng, H.-M., et al., *Aptamer-integrated DNA nanostructures for biosensing, bioimaging and cancer therapy.* Chemical Society Reviews, 2016. **45**(9): p. 2583–2602.
[17] Kherlopian, A.R., et al., *A review of imaging techniques for systems biology.* BMC Systems Biology, 2008. **2**(1): p. 1–18.
[18] Hahn, M.A., et al., *Nanoparticles as contrast agents for in-vivo bioimaging: current status and future perspectives.* Analytical and Bioanalytical Chemistry, 2011. **399**: p. 3–27.
[19] Patra, J.K., et al., *Nano based drug delivery systems: recent developments and future prospects.* Journal of Nanobiotechnology, 2018. **16**(1): p. 1–33.
[20] Yao, J., M. Yang, and Y. Duan, *Chemistry, biology, and medicine of fluorescent nanomaterials and related systems: new insights into biosensing, bioimaging, genomics, diagnostics, and therapy.* Chemical Reviews, 2014. **114**(12): p. 6130–6178.
[21] Holzinger, M., A. Le Goff, and S. Cosnier, *Nanomaterials for biosensing applications: a review.* Frontiers in Chemistry, 2014. **2**: p. 63.
[22] Diaz, D., A. Care, and A. Sunna, *Bioengineering strategies for protein-based nanoparticles.* Genes, 2018. **9**(7): p. 370.
[23] Bhaskar, S. and S. Lim, *Engineering protein nanocages as carriers for biomedical applications.* NPG Asia Materials, 2017. **9**(4): p. e371–e371.
[24] Truffi, M., et al., *Ferritin nanocages: a biological platform for drug delivery, imaging and theranostics in cancer.* Pharmacological Research, 2016. **107**: p. 57–65.
[25] Gill, S., R. Catchpole, and P. Forterre, *Extracellular membrane vesicles in the three domains of life and beyond.* FEMS Microbiology Reviews, 2019. **43**(3): p. 273–303.
[26] Hartzell, E.J., et al., *Modular hepatitis B virus-like particle platform for biosensing and drug delivery.* Acs Nano, 2020. **14**(10): p. 12642–12651.
[27] Van Niel, G., G. d'Angelo, and G. Raposo, *Shedding light on the cell biology of extracellular vesicles.* Nature Reviews Molecular Cell Biology, 2018. **19**(4): p. 213–228.
[28] Stephanopoulos, N. and M.B. Francis, *Choosing an effective protein bioconjugation strategy.* Nature Chemical Biology, 2011. **7**(12): p. 876–884.
[29] Lieser, R.M., et al., *Site-specific bioconjugation approaches for enhanced delivery of protein therapeutics and protein drug carriers.* Bioconjugate Chemistry, 2020. **31**(10): p. 2272–2282.
[30] Schoonen, L. and J.C. van Hest, *Functionalization of protein-based nanocages for drug delivery applications.* Nanoscale, 2014. **6**(13): p. 7124–7141.
[31] Rohovie, M.J., M. Nagasawa, and J.R. Swartz, *Virus-like particles: next-generation nanoparticles for targeted therapeutic delivery.* Bioengineering & Translational Medicine, 2017. **2**(1): p. 43–57.
[32] de Jong, O.G., et al., *Drug delivery with extracellular vesicles: from imagination to innovation.* Accounts of Chemical Research, 2019. **52**(7): p. 1761–1770.
[33] Cao, X., D. Huang, and Y.S. Zhang, Bionanomaterials as imaging contrast agents, in *Theranostic Bionanomaterials.* 2019, Elsevier. p. 401–421.
[34] Wu, M. and J. Shu, *Multimodal molecular imaging: current status and future directions.* Contrast Media & Molecular Imaging, 2018. **2018**.

[35] Singh, R.P. and K.R. Singh, *Bionanomaterials: fundamentals and biomedical applications*. 2021, IOP Publishing.
[36] Thubagere, A.J., et al., *A cargo-sorting DNA robot*. Science, 2017. **357**(6356): p. eaan6558.
[37] Wang, T., et al., *Targeted polymer-based antibiotic delivery system: a promising option for treating bacterial infections via macromolecular approaches*. Progress in Polymer Science, 2021. **116**: p. 101389.
[38] Carvalho, M., R. Reis, and J.M. Oliveira, *Dendrimer nanoparticles for colorectal cancer applications*. Journal of Materials Chemistry B, 2020. **8**(6): p. 1128–1138.
[39] Song, C., et al., *Superstructured poly (amidoamine) dendrimer-based nanoconstructs as platforms for cancer nanomedicine: a concise review*. Coordination Chemistry Reviews, 2020. **421**: p. 213463.
[40] Shirshahi, V. and M. Soltani, *Solid silica nanoparticles: applications in molecular imaging*. Contrast Media & Molecular Imaging, 2015. **10**(1): p. 1–17.
[41] Heuer-Jungemann, A., et al., *The role of ligands in the chemical synthesis and applications of inorganic nanoparticles*. Chemical Reviews, 2019. **119**(8): p. 4819–4880.
[42] Grodzinski, P., et al., *Integrating nanotechnology into cancer care*. 2019, ACS Publications.
[43] Hartshorn, C.M., et al., *Nanotechnology strategies to advance outcomes in clinical cancer care*. ACS Nano, 2018. **12**(1): p. 24–43.
[44] Bhatia, S.N., et al., *Cancer nanomedicine*. Nature Reviews Cancer, 2022. **22**(10): p. 550–556.
[45] de Lázaro, I. and D.J. Mooney, *Obstacles and opportunities in a forward vision for cancer nanomedicine*. Nature materials, 2021. **20**(11): p. 1469–1479.
[46] Malik, P., et al., *Nanobiosensors: concepts and variations*. International Scholarly Research Notices, 2013. **2013**.
[47] Li, M., et al., *Electrochemical and optical biosensors based on nanomaterials and nanostructures: a review*. Front Biosci (Schol Ed), 2011. **3**: p. 1308–1331.
[48] Zhou, Y., C.-W. Chiu, and H. Liang, *Interfacial structures and properties of organic materials for biosensors: an overview*. Sensors, 2012. **12**(11): p. 15036–15062.
[49] Guo, X., *Single-molecule electrical biosensors based on single-walled carbon nanotubes*. Advanced Materials, 2013. **25**(25): p. 3397–3408.
[50] Ko, P.J., et al., *Porous silicon platform for optical detection of functionalized magnetic particles biosensing*. Journal of Nanoscience and Nanotechnology, 2013. **13**(4): p. 2451–2460.
[51] Senveli, S.U. and O. Tigli, *Biosensors in the small scale: methods and technology trends*. IET Nanobiotechnology, 2013. **7**(1): p. 7–21.
[52] Valentini, F., et al., *Single walled carbon nanotubes/polypyrrole—GOx composite films to modify gold microelectrodes for glucose biosensors: study of the extended linearity*. Biosensors and Bioelectronics, 2013. **43**: p. 75–78.
[53] Su, L., et al., *Microbial biosensors: a review*. Biosensors and Bioelectronics, 2011. **26**(5): p. 1788–1799.
[54] Nie, S., et al., *Nanotechnology applications in cancer*. Annual Review of Biomedical Engineering, 2007. **9**: p. 257–288.
[55] Li, Y., H.J. Schluesener, and S. Xu, *Gold nanoparticle-based biosensors*. Gold Bulletin, 2010. **43**: p. 29–41.
[56] Biju, V., *Chemical modifications and bioconjugate reactions of nanomaterials for sensing, imaging, drug delivery and therapy*. Chemical Society Reviews, 2014. **43**(3): p. 744–764.
[57] Cai, H., et al., *Colloid Au-enhanced DNA immobilization for the electrochemical detection of sequence-specific DNA*. Journal of Electroanalytical Chemistry, 2001. **510**(1–2): p. 78–85.

[58] Crumbliss, A., et al., *Colloidal gold as a biocompatible immobilization matrix suitable for the fabrication of enzyme electrodes by electrodeposition.* Biotechnology and Bioengineering, 1992. **40**(4): p. 483–490.
[59] Xu, X., S. Liu, and H. Ju, *A novel hydrogen peroxide sensor via the direct electrochemistry of horseradish peroxidase immobilized on colloidal gold modified screen-printed electrode.* Sensors, 2003. **3**(9): p. 350–360.
[60] Zhao, J., et al., *A xanthine oxidase/colloidal gold enzyme electrode for amperometric biosensor applications.* Biosensors and Bioelectronics, 1996. **11**(5): p. 493–502.
[61] Hao, E., G.C. Schatz, and J.T. Hupp, *Synthesis and optical properties of anisotropic metal nanoparticles.* Journal of Fluorescence, 2004. **14**: p. 331–341.
[62] Mulvaney, P., *Surface plasmon spectroscopy of nanosized metal particles.* Langmuir, 1996. **12**(3): p. 788–800.
[63] Kelly, K.L., et al., *The Optical Properties of Metal Nanoparticles: the Influence of Size, Shape, and Dielectric Environment.* 2003, ACS Publications. p. 668–677.
[64] Reynolds, R.A., C.A. Mirkin, and R.L. Letsinger, *Homogeneous, nanoparticle-based quantitative colorimetric detection of oligonucleotides.* Journal of the American Chemical Society, 2000. **122**(15): p. 3795–3796.
[65] Oldenburg, S.J., et al., *Base pair mismatch recognition using plasmon resonant particle labels.* Analytical Biochemistry, 2002. **309**(1): p. 109–116.
[66] Liu, J. and Y. Lu, *Colorimetric biosensors based on DNAzyme-assembled gold nanoparticles.* Journal of Fluorescence, 2004. **14**: p. 343–354.
[67] Xu, W., et al., *Ultrasensitive and selective colorimetric DNA detection by nicking endonuclease assisted nanoparticle amplification.* Angewandte Chemie, 2009. **121**(37): p. 6981–6984.
[68] Xu, Q., et al., *Direct electrochemistry of horseradish peroxidase based on biocompatible carboxymethyl chitosan—gold nanoparticle nanocomposite.* Biosensors and Bioelectronics, 2006. **22**(5): p. 768–773.
[69] Wijaya, E., et al., *Surface plasmon resonance-based biosensors: from the development of different SPR structures to novel surface functionalization strategies.* Current Opinion in Solid State and Materials Science, 2011. **15**(5): p. 208–224.
[70] Zeng, S., et al., *Size dependence of Au NP-enhanced surface plasmon resonance based on differential phase measurement.* Sensors and Actuators B: Chemical, 2013. **176**: p. 1128–1133.
[71] Wu, K., et al., *Magnetic nanoparticles in nanomedicine: a review of recent advances.* Nanotechnology, 2019. **30**(50): p. 502003.
[72] Katz, E., *Synthesis, properties and applications of magnetic nanoparticles and nanowires—A brief introduction.* Magnetochemistry, 2019. **5**(4): p. 61.
[73] Mohammed, L., et al., *Magnetic nanoparticles for environmental and biomedical applications: a review.* Particuology, 2017. **30**: p. 1–14.
[74] Bishop, K.J., et al., *Nanoscale forces and their uses in self-assembly.* Small, 2009. **5**(14): p. 1600–1630.
[75] Haun, J.B., et al., *Magnetic nanoparticle biosensors.* Wiley Interdisciplinary Reviews: Nanomedicine and Nanobiotechnology, 2010. **2**(3): p. 291–304.
[76] Li, K., et al., *Fe2O3@ Au core/shell nanoparticle-based electrochemical DNA biosensor for Escherichia coli detection.* Talanta, 2011. **84**(3): p. 607–613.
[77] He, X., et al., *Plasmid DNA isolation using amino-silica coated magnetic nanoparticles (ASMNPs).* Talanta, 2007. **73**(4): p. 764–769.
[78] Min, J.H., et al., *Isolation of DNA using magnetic nanoparticles coated with dimercaptosuccinic acid.* Analytical Biochemistry, 2014. **447**: p. 114–118.
[79] Mejri, M., A. Tlili, and A. Abdelghani, *Magnetic nanoparticles immobilization and functionalization for biosensor applications.* International Journal of Electrochemistry, 2011. **2011**.

[80] Bi, S., et al., *Gold nanolabels for new enhanced chemiluminescence immunoassay of alpha-fetoprotein based on magnetic beads.* Chemistry—A European Journal, 2009. **15**(18): p. 4704–4709.

[81] Konry, T., et al., *Particles and microfluidics merged: perspectives of highly sensitive diagnostic detection.* Microchimica Acta, 2012. **176**(3): p. 251–269.

[82] Wang, S.X. and G. Li, *Advances in giant magnetoresistance biosensors with magnetic nanoparticle tags: review and outlook.* IEEE Transactions on Magnetics, 2008. **44**(7): p. 1687–1702.

[83] Tamanaha, C., et al., *Magnetic labeling, detection, and system integration.* Biosensors and Bioelectronics, 2008. **24**(1): p. 1–13.

[84] Jain, R.K., *Normalizing tumor microenvironment to treat cancer: bench to bedside to biomarkers.* Journal of Clinical Oncology, 2013. **31**(17): p. 2205.

[85] Murray, C., D.J. Norris, and M.G. Bawendi, *Synthesis and characterization of nearly monodisperse CdE (E= sulfur, selenium, tellurium) semiconductor nanocrystallites.* Journal of the American Chemical Society, 1993. **115**(19): p. 8706–8715.

[86] Weller, H., *Colloidal semiconductor q-particles: chemistry in the transition region between solid state and molecules.* Angewandte Chemie International Edition in English, 1993. **32**(1): p. 41–53.

[87] Khatun, Z., et al., *Imaging of the GI tract by QDs loaded heparin—deoxycholic acid (DOCA) nanoparticles.* Carbohydrate Polymers, 2012. **90**(4): p. 1461–1468.

[88] Khatun, Z., et al., *Oral delivery of near-infrared quantum dot loaded micelles for noninvasive biomedical imaging.* ACS Applied Materials & Interfaces, 2012. **4**(8): p. 3880–3887.

[89] Kim, J.S., et al., *In vivo NIR imaging with CdTe/CdSe quantum dots entrapped in PLGA nanospheres.* Journal of Colloid and Interface Science, 2011. **353**(2): p. 363–371.

[90] Geißler, D., et al., *Quantum dot biosensors for ultrasensitive multiplexed diagnostics.* Angewandte Chemie International Edition, 2010. **49**(8): p. 1396–1401.

[91] Giepmans, B.N., et al., *The fluorescent toolbox for assessing protein location and function.* Science, 2006. **312**(5771): p. 217–224.

[92] Xie, H.-Y., et al., *Luminescent CdSe-ZnS quantum dots as selective Cu2+ probe.* Spectrochimica Acta Part A: Molecular and Biomolecular Spectroscopy, 2004. **60**(11): p. 2527–2530.

[93] Yeh, H.-C., Y.-P. Ho, and T.-H. Wang, *Quantum dot—mediated biosensing assays for specific nucleic acid detection.* Nanomedicine: Nanotechnology, Biology and Medicine, 2005. **1**(2): p. 115–121.

[94] Tomasulo, M., et al., *pH-sensitive ligand for luminescent quantum dots.* Langmuir, 2006. **22**(24): p. 10284–10290.

[95] Xi, F., et al., *One-step construction of biosensor based on chitosan—ionic liquid—horseradish peroxidase biocomposite formed by electrodeposition.* Biosensors and Bioelectronics, 2008. **24**(1): p. 29–34.

[96] Ruan, C., et al., *One-pot preparation of glucose biosensor based on polydopamine—graphene composite film modified enzyme electrode.* Sensors and Actuators B: Chemical, 2013. **177**: p. 826–832.

[97] Kuila, T., et al., *Recent advances in graphene-based biosensors.* Biosensors and Bioelectronics, 2011. **26**(12): p. 4637–4648.

[98] Peng, J., et al., *Graphene quantum dots derived from carbon fibers.* Nano Letters, 2012. **12**(2): p. 844–849.

[99] Nurunnabi, M., et al., *Near infra-red photoluminescent graphene nanoparticles greatly expand their use in noninvasive biomedical imaging.* Chemical Communications, 2013. **49**(44): p. 5079–5081.

[100] Gao, L., H. Zhang, and H. Cui, *A general strategy to prepare homogeneous and reagentless GO/lucigenin&enzyme biosensors for detection of small biomolecules.* Biosensors and Bioelectronics, 2014. **57**: p. 65–70.

[101] Nurunnabi, M., et al., *Surface coating of graphene quantum dots using mussel-inspired polydopamine for biomedical optical imaging*. ACS Applied Materials & Interfaces, 2013. **5**(16): p. 8246–8253.
[102] Sun, H., et al., *Recent advances in graphene quantum dots for sensing*. Materials Today, 2013. **16**(11): p. 433–442.
[103] Wang, Z., et al., *Direct electrochemical reduction of single-layer graphene oxide and subsequent functionalization with glucose oxidase*. The Journal of Physical Chemistry C, 2009. **113**(32): p. 14071–14075.
[104] Yin, H., et al., *Electrocatalytic oxidation behavior of guanosine at graphene, chitosan and Fe3O4 nanoparticles modified glassy carbon electrode and its determination*. Talanta, 2010. **82**(4): p. 1193–1199.
[105] Gupta, V.K., et al., *A novel glucose biosensor platform based on Ag@ AuNPs modified graphene oxide nanocomposite and SERS application*. Journal of Colloid and Interface Science, 2013. **406**: p. 231–237.
[106] Gholivand, M.B. and M. Khodadadian, *Amperometric cholesterol biosensor based on the direct electrochemistry of cholesterol oxidase and catalase on a graphene/ionic liquid-modified glassy carbon electrode*. Biosensors and Bioelectronics, 2014. **53**: p. 472–478.
[107] Yola, M.L., T. Eren, and N. Atar, *A novel and sensitive electrochemical DNA biosensor based on Fe@ Au nanoparticles decorated graphene oxide*. Electrochimica Acta, 2014. **125**: p. 38–47.
[108] Valentini, F., M. Carbone, and G. Palleschi, *Carbon nanostructured materials for applications in nano-medicine, cultural heritage, and electrochemical biosensors*. Analytical and Bioanalytical Chemistry, 2013. **405**: p. 451–465.
[109] Davis, J.J., et al., *Chemical and biochemical sensing with modified single walled carbon nanotubes*. Chemistry—A European Journal, 2003. **9**(16): p. 3732–3739.
[110] Azamian, B.R., et al., *Bioelectrochemical single-walled carbon nanotubes*. Journal of the American Chemical Society, 2002. **124**(43): p. 12664–12665.
[111] Cai, C. and J. Chen, *Direct electron transfer of glucose oxidase promoted by carbon nanotubes*. Analytical Biochemistry, 2004. **332**(1): p. 75–83.
[112] Ménard-Moyon, C., et al., *Functionalized carbon nanotubes for probing and modulating molecular functions*. Chemistry & Biology, 2010. **17**(2): p. 107–115.
[113] Wang, J., *Carbon-nanotube based electrochemical biosensors: a review*. Electroanalysis: An International Journal Devoted to Fundamental and Practical Aspects of Electroanalysis, 2005. **17**(1): p. 7–14.
[114] Gruner, G., *Carbon nanotube transistors for biosensing applications*. Analytical and Bioanalytical Chemistry, 2006. **384**: p. 322–335.
[115] Besteman, K., et al., *Enzyme-coated carbon nanotubes as single-molecule biosensors*. Nano Letters, 2003. **3**(6): p. 727–730.
[116] Gigliotti, B., et al., *Sequence-independent helical wrapping of single-walled carbon nanotubes by long genomic DNA*. Nano Letters, 2006. **6**(2): p. 159–164.
[117] Tang, X., et al., *Carbon nanotube DNA sensor and sensing mechanism*. Nano Letters, 2006. **6**(8): p. 1632–1636.
[118] Li, C., et al., *Complementary detection of prostate-specific antigen using In2O3 nanowires and carbon nanotubes*. Journal of the American Chemical Society, 2005. **127**(36): p. 12484–12485.

Section C

Medical Applications

7 Hydroxyapatite-Based Nanocomposites

Synthesis, Optimization, and Functionalization for Medical Applications

Is Fatimah, Ganjar Fadillah, Ika Yanti, Suresh Sagadevan, and Ruey-an Doong

7.1 INTRODUCTION

Hydroxyapatite (HA), a compound with unit cell formula $Ca_{10}(PO_4)_6(OH)_2$ and general formula $Ca_5(PO_4)_3OH$, is a member of the apatite family that is a component of biological systems. HA is a major inorganic constituent of normal (bone, teeth, fish enameled, and shells of some species) and deposits in the body, such as dental and urinary stones. HA is the main component of human bones, as it is about 60–70% wt. of the human skeleton; 43–60% wt. of cartilage, teeth, tooth enamel, and other tissues; and 90% wt. of the inorganic bone matrix. Therefore, HA is one of the most important biomaterials in biomedical applications. The name "apatite" itself comes from the Greek word απατω, which means to deceive. These minerals exist in different forms and are often mistaken for valuable minerals such as aquamarine or amethyst. The characteristic differences occur in the apatite crystal structure due to ion exchange at the M-site, B-site, and A-site of the general structure of $M_{10}B_6A_2$, where M is a bivalent cation; B is the trivalent anion XO_4; and A is a monovalent anion and can be replaced by fluoride, chloride, or carbide. The structure is depicted in Figure 7.1.

As seen in Figure 7.1, Ca ions are found in the bottom, middle, and top quarter and three-quarter planes, so that the total Ca = = 2 (0.50) + {2 (0.00) + 2 (1.00))} / 2 + {3 (0.25) + 3 (0.75)} × 4/4 = 10. PO_4^{2-} can be found at the center and on the line, so PO_4^{2-} = 2 (middle) + 8 (line) / 2 = 6. OH^{2-} lies in the one-quarter and three-quarter positions, so OH^{2-} = {4 (0.25) + 4 (0.75)} / 4 = 2. As a result, HA is described as $Ca_{10}(PO4)_6(OH)_2$. HA is the hydroxyl-final member of the complex apatite group.

Other forms of calcium orthophosphate compounds, solubility, and stability besides HA can be seen in Table 7.1.

Because of its chemical and physical close similarities and biocompatibility, HA is widely used as a substitute for bone or artificial bones and teeth. In

DOI: 10.1201/9781003425427-10

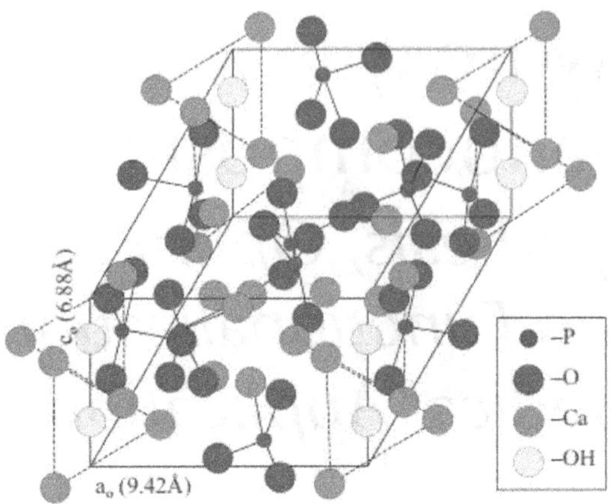

FIGURE 7.1 Structure of HA.

orthopedic and dental applications, HA is utilized as a biomaterial for substituting, repairing, and drug delivery systems. HA is a less water-soluble calcium phosphate with a Ca/P ratio of 1.67 and has chemical stability, as the structure remains unchanged in the pH range between 4 and 12. At room temperature, pure HA is a white powder, whereas natural HA can also be brown, yellow, or green, proportional to the discoloration of dental fluorosis. The chemical structure of HA gives rise to pores (porosity), and the mechanical properties of HA depend on the density, porosity, sinterability, phase composition, crystal size, and so on. The compressive strength, malleability, and tensile strength of HA lie in the range of 120–150, 38–250, and 38–300 MPa, respectively. Young's modulus values of dense HA vary from 35 to 120 GPa, depending on the pore structure and purity of the material, and in addition, the hardness of HA, referred to as the Vicker hardness, ranges from 3–7 GPa [1].

In terms of material chemistry, the physicochemical properties of HA bioceramics are highly dependent on the microstructure, purity, crystallinity, sintering ability, density, and grain density. This property can be influenced and controlled during synthesis, as well as a characteristic that allows other applications of HA apart from biomedical applications, for example, in adsorption technology and the environment. The latest development of HA synthesis presents various forms of HA, including one-dimensional (1D) forms, including rod-like, wire-like, 2-dimensional (2D sheet-like particles) and other HA composite forms. Many composite forms are chosen to arrange certain characteristics of the material, for example, nanocomposites of metal or metal oxide-doped HA as functionalized materials for specific purposes. As an example, silver nanoparticle (Ag NP)–doped HA nanocomposites were developed and reported as antibacterial biomaterials for wide applications such as in medical surgery and antibacterial ceramics technology. Therefore, synthesis and characterization become important aspects of HA and functionalized HA preparation that are also close

TABLE 7.1
The Calcium Orthophosphate Compounds, Solubility, Stability, and pH

Ca/P	Compound	Formula	Solubility at 25°C (–log Ks)	Solubility at 25°C (g/L)	pH Stability at 25°C
2.00	Tetracalcium phosphate (TTCP or TetCP), mineral hilgenstockite	$Ca_4(PO_4)_2O$	38–44	~0.0007	
1.67	Oxyapatite (OA, OAp, or OXA)	$Ca_{10}(PO_4)_6O$	~69	~0.087	
1.67	Hydroxyapatite (HA, HAp, or OHAp)	$Ca_{10}(PO_4)_6(OH)_2$	116.8	~0.0003	9.5–12
	Fluorapatite (FA or FAp)	$Ca_{10}(PO_4)_6F_2$	120.0	~0.0002	7–12
1.5–1.67	Calcium-deficient hydroxyapatite (CDHA or Ca-def HA)	Ca_{10}-$x(HPO_4)x(PO_4)_6$-$x(OH)_2$-x $(0<x)$	~85	~0.0094	6.5–9.5
1.5	α-tricalcium phosphate (α-TCP)	α-$Ca_3(PO_4)_2$	25.5	~0.0025	
1.5	β-tricalcium phosphate (β-TCP)	β-$Ca_3(PO_4)_2$	96.6	~0.0025	
1.33	Octacalcium phosphate (OCP)	$Ca_8(HPO_4)_2(PO_4)_4{\cdot}5H_2O$	96.6	~0.0081	5.5–7.0
1.2–2.2	Amorphous calcium phosphates (ACPs)	$CaxHy(PO_4)z{\cdot}nH_2O$, n = 3–4.5; 15–20% H2O			~5–12
1.0	Dicalcium phosphate dihydrate (DCPD)	$CaHPO_4.H_2O$	6.59	~0.088	2.0–6.0
1.0	Dicalcium phosphate anhydrous (DCPA)	$CaHPO_4$	6.90	~0.048	
0.5	Monocalcium phosphate monohydrate (MCPM)	$Ca(H_2PO_4)_2.\ H_2O$	1.14	~18	0.0–2.0
0.5	Monocalcium phosphate anhydrous (MCPA)	$Ca(H_2PO4)_2$	1.14	~17	

related to their applications. Physicochemical characterization utilizing instrumental analysis plays an important role in material development. In addition, future perspectives and innovative outlooks on HA and functionalized HA such as utilization of waste material as a resource are interesting to discuss. Figure 7.2 represents the aspect of HA functionalization for potential biomedical applications discussed in this chapter.

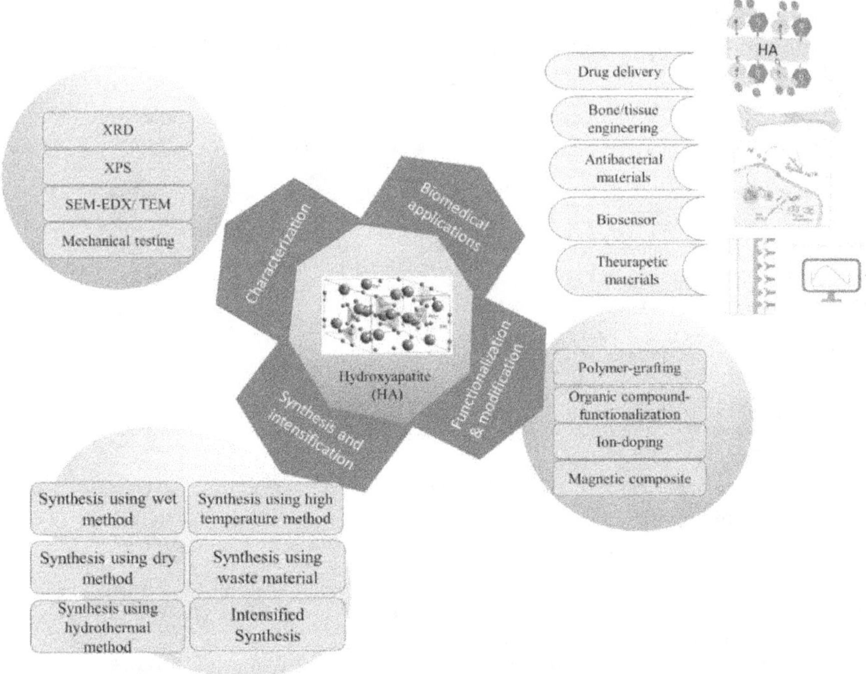

FIGURE 7.2 Schematic representation of synthesis, characterization, functionalization, and applications of HA-based nanocomposites.

7.2 SYNTHESIS OF HA

Various techniques have been reported for the preparation of HA either in bulk or nanoparticle forms. From various papers, it is concluded that the diverse characteristics and properties, such as porosity, surface area, crystal defects, and the affinity for organic matter found in physiological media, depend on the synthesis technique. Generally, HA can be synthesized by a variety of techniques, which can broadly be grouped into six sets of methods, as follows.

7.2.1 Dry Method

In this method, the reaction between calcium and the phosphate precursor is conducted at high temperatures without any solvent. The process is considered a low-cost process with high reproducibility despite the risk of contamination during milling. As a precursor, calcium and phosphorous compounds can be used as the starting materials, such as calcium hydroxide ($Ca(OH)_2$), monocalcium phosphate monohydrate ($Ca(H_2PO_4)_2{\cdot}H_2O$), calcium carbonate ($CaCO_3$), calcium pyrophosphate ($Ca_2P_2O_7$), dicalcium phosphate dihydrate ($CaHPO_4{\cdot}2H_2O$), dicalcium phosphate anhydrous ($CaHPO_4$), and calcium oxide (CaO). Even though there is no excess liquid as a solvent, a small amount of mixed-in acetone and water is usually added to homogenize the reactant before heat treatment. Several parameters, such as time of milling, rotation speed, and temperature, are recognized as the influencing factors for the characteristics of HA, including crystallinity and specific surface area.

7.2.2 Wet Method

In this method, chemical precipitation at low temperature is the main reaction, with varied possible specific routes, such as co-precipitation, the sol-gel route, and hydrolysis. However, the basic method is the wet method, which is usually performed by reacting calcium oxide or $Ca(OH)_2$ with phosphoric acid or ammonium diphosphate, where, in the precipitation stage, the calcium-to-phosphate molar ratio is assumed to be 1.67 via the following equation:

$$CaO_{(s)} + H_2O_{(l)} \rightarrow Ca(OH)_2\,(aq)$$
$$10Ca(OH)_2 + 6H_3PO_4 \rightarrow Ca_{10}(PO_4)_6(OH)_2 + 18H_2O$$

This chemical precipitation approach has proven popular due to easiness, speed, and economic advantages.

The simple method for HA by the wet method was preparing $(NH_4)_2HPO_4$ solution as a P source and $Ca(NO_3)_2{\cdot}4H_2O$ as Ca precursors. The slow addition of $(NH_4)_2HPO_4$ solution into $Ca(NO_3)_2{\cdot}4H_2O$ under stirring at room temperature is applied before the solution is further set up at pH 9–11 by the addition of a base solution (NaOH or NH_4OH). The Ca/P ratio of 1.67 needs to be kept at 1.67 while white precipitate is obtained from these steps, which then could be sintered at optimum temperature.

As with other precursors, the precipitation rate will strongly influence the crystallinity and applicability. High purity of HA is reported to be derived by the sol-gel method. For this mechanism, Ca precursors such as calcium diethoxide ($Ca(OEt)_2$), calcium nitrate, and calcium propionate were used; meanwhile, as P precursors, triethyl phosphate ($PO(OEt)_3$, phenyldichlorophosphite ($C_6H_5PCl_2$), and phosphonoacetic acid ($HOOCCH_2PO(OH)_2$) can be chosen. Research work on pure HA synthesis was reported by using ($Ca(OEt)_2$) and ($PO(OEt)_3$) in ethanol solvent by the sol-gel mechanism and followed by sintering at 600°C. The sol-gel reaction can be represented as follows:

$$P(OEt)_{3\text{-}x}\,(OH)_x + Ca(NO_3)_{2\text{-}x}\,(OEt)_y \dashrightarrow (OEt)_{y'}(NO_3)_{2\text{-}y'}\text{-}Ca\text{-}O\text{-}HPO(OEt)_{3\text{-}x'} + C_2H_5OH + H_2O.$$
$$P(OEt)_{3\text{-}x}\,(OH)_x + Ca^{2+} + NO_3^{-} \dashrightarrow (NO_3)^{-1}OH)\text{-}Ca\text{-}O\text{-}PHO(OEt)_{3\text{-}x} + H^+ + C_2H_5OH + H_2O.$$

Referring to the mechanism, the proton will be released from step 2, leading to a decreased pH level of the sol. A further reaction step is solvent evaporation, which accelerates polymerization via dehydration/condensation and forms dry gels[2]. Another hydrous sol-gel mechanism was utilized with calcium nitrate ($Ca(NO_3)_2{\cdot}4H_2O$) and phenyldichlorophosphite ($C_6H_5PCl_2$) as precursors and a calcination temperature of 900°C and the use of using calcium nitrate and phosphonoacetic acid ($HOOCCH_2PO(OH)_2$) in an aqueous solution and calcination at 700°C. Generally, higher temperatures will lead to increasing crystallinity [3].

Such optimization of synthesis parameters for the characteristics and morphology of HA have been studied, including the effects of additional surfactant, solvent, temperature calcination/sintering, and the use of different reagents. From some investigations, the presence of cationic or non-ionic surfactants acts as a soft template to govern the nucleation and crystal growth of HA. This affects the particle size and form, as well as the effect of a kind of precursor or different reagent. From research conducted by Salarian, homogeneous nano-rod morphology and size were successfully prepared using cetiltrimethyl ammonium bromide (CTAB) and polyethylene glycol (PEG) under hydrothermal conditions. Irregular-shaped and aggregated nanocrystals were obtained when the surfactant was not utilized, mainly caused by homogeneous nucleation followed by agglomeration [4]. Figure 7.3 represents the role of surfactant in determining the particle morphology and size.

Surfactant molecules are characterized as having two sides with different properties: a hydrophilic side that comes from positive or negative charges contained in the structure having strong affinity for water molecules, and on the other hand, the hydrocarbon chain has hydrophobic properties with less affinity for water molecules. At a certain concentration exceeding the critical micelle concentration (CMC) in water, surfactant molecules tend to form micelles by a self-assembly mechanism, creating the structure composed of hydrocarbon chains inside, and the surface of micelles has a positive charge. This conformation results in the strong electrostatic attraction of the micelle surface when mixed with the phosphate solution of HA precursor. When CTMA is used as a surfactant, the CTA^+ $-PO_4^{3-}$ micelles act as nucleation centers surrounded by alkyl hydrocarbons. The CTA^+ and PO_4^{3-} bonding is in the form a tetrahedral structure, which is stereochemically in confirmation and able to interact with each other to create a homogeneous form and particle size of precipitated HA [5].

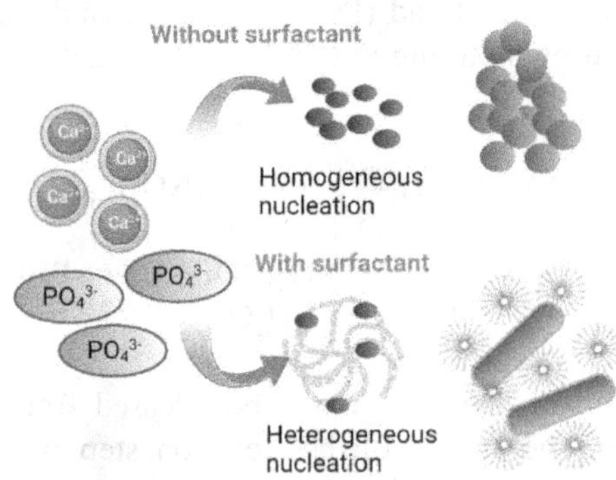

FIGURE 7.3 Schematic representation of precipitation during HA synthesis with and without surfactant.

7.2.3 Hydrothermal Method

The hydrothermal method is based on the use of high-temperature and high-voltage aqueous solutions. The reaction system is subjected to a high temperature and pressure conditions that are produced by the vapor of the solvent. The combined effect between high pressure, temperature, and the presence of calcium carbonate favors the nucleation processes by fast calcium ion diffusion into dicalcium diphosphate to transform it into HA [6]. The system improves the crystallinity and purity of the product significantly. However, high temperature and pressure typically mean obtaining HA crystals with irregular morphology. Control of the morphology can be obtained by the addition of polymer [6] or surfactant and intensifying conditions such as utilizing a microemulsion system.

7.2.4 High-Temperature Method

In this kind of method, HA can be synthesized by combustion and pyrolysis processes. Generally, the high-temperature process aims to produce fully dense ceramic bodies with high mechanical strength. A mixture of precursors in deionized water is usually heated at a high temperature (500–600°C); then, in order to break down the obtained agglomerates, a milling procedure of 5–8 h is conducted[7]. Intensification of the combustion mechanism can also be proceeded by microwave irradiation [8]. In general, the mechanical behavior of HA is controlled by the sintering process for a tailored microstructure, and the porosity can be controlled by reducing the heating rate, and shrinkage occurs in synchrony with the densification rate [7]. Some investigations on a combination of synthesis methods are also recorded. As examples, the microwave-assisted and hydrothermal-assisted sol-gel method were reported as innovative fast processes using the sol-gel mechanism [9].

Microwave irradiation to the sol-gel synthesis is a combined method that excludes the aging procedure on sol-gel preparation; meanwhile the conventional method involves a 48-h aging period, compared with other combinations such as microwave and hydrothermal, microwave and co-precipitation, as well as the use of ultrasound irradiation. A basic principle of the intensification using microwave and ultrasound is the increasing effectiveness of molecular collision during the process, referring to absorbed energy by the reactants. In intensifying the combustion method, for example, microwave irradiation reduces the optimum temperature of the method. Table 7.2 lists some research on several effects of synthesis parameters on the characteristics of HA.

7.3 FUNCTIONALIZATION AND MODIFICATION OF HA

Functionalization and modification of HA are usually intended for a specified material purpose, including as a bioactive material or drug delivery agent for biomedical applications. In general, HA functionalization and modification could be divided into two main categories: a) organic and/or polymer-functionalized HA and b) metal/metal oxide modified HA.

TABLE 7.2
Research on Several Effects of HA Synthesis Parameters

Synthesis Method	Parameter of Synthesis	Morphology	Particle Size	Specific Remark	Reference
Dry method	Calcination temperature	–	11–249 nm	Increasing temperature tends to form increasing crystallite size	[10]
Coprecipitation	Calcium precursors	Mixed of short rod-like to rice-like forms	15–30 nm	From varied nitrate, hydroxide, and carbonate, only calcium nitrate gave the single phase of HA without any other phase	[11]
Coprecipitation	pH	Morphology of HA depends on the pH of co-precipitation	Depending on pH	The particles are in spherical shapes at pH = 5, with sizes at 50–90 nm; at pH 6: HA particles are in filament form with size of 100–150 nm; at pH =7: particles are in needle shape, 5–10 nm in diameter and 150–200 nm in length; at pH 9: HA particles are in nanorods with dimensions of 20–30 nm in diameter and 200–300 nm in length; at pH = 12: particles are in rice-like shapes, with the size ranging from 10–20 nm in length and not exceeding 60 nm	[12]
Coprecipitation	Surfactant	Rod-like nanoparticles		The presence of CTAB and PEG resulted a uniform particle size	[4]
Coprecipitation	Solvent effect	Morphology depends on solvent	3–103 nm	Ethanol and tetrahydrofuran (THF) gave rod-like and spherical forms; meanwhile water gave irregular form	[13]
Hydrothermal	Solvent effect	Rod-like nanoparticles	16.77–77 nm	The composition of the isopropanol–water mixture does not significantly affect the particle size distribution	[14]
Hydrothermal	Polymer addition	Rod-like nanoparticles	≥20 nm	Nanoporous HA was obtained by the addition of novel polymer poly-ethylene oxide (PEO) in hydrothermal synthesis using $Ca(OH)_2$ and $NaH_2PO_4.2H_2O$ precursors. The higher PEO acts to decrease HA particle size and increase Young's modulus	[15]
Hydrothermal	Time of hydrothermal treatment	Nanorod	200–500 nm	Varied time of hydrothermal treatment (24–72 hrs) has no effect on the particle morphology or size	[6]

7.3.1 Polymer-Functionalized HA

Biocompatibility is an important feature that is required in tissue engineering and drug delivery systems. In order to provide the desired properties, the combination of bioresorbable polymers and HA over a self-assembly of nanocomposite with tunable mechanical, thermal, and electrical properties become important. Grafting polymers on the HA surface is the most common method to enhance the dispersion of HA. Moreover, the designed polymer-grafted HA is also attempted to enhance protein absorption in tissue engineering. Various types of polymers such as poly(methyl methacrylate) (PMMA), poly(lactic acid-co-glycolic acid) (PLGA), poly lactic acid (PLA), poly(N-isopropylacrylamide) (PNIPAM),), ε-caprolactone (CL), and poly(γ-benzyl-L-glutamate) (PBLG) were reported. Some polymers/HA with applications are listed in Table 7.3. In situ polymerization in the grafting mechanism could be performed, including enzyme-assisted polymerization in preparation of PLGA/HA. Other non-enzymatic polymerization such as the mechanism of atom transfer radical polymerization (ATRP) could be applied instead of reversible addition—fragmentation chain transfer (RAFT) polymerizations [16].

Figure 7.4 represents the mechanism of the grafting process in PMMA/HA preparation. Hydroxyl groups are available on the HA surface and are first functionalized

TABLE 7.3
Polymer/HA Nanocomposites and Their Applications

Polymer	Remark	Application	Reference
Poly(lactic acid-co-glycolic acid) (PLGA)	PLGA/HA *in situ* polymerization of glycolide in porous HAp disks and L-lactide using lipase *MM*, derived from *Mucor miehei*	Nanocomposites have better mechanical strength to support cell adhesion and proliferation properties	[17]
PNIPAM/HA	The nanocomposite can be prepared by physical crosslinking of *N*-isopropylacrylamide with a nano-HA, electrochemical, or co-precipitation method	Thermos-responsive bone imitation, non-hemolytic biomaterial, injectable hydrogel	[18–20]
Alginic acid/HA	Nanocomposites can be prepared by wet chemical precipitation and freeze-drying methods	Protein absorption improvement	
Poly-ε-caprolactone/HA	Mechanical method	Provides porosity for bone-imitation applications	[21]
Poly(2,2-dimethyl trimethylene carbonate) s (PDTC-COOH)/HA	Mechanical method	Improvement of mechanical properties	

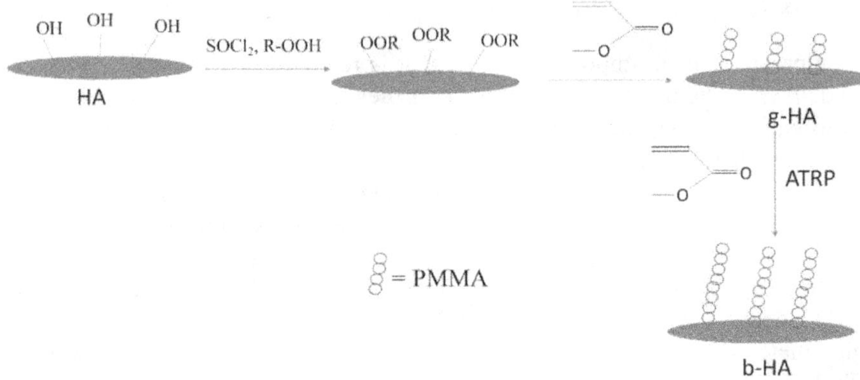

FIGURE 7.4 Schematic representation of polymer grafting by ATRP mechanism.

with peroxide groups for reverse ATRP. Then, an initiator of bromide groups is used to initiate MMA via ATRP continuously. Reverse ATRP of MMA on the surface of ROO-HAP nanoparticles was conducted in cyclohexanone using $CuCl_2$ and PMDETA as the catalytic system. By these grafting processes, PMMA/HA has been reported to have good physical characteristics of bioaffinity, biocompatibility, compressive strength, and osteo-conductivity that fit with dental implant fixtures. A similar effect was reported from the PLGA/HA that increased the mechanical strength by more than twice. Higher grafting efficiency was achieved by intensive surface modification by optimizing the effectiveness of the molecular weight of PLGA. The grafting efficiency itself was monitored by thermal gravimetry analysis (TGA) [22]. Polymer modification to HA was also attempted to provide bioceramic nanoparticles with the capability to support proteins related to bone repair and regeneration. The absorption capability of collagen and albumin in osteoblastic cell culture and suppresses apoptosis. Cellulose/HA, alginate/HA, and PLA/HA are examples of this purpose[23]. Within this scheme, a biodegradable polymer scaffold provides several benefits for bone tissue engineering, such as better growth function and an enhanced environment for cell seeding and survival.

Besides chemical methods, mechanical methods could be applied in polymer/HA preparation. The method is based on mixing and stirring inorganic filler and the polymer solution at a certain temperature and pressure. Such a method was reported for PCL/HA [21], poly(2,2-dimethyl trimethylene carbonate)s/HA, and PLA/HA[24]. The method is fast, simple, and universal; however, it has some drawbacks, such as low strength characteristics of the materials as a consequence of the irregular distribution of the filler in the polymer matrix. In order to optimize polymer modification, organic functionalization to the HA surface could be conducted to form a highly bioactive scaffold.

7.3.2 Organic-Functionalized HA

Surface functionalization of the HA surface is generally performed for drug release applications, especially to enhance drug adsorption affinity. As an example, organic

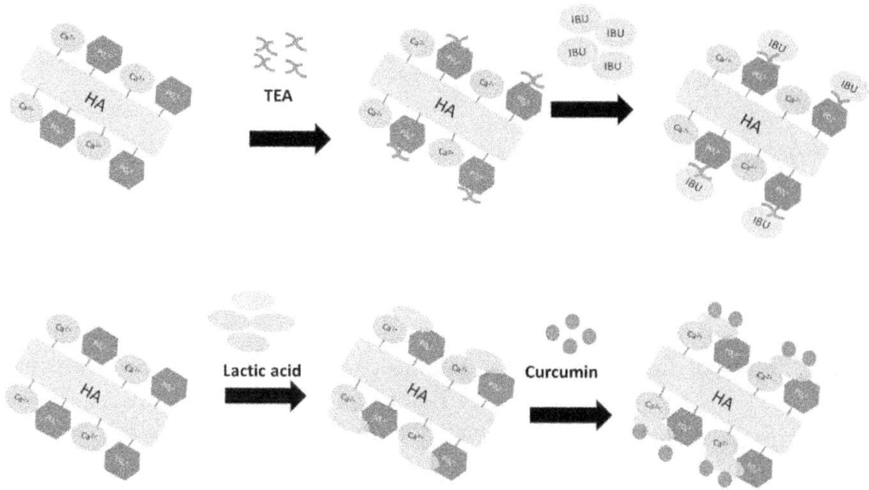

FIGURE 7.5 Schematic representation of organic-modified HA for drug delivery applications (above: TEA-HA as ibuprofen (IBU) delivery; below: lactic acid-HA as curcumin delivery system) [26].

acids consisting of lactic acid, tartaric acid, and citric acid were used to create a binding site to curcumin as a bioactive compound. The surface modification is based on the electrostatic interaction of opposite charges between both particles[25]. The positive and negative charges of HA surface coming from Ca^{2+} and PO_4^{3-} have the capability to bind with organic acids with dipoles. Furthermore, the available hydrophobic sites of organic acid could conduct covalent binding to curcumin. A similar mechanism was represented in triethylamine-modified HA for preparing ibuprofen-loaded and alginate-HA for a sustained release of ciprofloxacin [26]. Surface functionalization of HA by TEA is referred to as hydrogen bonding formation of nitrogen in TEA and protonated phosphate, which is facilitated by ethanol solvent. Figure 7.5 is a schematic representation of organic-modified HA for drug delivery applications.

Surface modification of HA by using furosemide (Fur), 4-chloro-2-[(2-furanylmethyl)-amino]-5-sulfamoylbenzoic acid, is a hybrid material to develop efficacy in drug delivery systems. Furosemide is a high ceiling loop diuretic drug to treat hypertension and edema linked to congestive heart failure, liver cirrhosis, and renal disease. It has characteristics of low solubility and permeability values, so it is almost insoluble in water, leading to poor oral bioavailability. The development of oral formulations that minimize the dose size and effect could be designed with inorganic–organic hybrid materials, including the use of HA. Fur/HA showed greatly improved solubility, wettability, and dissolution rate in a simulated gastro-intestinal environment, representing improved bioavailability [27]. Chemical binding of the active compounds to HA and synergistic interaction to improve drug bioactivity are the main points for organic-modified HA. Epigallocatechin-3-gallate (EGCG)-HA is a hybrid material that fits this consideration. As EGCA is active to inhibit tumor growth, the resulting conjugate (HA-EGCG) exhibited significantly enhanced anticancer activity compared to free EGCG [28]. In more complex material design, a multimaterial composite of drug-loaded alginate/hydroxyapatite/collagen was prepared for infected bone defect applications.

7.3.3 Ion-Doped HA

Ion-doped HA is an existing modified HA for many applications. For example, Ag-doped and Zn-doped HA are popular materials in antibacterial bioceramics. The basic principle of ion doping is based on the crystal structure of hydroxyapatite, in which the Ca^{2+}, PO_4^{3-} and OH^- in the structure could be replaced by other ions. The strength, atomic diameter, and reduction potential of the ions determine the exchangeability of the structure. For example, OH^- is easily replaced by a small amount of F^- and CO_3^{2-}. In the case of F-doping, one F^- ion replaces an OH^- ion, forming two hydrogen bonds with two neighboring OH^- ions. This inhibits the motion of OH^- ions, resulting in ordered configuration in F-HA with enhanced durability. Figure 7.6 describes the A position as anionic sites for substitution, such as PO_4^{3-}, which can be replaced by CO_3^{2-} and SiO_4^{4-}, and at B position, where the cationic site Ca^{2+} can be replaced by other cations. The existence of these ions may affect the mechanical properties, crystallization, and biological activity of apatite, degradation, and furthermore the physiological function of hard tissue in the organism [12,17–20]. In order to obtain biomimetic materials like natural apatite in composition, structure, and function, ion-doped hydroxyapatite has been extensively investigated. Identification of the structural changes during F- and CO_3^{2-} doping could be described by x-ray diffraction analysis, especially the evaluation of lattice parameter analysis. The lattice parameters of the a-axis decrease concomitantly with increased F-substitution; meanwhile, the c-axis increases slightly [29,30]. In the case of Mg^{2+} doping, as the ionic radius of Mg^{2+} is smaller than that of Ca^{2+}, the lattice parameters and unit cell volume decrease, which refers to unit cell distortion. In contrast, the

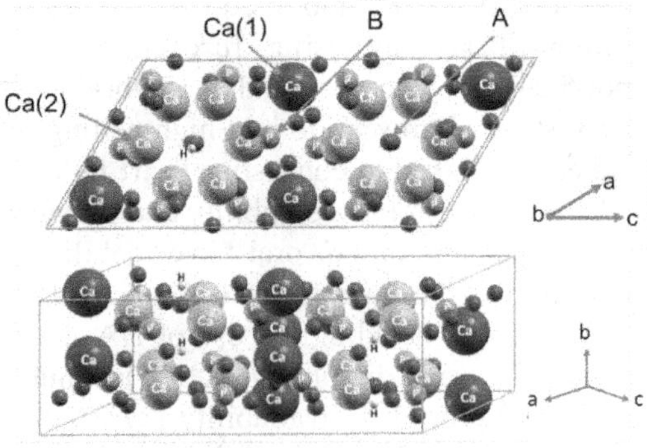

FIGURE 7.6 Incorporation of foreign cations in Ca(1) and Ca(2) positions and anions in A and B positions. Color coding: Ca in Ca(1) position—magenta; Ca in Ca(2) position—cyan; P—orange; O—red; H—grey. [32].

larger ionic radius of Sr^{2+} with respect to Ca^{2+} in Sr-doped HA, the expanded unit cell of HA, is confirmed by the increase in parameters a, b, c, and the volume of the unit cell [31].

A diverse range of methods is available for ion-doped HA preparation. Similar to the preparation of HA and polymer/HA, there are dry methods, wet methods, hydrothermal methods, high temperatures, and combined methods that can be chosen. In addition, intensification of the procedure has also gained attention. The simple precipitation method using (Fe doping/Ca^+ doping ion) molar ratio is usually chosen. In addition, intensification procedures, such as electrochemical, hydrothermal, and solvothermal methods, can also be used. For example, silver-doped and titanium-doped HA were prepared at [doping ion/Ca^+ doping ion] ranging from 0.1–0.9, so the formula of $Ca_{10-x}Ag_x(PO_4)_6(OH)_2$ and $Ca_{10-x}Ti_x(PO_4)_6(OH)_2$ from Ag and Ti doping, respectively. Both Ag- and Ti-doped HA showed clear antibacterial activity, which was associated with the [Ag/(Ca + Ag)] and [Ti/(Ca + Ti)] ratios. The significant effect on antibacterial activity is reflected by the biocidal effect of ion-doped HA materials, even though the Ag or Ti content is less than 2%.

The simple method for Ag- and Ti-doped HA by the wet method is similar to a wet method for HA preparation. In particular, $(NH_4)_2HPO_4$ dissolved in water or ethanol: water was used as the P source, and a mixture of $Ca(NO_3)_2{\cdot}4H_2O$ and $AgNO_3$ dissolved in the same solvent was used as the Ca and Ag precursors. A solution containing P was added dropwise to the solution containing Ca and Ag under continuous stirring at 12 h at 100°C, and the [Ca+Ag]/P ratio should be set at 1.67. During the stirring, pH should be maintained at 10 by adding ammonium hydroxide. A similar procedure was utilized for Zn^{2+}, Te^{2+}, Co^{2+}, Sr^{2+}, Mg^{2+}, and Li^{2+}. Table 7.4 lists important notes from ion-doped HA prepared by the precipitation method.

The accessible ionic form of dopants generally gives ion-doped HAs antibacterial and antifungal active properties. Antibacterial activity testing presents this by inhibition zone for several bacteria such as *Escherichia coli*, *Staphylococcus aureus*, *Kliebsiella pneumonia*, and *Streptococcus pyogenes*. The mechanism of antibacterial and antifungal activity is associated with the capability of dopants to penetrate the cell wall and create ionic conditions causing cell lysis or prevent ATP and DNA replication, leading to the death of bacteria. As the ions enter the cytoplasm, there is an increase in permeability, which causes bacterial death. Another mechanism is that the $dopant^+$ ions penetrate the cell membrane and react directly with the -SH (thiol) group and other possible ligands of cell proteins. The interaction causes the release of ATP synthesis during cellular respiration and proton release, and the cell metabolism system is disrupted. Figure 7.7 gives a schematic representation of antibacterial activity. Similarly, the fungi cells will be interfered with as the dopant ion enters the cell, being diffused and leading them to produce reactive oxygen species (ROS). The presence of ROS induces oxidative stress of the cell, which is easily followed by cell disruption.

The progression of ion-doped HA is the use of the nanoparticle form of the dopant for more effective activity [40]. By adopting green chemistry principles, green synthesized nanoparticles (NPs) using plant extracts are a new methodology. Using this scheme, biosynthesized Au NP-doped HA and Ag NP-doped HA were reported to have high antibacterial activity [41].

TABLE 7.4
Remarks on Ion-Doped HA Preparation

Ion Doped	Optimum Temperature/ Time of Stirring	Remark	Reference
Zn^{2+}	100°C, 24 h	The increasing Zn amount leads to a decrease of the crystallite size of HA.	[33]
Ta^{5+}	Room temperature, 24 h	The rapid and intensive crystallization of HA upon thermal annealing was obtained by the incorporation of tantalum into the HA.	[34]
Se^{2+}	Room temperature, 24 h	The increasing Se^{2+} dopant caused decreasing crystallinity. Incorporated Se^{2+} tends to preferentially occupy the CaII position.	[35]
Sr^{2+}	Room temperature, 24 h	Incorporation of Sr^{2+} leading to the expanded unit cell of HA is confirmed by the increasing of parameters a, b, and c and the volume of unit cells.	[31]
Se^{2+}	Room temperature, 24 h	The width of the crystal elongated as the selenium doping increased. The thermal stability of the products depends on the selenium content.	[36]
Fe^{2+}	70°C, 24 h	The higher Fe concentration tends to change the particle's form from spherical to needle shape.	[37]
Li^{+}	Room temperature, 48 h	A lower degradation rate of LiHA scaffolds compared to HA scaffolds was obtained.	[38]
Mn^{2+}	Room temperature, 8 h	Mn-HA has capability as an adjuvant for photo-therapy for cancer.	[39]

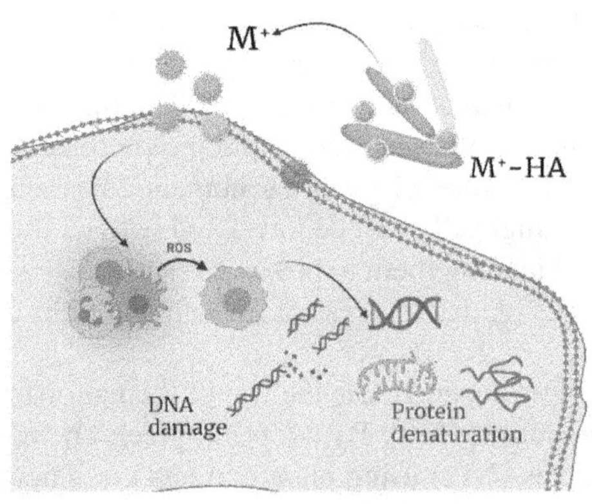

FIGURE 7.7 Mechanism of antibacterial by ion-doped HA.

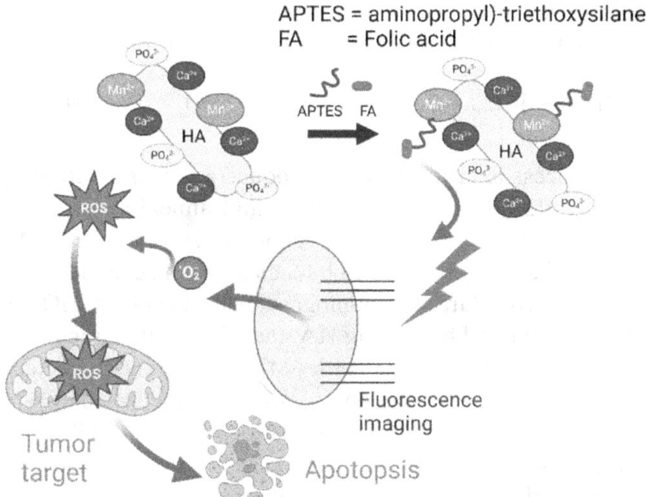

FIGURE 7.8 Schematic representation of the photodynamic therapy mechanism.

The capability of conducting a reduction-oxidation reaction of the dopant ion has potential for sensor and therapeutic applications. Cu-doped HA in exhibits good response and sensitivity for detecting uric acid with the lowest stable detection limit, ~0.5 μM. The reduction-oxidation state of Cu/Cu^{2+} is the electrochemical state as the non-enzymatic identification of uric acid. From the study, it is found that electrochemical activities and non-enzymatic selectivity depend on the surface character of Cu-HA [42]. The electrochemical state of Mn in Mn-HA has the potential to be an adjuvant for cancer therapy via photodynamic therapy (PDT). Mn-HA was linked with the counter folic acid and IR-783 fluorescence dye that were chosen as fluorescence sites. By photon interaction, doped Mn allows the release of Mn^{2+} ions, which triggers the production of toxic hydroxyl radicals (•OH) via Fenton or Fenton-like reactions to decompose H_2O_2 at tumor sites. Figure 7.8 is a schematic representation of the photodynamic therapy mechanism.

7.3.4 Metal or Metal Oxide/HA Nanocomposites

Besides ion doping to HA structure, HA-based nanocomposites are also prepared by a combination of metal or metal oxides. Generally, a nanocomposite is composed of the mixing of prepared HA and metal or metal oxide nanoparticles, followed by either wet, hydrothermal, or high-temperature blending methods. In principle, the difference between this scheme with ion-doped HA is that the metal or metal oxide is not positioned at the crystallite structure of HA. In addition, the percentage of combined metal/metal oxide is usually higher.

As an example, TiO_2/HA can be prepared by mechanochemical, hydrothermal, and electrochemical deposition methods. Mechanochemical synthesis of TiO_2/HA was successfully prepared by using CaO, $CaHPO_4$, and metallic Ti as reported precursors. The mechanochemical reaction was performed by zirconia ball milling under a highly purified argon gas atmosphere at ball ratio and rotational speed of 1:20 and 600 rpm, respectively, so the synthesis reaction was as follows:

$$6CaHPO_4 + 4CaO + Ti \longrightarrow Ca_{10}(PO_4)_6(OH)_2 + Ti + 2H_2O \text{ [43].}$$

Although the process is fast, it tends to produce a low-crystallinity powder. This low crystallinity could be improved by a thermal annealing process at 650°C. In principle, the reaction condition should be set at a Ca/P ratio of 1.67. Morphological evaluation revealed that milling up to 15 h tends to produce agglomeration [43].

By hydrothermal procedure, the homogeneously dispersed TiO_2 in a TiO_2/HA nanocomposite was obtained by mixing HA and TiO_2 as the precursors and reacted under autoclave at 150–250°C for 12–48 h [44]. A different characteristic of nanocomposites compared to Ti-HA nanocomposites can be identified by x-ray diffraction (XRD) analysis, as TiO_2/HA will be a combination of HA and TiO_2 phase reflection. However, such specific surface area and optical properties of the nanocomposite are also crucial factors affecting bioactivity as an antibacterial agent or sensor material [45]. More controllable crystallinity of nanocomposite was reported by using $Ti(OH)_4$ as a precursor instead of TiO_2 [46].

Electrochemical deposition is the third method usually chosen for creating a uniform surface of the metal oxide, usually called TiO_2-coated HA. This method was applied for ZnO/HA [47], ZrO_2/HA[47], and $CoFe_2O_4$ [48]. Electrodeposition of HA was conducted by chronoamperometry at optimum potential and pH using electrodes. For example, in the preparation of ZnO/HA, a stainless-steel sample was used as a working electrode, a platinum sheet was used as a counter electrode, and a saturated calomel electrode (SCE) was utilized as the reference electrode. Excellent performance of the nanocomposite as a sensor and antibacterial or antifungal agent was reported as function of homogeneous surface dispersion.

7.3.5 Magnetic HA Nanocomposites

The use of magnetism in medicine was introduced by Freeman et al. in the 1960s, and after that, quick progression of the use of magnetic nanoparticles for therapeutic applications has considerably surpassed expectations. Basically, magnetic materials are classified into five major groups: antiferromagnetic, diamagnetic, paramagnetic, ferromagnetic, and ferrimagnetic. Magnetic targeting for drug delivery and theragnostic applications is based on two key aspects: a magnetically responsive carrier and a magnetic field gradient (magnetic force) that is able to attract or place magnetically responsive carriers inside a body system. By this, magnetic nanoparticles have the capability to specifically target and easily recover excess or unused drugs. In addition, magnetic nanoparticles facilitate numerous therapeutic applications such as detoxifying biological fluids, magnetic resonance imaging (MRI), controlled and specific targeted drug delivery, cancer therapy, magnetic fluid hyperthermia, photocatalysis, and sensors. Among various magnetic material categories, iron oxide–based magnetic composites with HA are widely studied for their application as inorganic nanocarriers together with a polymer sheath around them. This is due to the superior characteristic of iron oxide compared with other magnetic compounds, such as high magnetic saturation, less toxicity, and easy fabrication.

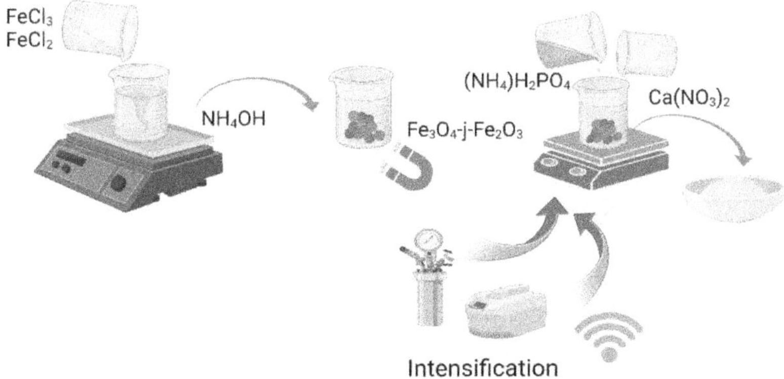

FIGURE 7.9 Example scheme of magnetic-HA synthesis by the chemical precipitation method with possible intensification.

Several approaches are employed in order to prepare iron oxide magnetic-HA. Chemical precipitation is the method that obtains a homogeneous nanocomposite and controlled structure. In addition, modifications and intensifications, such as utilizing microwaves, ultrasound, or hydrothermal treatments, are performed to provide better character, faster reactions, lower energy inputs, and so on. The simple method within this scheme is by mixing co-precipitated iron oxide into a mixture of Ca and P precursors. Magnetic compounds of γ-Fe_2O_3 or Fe_3O_4 could be obtained by the mixing of $FeCl_3$ and $FeCl_2$ precursors and NH_4OH. Figure 7.9 depicts an example scheme of the steps. The pH of the solution is key factor of this process, and it could be maintained by an ammonia solution. Furthermore, the concentration of Fe in the nanocomposite determines the magnetism [49].

7.4 CHARACTERIZATION OF HA-BASED NANOCOMPOSITES

Characterization of prepared nanocomposites plays an important role in synthesis. For solid nanocomposites, physicochemical characterizations are usually performed by instrumentation such as x-ray diffraction, scanning electron microscope-energy dispersive-x-ray (SEM-EDX), transmission electron microscopy (TEM), Fourier transform infrared (FTIR), and x-ray photoelectron spectroscopy (XPS).

7.4.1 XRD Analysis

XRD is one of the tools used to identify the structure, crystal phase, crystallite size, lattice parameters, and degree of crystallization of a material. It is non-destructive analysis in which the sample form could be powder, solids, thin layers, or ribbons. In principle, XRD analysis utilizes monochromatic X-rays impinging on a solid. Furthermore, the rays will be scattered in all directions, but the regular location of the atoms in the crystal, and in certain directions, the waves will interfere constructively, and in other directions, will interfere destructively. Requirements such that the planes that are parallel when scattered by the atoms will interfere constructively have a different range of $n\lambda$ path distances, where the difference between the distances

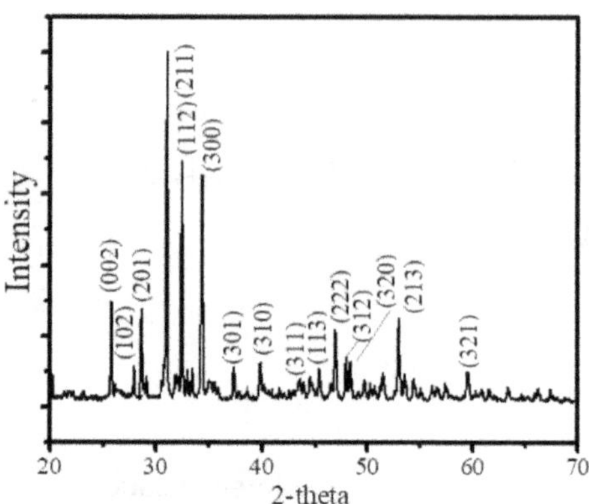

FIGURE 7.10 XRD pattern of HA (Fatimah, unpublished material).

between two files is 2D sin θ and meets Bragg's law (Eq. 1):

$$n\lambda = 2d \sin \theta \tag{1}$$

where n is the diffraction order, d is the grid distance, λ is the wavelength of the x-ray, and θ is the diffraction angle.

The resulting data, called a diffractogram, is the intensity of the peak at recorded at a 2θ angle. Characteristic peaks should be compared to the Joint Committee Powder Diffraction Standard (JCPDS) or ICCD to obtain information on whether the orientation of the crystal matches a selected crystal. In this case, for HA nanocomposites, the standard is No: JCPDS 00–009–0432, which provides peaks at 2θ: 28.683 (102), 29.337 (210), 35.493 (211), 47.102 (222), 50.795 (321), and dan 54.347 (104), consistent with the International Centre for Diffraction Data, ICDD 9–432. Figure 7.10 is an example of the XRD pattern of HA.

As mentioned, doping to the HA structure will not change the pattern but change the lattice parameter. It is different from metal or metal oxide impregnation to HA that expresses additional peaks correlated with the formation of the phases. For example, TiO_2 anatase or rutile phases will appear from prepared TiO_2/HA. For this case, the crystallite sizes of TiO_2 in all samples are determined by using the Scherer equation (Eq. 2):

$$D = \frac{K\lambda}{\beta \cos\theta} \qquad \text{(Eq. 2)}$$

where K is the reflection constant, λ is the wavelength of XRD light, β is the full width and half maximum (FWHM) of the reflections, and θ is the reflection angle.

7.4.2 SEM-EDX Analysis

A scanning electron microscope is an electron microscope capable of showing the morphology or an image of the surface with high magnification. Meanwhile, energy dispersive x-ray is a technique for determining the atomic composition of materials

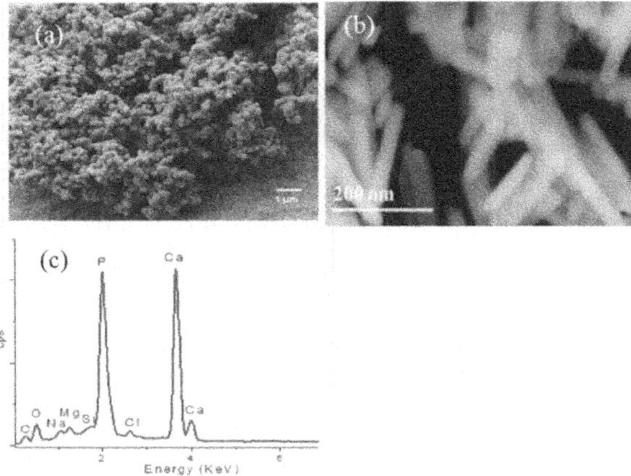

FIGURE 7.11 (a-b) SEM image of HA with spherical and nanorod forms; (c) EDX spectrum.

by measuring them based on the properties of x-rays emitted from the interaction with all components in the sample. SEM functions to observe surface and pore shapes are viewed on a larger and finer scale and equipped with an EDX whose function is to detect the percentage of elements and surfaces of the material being analyzed via electron conductors.

The basic principle of SEM is that the electron waves emitted by the e-gun and condensed on the condenser lens are comfortable as clear points by the objective of the lens. The high-energy electron beam striking the sample generates secondary electrons, which are then collected by the secondary detector. The results of the detected interactions will then be converted into an image of the sample surface. The resulting image consists of many points with many intensities on the surface of the cathode ray tube (CRT) as a topographical image. In addition, the results obtained will also be produced in the form of a graph or diagram on the EDX that shows the percentage of elements from the sample being analyzed. The working principle of EDX is to capture and analyze x-ray fluorescence rays coming out of the sample. The resulting x-ray energy is characteristic of the energy difference between the two higher-energy and lower-energy shells and the characteristics of the atomic structure, allowing compositions and specimens to be analyzed and measured.

Even though IR depends on various parameters in the synthesis, the surface morphology of HA is generally spherical-like to rod-like forms and sometime non-uniform (see Figure 7.11) [50]. The proof of HA as major component will be recognized from energy dispersive-spectra (EDS).

7.4.3 Fourier Transform Infrared

Fourier-transform infrared analysis is found to be the most analytical type of technique to identify the functional groups in solid materials and their change during preparation and modification. The spectra from the analysis result from the interactions between matter and electromagnetic fields in the infrared region ($3 \times 10^{-4} - 7.8$

× 10^{-7} m). As the functional group interacts with the IR spectrum, the bonding in a molecule will absorb the radiation and convert the energy into molecular vibrations. Molecules are excited to a higher vibrational state by absorbing IR radiation. The recorded spectrum represents the energy of vibration, rotation, and translation of each bonding. By using FTIR analysis, an HA sample will show functional groups as identification of PO_4^{3-}, ^{-}OH, CO_3^{2-}, and HPO_4^{2-} that are characteristic of HA. The PO_4^{3-} group will appear at absorption bands at 560 and 600 cm^{-1} together with the band at 1000–1100 cm^{-1}; meanwhile, CO_3^{2-} will be confirmed by intensive peaks at 1460 and 1530 cm^{-1}. The presence of ^{-}OH is usually also associated with the adsorbed water and will be exhibited by a wide band at the range of 3600 to 2600 cm^{-1}. Metal or metal oxide modification will affect the shift of bonding energy and will be presented by a shifted absorption peak.

7.4.4 Transmission Electron Microscopy

TEM operates on the same basic principles as SEM and other light microscope instruments, but the difference lies in using electrons instead of light. Optimal resolution will be gained by using electrons due to the small wavelength of electrons; therefore, the optimal resolution is attainable. By intensive interaction of electrons and sample, TEM has the capability to reveal the finest details of solid internal structures at very high-resolution images down to a level of several angstroms (~0.19 nm). The specific nanoform of HA will be clearly stated based on TEM. For example, an HA sample denoted as nanosphere, nanotube, or nanorod will be valid, as it is coming from TEM analysis. Figure 7.12(a) demonstrates a TEM image of an HA sample with a size ranging from

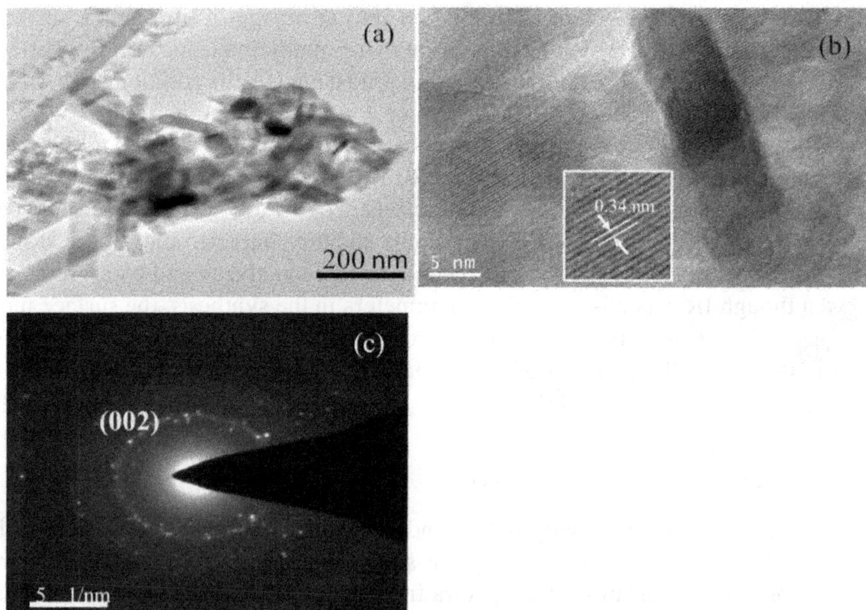

FIGURE 7.12 (a) TEM image, (b) HRTEM image, and (c) SAED pattern of HA [51].

60–200 nm in length. The high-resolution image from high-resolution TEM analysis (HRTEM) demonstrates an interplanar distance of 0.34 nm, which can be inferred from the reflections of the (210) plane [Figure 7.12(b)]. The confirmation of the nanocrystal structure could be reflected using the SAED pattern, which is confirmed by XRD analysis. From the figure, it is seen that the result is the close values of d-spacing associated with the lattice spacings of (300), (112), and (211) [Figure 7.12(c)].

7.5 FUTURE PERSPECTIVES AND INNOVATIVE CHALLENGES

Hydroxyapatite is a biomaterial that has been recognized as potential material for various biomedical applications. This chapter presents numerous modifications with many techniques that could be applied to get functionalized HA. In fact, the main resources for HA synthesis are calcium and phosphate, which can be obtained from many natural resources. Many papers have revealed the applicability of animal waste such as snail shells, chicken bones, fish bones, and so on as calcium sources [1]. Based on these, many innovations in functionalization and application of HA-based nanocomposites can be developed. In addition, modifiers and certain properties can be synthesized from the perspective of green chemistry. The use of safer chemicals, less energy, reusability, and a time-efficient process are some of the principles that are fundamental for selecting the synthesis procedure. As an example, metal or magnetic nanoparticles as HA modifiers could be synthesized by using the bio-reduction method.

REFERENCES

[1] Taji, L.S.; Wiyono, D.E.; Karisma, A.D.; Surono, A.; Ningrum, E.O. Hydroxyapatite based material: natural resources, synthesis methods, 3D print filament fabrication, and filament filler. *IPTEK J. Eng.* **2022**, *8*, 26–35.

[2] Liu, D.; Troczynski, T.; Tseng, W.J. Water-based sol-gel synthesis of hydroxyapatite: process development. *Biomaterials.* **2001**, *22*, 1721–1730.

[3] Takahashi, H.; Yashima, M.; Yoshimura, M. Synthesis of stoichiometric hydroxyapatite by a gel route from the aqueous solution of citric and phosphoneacetic acids. *Inorg. Chem.* **1995**, *32*, 829–835.

[4] Shiba, K.; Motozuka, S.; Yamaguchi, T.; Ogawa, N.; Otsuka, Y.; Ohnuma, K.; Kataoka, T.; Tagaya, M. Effect of cationic surfactant micelles on hydroxyapatite nanocrystal formation: an investigation into the inorganic–organic interfacial interactions. *Cryst. Growth Des.* **2016**, *16*, 1463–1471, doi:10.1021/acs.cgd.5b01599.

[5] Nosrati, H. Crystallographic study of hydrothermal synthesis of hydroxyapatite nano-rods using Brushite precursors. *J. Tissues Mater.* **2019**, *10*, 1–8, doi:10.22034/jtm.2019.199830.1022.

[6] Earl, J.S.; Wood, D.J.; Milne, S.J. Hydrothermal synthesis of hydroxyapatite. *J. Phys. Conf. Ser.* **2006**, *26*, 268–271, doi:10.1088/1742-6596/26/1/064.

[7] Canillas, M.; Rivero, R.; García-carrodeguas, R.; Barba, F.; Rodríguez, M.A. Processing of hydroxyapatite obtained by combustion synthesis. *Boletín la Soc. Española Cerámica y Vidr.* **2017**, *56*, 237–242, doi:10.1016/j.bsecv.2017.05.002.

[8] Lamkhao, S.; Phaya, M.; Jansakun, C.; Chandet, N. Synthesis of hydroxyapatite with antibacterial properties using a microwave-assisted combustion method. *Sci. Rep.* **2019**, 1–9, doi:10.1038/s41598-019-40488-8.

[9] Bilton, M.; Milne, S.J.; Brown, A.P. Comparison of hydrothermal and sol-gel synthesis of nano-particulate hydroxyapatite by characterisation at the bulk and particle level. *Open J. Inorg. Non Met. Mater.* **2012**, *2012*, 1–10.

[10] Guo, X.; Yan, H.; Zhao, S.; Zhang, L.; Li, Y.; Liang, X. Effect of calcining temperature on particle size of hydroxyapatite synthesized by solid-state reaction at room temperature. *Adv. POWDER Technol.* **2013**, doi:10.1016/j.apt.2013.03.002.

[11] Othman, R.; Mustafa, Z.; Wee, C.; Fauzi, A.; Noor, M. Effect of calcium precursors and pH on the precipitation of carbonated hydroxyapatite. *Procedia Chem.* **2016**, *19*, 539–545, doi:10.1016/j.proche.2016.03.050.

[12] Abd-elwahab, S.F.M.S.I.E.M.A.A.S.M. Effect of preparation conditions on the nanostructure of hydroxyapatite and brushite phases. *Appl. Nanosci.* **2016**, *6*, 991–1000, doi:10.1007/s13204-015-0509-4.

[13] Alobeedallah, H.; Ellis, J.L.; Rohanizadeh, R.; Gerard, H.; Coster, L. Preparation of nanostructured hydroxyapatite in organic solvents for clinical applications preparation of nanostructured hydroxyapatite in organic solvents for clinical applications. *Trends Biomater. Artif. Organs.* **2011**, *25*(1).

[14] Guo, X.; Xiao, P. Effects of solvents on properties of nanocrystalline hydroxyapatite produced from hydrothermal process. *J. Eur. Ceram.* **2006**, *26*, 3383–3391, doi:10.1016/j.jeurceramsoc.2005.09.111.

[15] Tp, G.; Um, R.; Rmg, R. Facile synthesis of hydroxyapatite nanoparticles by a polymer-assisted method: morphology, mechanical properties and formation mechanism. *J. Clin. Investig. Stud.* **2018**, *1*, 1–5, doi:10.15761/JCIS.1000101.

[16] Wang, Y.; Xiao, Y.; Huang, X.; Lang, M. Journal of colloid and interface science preparation of poly (methyl methacrylate) grafted hydroxyapatite nanoparticles via reverse ATRP. *J. Colloid Interface Sci.* **2011**, *360*, 415–421, doi:10.1016/j.jcis.2011.04.093.

[17] Takeoka, Y.; Hayashi, M.; Sugiyama, N.; Yoshizawa-fujita, M.; Aizawa, M.; Rikukawa, M. In situ preparation of poly (L-lactic acid-co-glycolic acid)/hydroxyapatite composites as artificial bone materials. **2015**, 164–170, doi:10.1038/pj.2014.121.

[18] Aparecida, C.; Vinicius, M.; Martins, S.; Helena, A.; Carlos, J.; Elena, M.; Antonio, A.; Queiroz, A. De Materials Science & Engineering C Electrochemical preparation and characterization of PNIPAM-HAp scaffolds for bone tissue engineering. *Mater. Sci. Eng. C.* **2017**, *81*, 156–166, doi:10.1016/j.msec.2017.07.048.

[19] Yu, Y.; Zhang, H.; Sun, H.; Xing, D.; Yao, F. Nano-hydroxyapatite formation via co-precipitation with chitosan-g-poly (N-isopropylacrylamide) in coil and globule states for tissue engineering application. *Front. Chem. Sci. Eng.* **2013**, *7*, 388–400, doi:10.1007/s11705-013-1355-0.

[20] Wang, S.; Zhang, Z.; Zhang, Q.; Li, L. Physical crosslinked poly (N-isopropylacrylamide)/Nano-hydroxyapatite thermosensitive composite hydrogels. *J. Inorg. Organomet. Polym. Mater.* **2018**, doi:10.1007/s10904-018-0893-9.

[21] Natu, M.V.; Gaspar, M.N.; Wan, C.; Chen, B.; Gou, M.; Shi, H. Synthesis of poly-ε-caprolactone/hydroxyapatite composite materials by in situ and mechanical mixing methods and investigation of their physico-chemical properties Synthesis of poly-ε-caprolactone/hydroxyapatite composite materials by in situ and mechanical mixing methods and investigation of their physico-chemical properties. *IOP Conf. Ser.: Mater. Sci. Eng.* **2019**, *597*(1), 012007, doi:10.1088/1757-899X/597/1/012007.

[22] Park, J.; Hwang, J.; Back, J.; Jang, S.; Kim, H.; Kim, P.; Shin, S.; Kim, T. High strength PLGA/hydroxyapatite composites with tunable surface structure using PLGA direct grafting method for orthopedic implants. *Compos. Part B.* **2019**, *178*, 107449, doi:10.1016/j.compositesb.2019.107449.

[23] Fu, L.; Qi, C.; Liu, Y.; Cao, W.; Ma, M. Sonochemical synthesis of cellulose/hydroxyapatite nanocomposites and their application in protein adsorption. *Sci. Rep.* **2018**, 1–12, doi:10.1038/s41598-018-25566-7.

[24] Kasuga, T.; Ota, Y.; Nogami, M.; Abe, Y. Preparation and mechanical properties of polylactic acid composites containing hydroxyapatite fibers. *Biomaterials.* **2001**, *22*, 19–23.
[25] Lee, W.; Loo, C.; Rohanizadeh, R. Materials Science & Engineering C Functionalizing the surface of hydroxyapatite drug carrier with carboxylic acid groups to modulate the loading and release of curcumin nanoparticles. *Mater. Sci. Eng. C.* **2019**, *99*, 929–939, doi:10.1016/j.msec.2019.02.030.
[26] Venkatasubbu, G.D.; Ramasamy, S.; Ramakrishnan, V.; Kumar, J. Hydroxyapatite-alginate nanocomposite as drug delivery matrix for sustained release of ciprofloxacin. *J. Biomed. Nanotechnol.* **2011**, *7*(6), 759–767, doi:10.1166/jbn.2011.1350.
[27] La, M.; Alessia, R.; Giovanna, R.; Friuli, V.; Maggi, L.; Bini, M. New emerging inorganic–organic systems for drug-delivery: hydroxyapatite @ furosemide hybrids. *J. Inorg. Organomet. Polym. Mater.* **2022**, *5*, doi:10.1007/s10904-022-02302-3.
[28] Ren, J.; Sun, L.; Xiao, C.; Zhou, S.; Liang, Q.; Sun, S. Smart materials in medicine chemical bonding of epigallocatechin-3-gallate to the surface of nano-hydroxyapatite to enhance its biological activity for anti-osteosarcoma. *Smart Mater. Med.* **2023**, *4*, 396–406, doi:10.1016/j.smaim.2022.12.003.
[29] Liao, J.G.; Li, Y.Q.; Duan, X.Z.; Liu Q. Synthesis and characterization of CO-3(2-) doping nano-hydroxyapatite. *Guang pu Xue Yu Guang Pu Fen Xi.* **2022**, *11*, 3011–3014.
[30] Horiuchi, N.; Endo, J.; Wada, N.; Nozaki, K.; Nakamura, M.; Nagai, A. Dielectric properties of fluorine substituted hydroxyapatite: the effect of the substitution on configuration of hydroxide ion chains. *J. Mater. Chem. B* **2015**, 13–18, doi:10.1039/C5TB00944H.
[31] Nagyné-kovács, T.; Studnicka, L.; Kincses, A.; Spengler, G.; Molnár, M.; Tolner, M.; Endre, I.; Szilágyi, I.M.; Pokol, G. Synthesis and characterization of Sr and Mg-doped hydroxyapatite by a simple precipitation method. *Ceram. Int.* **2018**, 0–1, doi:10.1016/j.ceramint.2018.09.096.
[32] Makshakova, O.N.; Gafurov, M.R. The mutual incorporation of Mg^{2+} and CO_3^{2-} into hydroxyapatite: a DFT study. *Materials (Basel).* **2022**, *15*(24), 9046.
[33] Andrew, E.; Idowu, A.; Abidemi, M.; Olateju, S. Heliyon synthesis and characterization of Zn-doped hydroxyapatite: scaffold application, antibacterial and bioactivity studies. *Heliyon* **2019**, *5*, e01716, doi:10.1016/j.heliyon.2019.e01716.
[34] Baradaran, S.; Nasiri-tabrizi, B.; Shirazi, F.S.; Saber-samandari, S.; Basirun, W.J. Wet chemistry approach to the preparation of tantalum-doped hydroxyapatite: Dopant content effects. *Ceram. Int.* **2017**, doi:10.1016/j.ceramint.2017.11.016.
[35] Wei, L.; Yang, H.; Hong, J.; He, Z.; Deng, C. Synthesis and structure properties of Se and Sr co-doped hydroxyapatite and their biocompatibility. *J. Mater. Sci.* **2019**, *54*, doi:10.1007/s10853-018-2951-7.
[36] Liu, Y.; Ma, J.; Zhang, S. Synthesis and thermal stability of selenium-doped hydroxyapatite with different substitutions. **2015**, *9*, 392–396, doi:10.1007/s11706-015-0313-9.
[37] Jose, S.; Senthilkumar, M.; Elayaraja, K.; Haris, M.; George, A.; Raj, A.D.; Sundaram, S.J.; Bashir, A.K.H.; Maaza, M.; Kaviyarasu, K.; et al. Preparation and characterization of Fe doped n-hydroxyapatite for biomedical application. *Surf. Interfaces.* **2021**, *25*, 101185, doi:10.1016/j.surfin.2021.101185.
[38] Wang, Y.; Yang, X.; Gu, Z.; Qin, H.; Li, L.; Liu, J.; Yu, X. In vitro study on the degradation of lithium-doped hydroxyapatite for bone tissue engineering scaffold. *Mater. Sci. Eng. C.* **2016**, *66*, 185–192, doi:10.1016/j.msec.2016.04.065.
[39] Park, S.; Choi, J.; Hoang, V.; Doan, M.; O, S.H. Biodegradable manganese-doped hydroxyapatite antitumor adjuvant as a promising photo-therapeutic for cancer treatment. *Front. Mol. Biosci.* **2022**, 1–15, doi:10.3389/fmolb.2022.1085458.

[40] Kumar, V.B.; Khajuria, D.K.; Karasik, D.; Gedanken, A. Silver and gold doped hydroxyapatite nanocomposites for enhanced bone regeneration. *Biomed. Mater.* **2019**, *8*, 055002.

[41] Fatimah, I.; Citradewi, P.W.; Yahya, A.; Nugroho, B.H.; Hidayat, H.; Purwiandono, G.; Sagadevan, S.; Ahmad, S.; Sheikh, I.; Ghazali, M.; et al. Biosynthesized gold nanoparticles-doped hydroxyapatite as antibacterial and antioxidant nanocomposite Biosynthesized gold nanoparticles-doped hydroxyapatite as antibacterial and antioxidant nanocomposite. *Mater. Res. Express.* **2021**, *8*, 115003.

[42] Chatterjee, T.; Chatterjee, P.; Kumar, C.A.; Pradhan, S.K.; Meikap, A.K. Template-free growth of copper-doped hydroxyapatite nanowhiskers and their use as uric acid electrochemical sensor. *Mater. Today Commun.* **2022**, *33*, 104870.

[43] Fahami, A.; Ebrahimi-kahrizsangi, R.; Nasiri-tabrizi, B. Mechanochemical synthesis of hydroxyapatite/titanium nanocomposite. *Solid State Sci.* **2011**, *13*, 135–141, doi:10.1016/j.solidstatesciences.2010.10.026.

[44] Noviyanti, A.R.; Asyiah, E.N.; Permana, M.D.; Dwiyanti, D.; Eddy, D.R. Preparation of hydroxyapatite-titanium dioxide composite from eggshell by hydrothermal method: characterization and antibacterial activity. *Crystals.* **2022**, *12*(11), 1599.

[45] Taha, S.; Begum, S.; Narwade, V.N.; Halge, D.I.; Dadge, J.W.; Mahabole, M.P.; Khairnar, R.S.; Bogle, K.A. Development of alcohol sensor using TiO_2-hydroxyapatite nano-composites. *Mater. Chem. Phys.* **2020**, *240*, 122228, doi:10.1016/j.matchemphys.2019.122228.

[46] Anmin, H.; Tong, L.; Ming, L.; Chengkang, C.; Huiqin, L.; Dali, M. Preparation of nanocrystals hydroxyapatite/TiO_2 compound by hydrothermal treatment. *Appl. Catal. B: Environ.* **2006**, *63*, 41–44, doi:10.1016/j.apcatb.2005.08.003.

[47] Geuli, O.; Lewinstein, I.; Mandler, D. Composition-tailoring of ZnO-hydroxyapatite nanocomposite as bioactive and antibacterial. *ACS Appl. Nano Mater. Nano Mater.* **2019**, *2*, 2946–2957, doi:10.1021/acsanm.9b00369.

[48] Abdel-hamid, Z.; Rashad, M.M.; Mahmoud, S.M.; Kandil, A.T. Electrochemical hydroxyapatite-cobalt ferrite nanocomposite coatings as well hyperthermia treatment of cancer. *Mater. Sci. Eng. C.* **2017**, *76*, 827–838, doi:10.1016/j.msec.2017.03.126.

[49] Biedrzycka, A.; Skwarek, E.; Hanna, U.M. Hydroxyapatite with magnetic core: synthesis methods, properties, adsorption and medical applications. *Adv. Colloid Interface Sci.* **2021**, *291*, 102401, doi:10.1016/j.cis.2021.102401.

[50] Rivera-Muñoz, E.M. Hydroxyapatite-based materials: synthesis and characterization. *Biomed. Eng. Front. Challeng.* **2011**, 978-953-51-4472-4.

[51] Fatimah, I.; Hidayat, H.; Purwiandono, G.; Khoirunisa, K.; Zahra, H.A. Green synthesis of antibacterial nanocomposite of silver nanoparticle-doped hydroxyapatite utilizing curcuma longa leaf extract and land snail (*Achatina fulica*) shell waste. *J. Funct. Biomater.* **2022**, *13*(2), 84.

8 Insights into 2D MXenes for Versatile Biomedical Applications

Zambaga Otgonbayar and Won-Chun Oh

8.1 INTRODUCTION

2D nanomaterials have recently been employed extensively in a variety of fascinating disciplines, including various fields [1–11] and biomedical procedures [12–15]. Early twentieth-century classical physicists predicted that, owing to thermal chassis oscillations, the stability of 2D accessories was nebulous at any fixed temperature. Nevertheless, the discovery of a 2D graphene monolayer in 2004 signaled a scientific breakthrough in the field of material wisdom. Physically, 2D accouterments are atomically thin crystalline solids clicked by covalent and van der Waals (vdW) cling [16]. In contrast to their 3D counterparts, 2D nanomaterials consist of layers with densities ranging from minuscule to nanoscale. However, the use of old 2D nanomaterials in academic and research settings remains significant. Only recently have new accessories with significantly improved physical and chemical components been developed that are perfect for active exploration fields. Electronic engineers are developing energy storage devices and detectors, and in recent investigations, they have been used to study bacterial cells and dead cancer cells [17, 18]. MXenes are a branch of the 2D material family first discovered by Prof. Yury Gogotsi and Michel W. Barsoum with collaborators at Drexel University in 2011. Because of their infinitesimal subcaste consistency and excellent colorful parcels, such as semiconductor parcels, face hydrophilic parcels, glamorous parcels, and rich face functional groups, MXenes have attracted significant attention in a variety of colorful exploration fields. The MXene was etched from its parent MAX phase. Its formula is $M_{n+1}X_nT_x$, where "M" stands for the transition essence, "X" for C and/or N rudiments, "T_x" for face termination groups such as O, F, and OH, and $n = 1$ to 3. The basic elements of group IIIA/IVA are denoted by $Ti_3C_2T_x$, the first MXene created by etching Al layer from Ti_3AlC_2 substrate [19]. The family of MXenes has grown as a result of the variety of elemental compositions that have led to the synthesis of more than 30 different types of MXene compositions through trials and theoretical predictions. This is due to their abundant functional groups, exceptional electrical and optical properties, extraordinary mechanical and glamorous parcels, and other properties. Numerous researchers have focused on the biomedical applications of 2D materials, including various types of materials. Among them, MXenes retain many functional groups on their faces in comparison with other 2D materials, making revisions more diverse.

DOI: 10.1201/9781003425427-11

Additionally, MXenes have hydrophilicity, changeable compositions, and minuscule layers; the combination of these properties makes them ideal for biological applications. Additionally, the cost-effective scale-up of products of MXenes may be a strength. To date, the primary functions of MXenes in the biomedical arena include bioimaging, antimicrobial technology, biosensing, medication distribution, and therapeutic clothing (Figure 8.1).

For example, MXenes are considered for tumor ablation, a photothermal treatment (PT) that can penetrate deep within the body, target cancer cells, and have a lower adverse effect on the human body. This treatment uses extreme light immersion and near-infrared (NIR) light conversion. Modified MXenes also have the ability to carry and release medications where needed. The effectiveness of cancer treatment has been considerably improved by the combination of PT, CT, and real-time bioimage monitoring. Additionally, MXenes are projected to be the sought-after tools for producing biocompatible materials that can quickly and fluidly describe natural events.

Typically, transition-state carbides and nitrides are used to produce MXene. Similar to graphene, MXenes are typically produced by selectively etching certain layers from their 3D parent compounds, known as MAX phase; this method is referred to as a top-down approach. Ternary carbides or nitrides with the general formula $M_{n+1}AX_n$ serve as 3D precursors for MXenes and are known as MAX phases. Figure 8.2 illustrates these characteristics; as shown, the A-layers are extensively erased by strong acid etching to generate layered $M_{n+1}X_n$. Owing to their high face energy, MXenes typically end their shells with fluorine (-F), hydroxide (-OH), and oxygen (-O) groups during the drawing process. As a result, $M_{n+1}X_nT_x$, where T_x is the face functional group, is the final chemical formula of MXene. Ti_3C_2 and Ti_2C

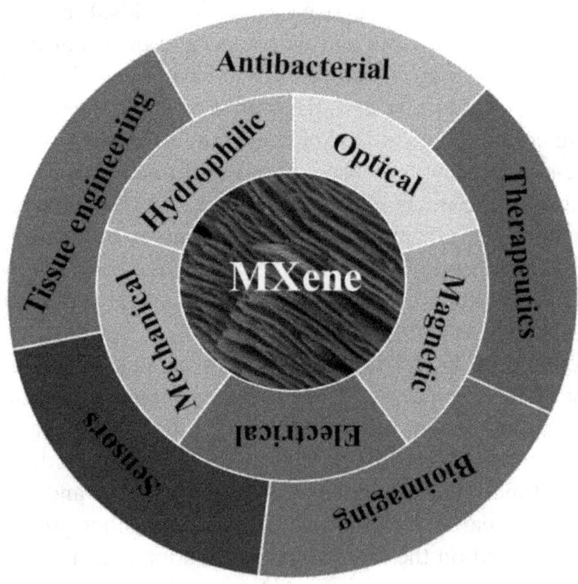

FIGURE 8.1 MXene characteristics and biological uses [20].

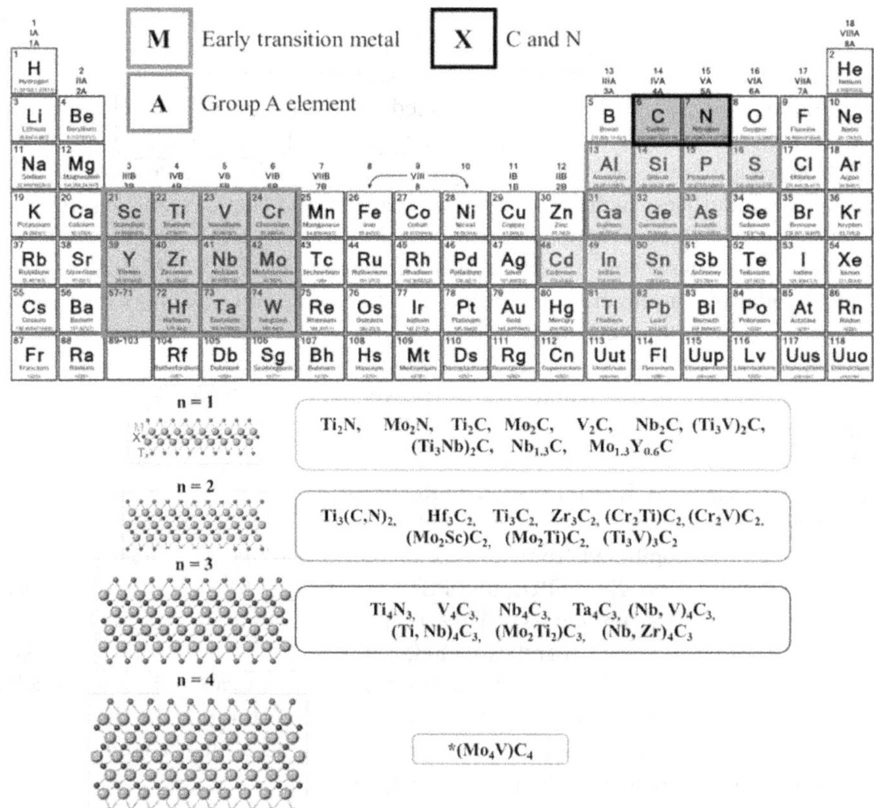

FIGURE 8.2 MAX phase and MXene general elemental makeup is as follows: M: earliest transition metal. A: Element of Group A. X: C or N, surface functional group (T_x) [21].

groups now contain the most useful MXenes for biomedicine and biotechnology. This gives rise to the possibility of using different Ti- and C-less M- and A-group bases.

MXenes have been increasingly used in the biomedical sector owing to their fascinating physicochemical properties. They have been developed for various biomedical applications, including biosensing, bioimaging, therapeutic diagnostics, implants, and antibacterial agents [22]. The expanding library of MXenes with tunable properties paves the way for high-performance, application-specific MXene-based sensing and therapeutic platforms [23, 24]. MXene functional groups also contribute to their hardness and rigidity, which are crucial for bi-electronic bias in thin film conformation [25]. Because MXene nanosheets have a large face area, they can be used for the distribution and MXene functional groups also contribute to their hardness and rigidity, which are crucial for bi-electronic bias in thin-film conformations [25]. As MXene nanosheets have a large face area, they can be used for the distribution and loading of medications for diagnostic procedures and synergistic complaint treatment

[22, 26]. The presence of hydrophilic functional groups is crucial for the administration of medicines because their modification or functionalization improves their biocompatibility with the living cells of Atkins.

With the keywords "MXene" and "Biomedical", 121 papers on MXenes for biomedical procedures were reported on lens.org for the past six months (Figure 8.3) [27]. Comparing biomedical engineering's volume of publications to those of other well-known industries, such as electronics, catalysts, and energy storage, shows that MXenes are still in short supply in biomedical operations. Although research centered on MXenes has been published since the beginning of 2012, MXene research on biomedical operations has only been published three times. This is because the first three publications were devoted to initial studies of MXenes, such as characterizing MXene structures to identify their parcels. A thorough study is necessary because 2D MXene research and studies are expanding to increase research interest for application in biomedicine and biotechnology.

MXene articles published between January 2015 and December 2020 were analyzed using a rigorous literature search. Relevant exploratory papers were named using lens.org based on the previously chosen criteria. The metadata obtained from lens.org was combined from various sources, including Microsoft Academic, PubMed, Crossref, ORCID, CORE, and PubMed Central. To support the suggested themes and uphold the uniqueness of the review, pertinent and recent studies are required. Original literature searches included searching for articles using the keywords MXene, biomedical operation, biosensors, cancer therapy, medication delivery, and antimicrobial effort. Table 8.1 summarizes the findings of the literature search. The overall review was gathered and considered with pertinent examples after examining and assessing the pertinent papers.

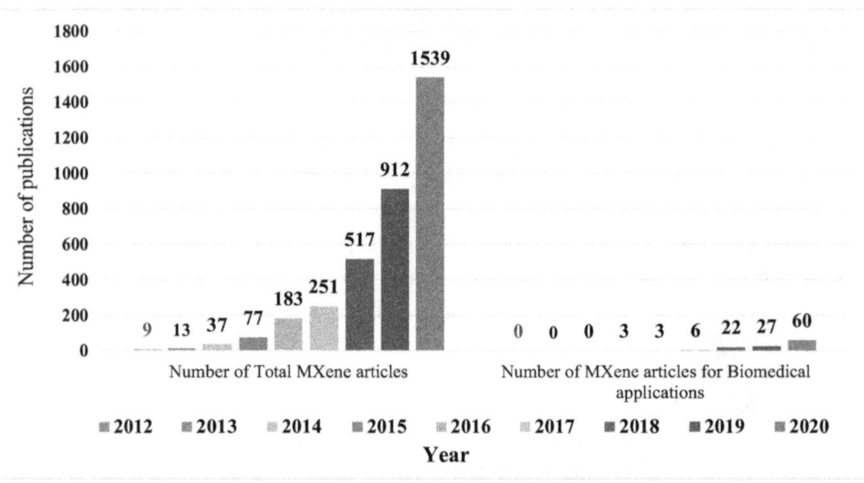

FIGURE 8.3 "MXene" and "Biomedical" keywords were used to find the number of journal papers on MXenes for biomedical applications [27].

TABLE 8.1
Results of a Literature Search Gathered from lens.org [27]

Search Terms	Journal Article Output by Year in Terms of Number						Total Articles
	2015	2016	2017	2018	2019	2020	
"MXene", "Biomedical"	3	3	6	22	27	60	121
"MXene", "Biosensors"	2	2	1	7	17	30	59
"MXene", "Cancer theranostics"	—	—	1	3	2	3	9
"MXene", "Drug delivery"	—	—	2	4	1	9	16
"MXene", "Antimicrobial"	1	—	—	7	2	8	18

8.2 SYNTHESIS OF MXENES

Either top-down or bottom-up techniques can be used to synthesize MXenes. The key to determining the overall physical and chemical components, such as the size, shape, and functioning of the material, is to choose the appropriate technique [28]. To improve the biocompatibility of the material or reduce its cytotoxicity for use in biomedical procedures, MXene faces can be further modified after conflation (in any bottom-up or top-down direction).

8.2.1 Top-Down Synthesis Method

The top-down fabrication method, which starts with the MAX phase and utilizes the acid exfoliation of concentrated A solids, is a widely used pharmaceutical system for MXene layers (Figure 8.4). Because the MXene layered structure resembles that of its 3D equivalent, this system is regarded as a classic. The MAX phase often undergoes two-way etching or fractionalization, and the molten MXene layers

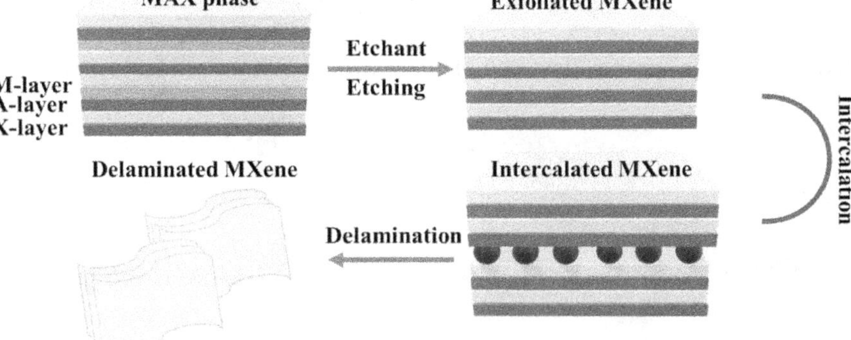

FIGURE 8.4 Schematics diagram of the top-down synthesis of MAX and MXene [31].

delaminate [29, 30]. One problem with this technique is that excessive amounts of "A-element" layers are required because this layer readily evaporates and smells into a vacuum/argon (Ar) atmosphere at high temperatures [28]. Another popular method for MAX-phase condensation is pressure-free sintering [29]. Compared to hot-pressed Ti_3AlC_2, low-pressure sintering has been reported to produce a Ti_3AlC_2 MAX phase that is well known.

The initial stage in the synthesis of the MAX phase is to transfer the 3D-MAX phase to Ti_3AlC_2 using a strong flux, often HF. Fluoride ions (F^-) must be present in large concentrations to etch the A element of the MAX common phase because they have a significant affinity for it (or Al). In addition, as illustrated in Figure 8.5 [32], the HF treatment of the MAX phase produced various surface finishes with -OH, -O, and -F group chemical equations.

$$M_{n+1}AlX_n(s) + 3\ HF\ (aq) \rightarrow M_{n+1}\ X_n(s) + AlF_3(aq) + 3/2H_2(g)\ [1]$$

$$M_{n+1}X_n(s) + 2\ HF\ (aq) \rightarrow M_{n+1}X_nF_2(s) + 2\ H_2(g)\ [2]$$

$$M_{n+1}X_n(s) + 2\ H_2O\ (aq) \rightarrow M_{n+1}X_n(OH)_2(s) + H_2(g)\ [3]$$

In earlier research, which is included in Table 8.2, the HF of MAX phases was reported using various HF concentrations, immersion periods, and temperatures [33, 34]. In their initial study, Nagib et al. achieved complete MAX phase separation by soaking Ti_3AlC_2 particles in 50% HF concentration for 2 h at room temperature. Other examples include the production of Ti_3C_2 MXene by stirring in 48% aqueous HF for 24 h at room temperature [22] and the production of multilayered Ti_3C_2 nanorods by swirling in 40% HF for 2 days at 45°C.

The in situ synthesis of HF, which occurs when HF is combined with an acid (such as hydrochloric acid or HCl) and a fluoride salt, is another typical industrial strategy [24]. The safety of aluminum (Al) is improved by this procedure, which also prevents the negative consequences of high HF concentrations. Ti_3C_2 and Ti_2C nanosheets can also be created by combining various fluoride salts with HCl at the right temperature [35]. In a different study, Ti_3AlC_2 films were etched using only ammonium bifluoride (NH_4HF_2) without the need for a strong acid [36]. In contrast to Ti_3C_2-gravitated HF, they discovered that MXene layers intercalated simultaneously with ammonium (NH_4^+) species in the pot flow and created layers with greater lattice parameters

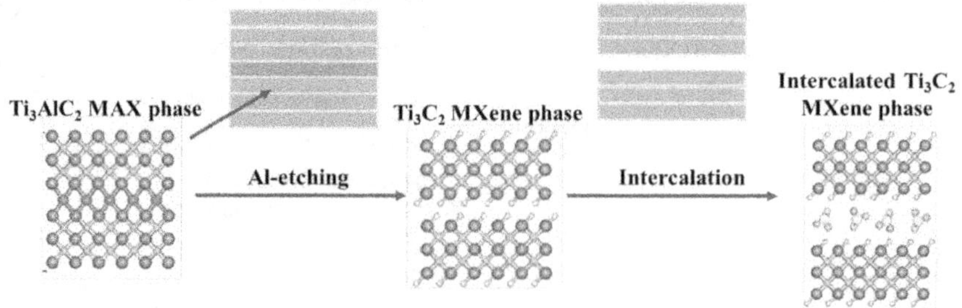

FIGURE 8.5 The intercalation and the etching process [32].

TABLE 8.2
Handling and Storage Conditions for MXene Synthesis (Bottom-Up Approach), as Reported in Several Major Studies

MAX Phase (S)	Etchant (Conditions)	Intercalant (Conditions)	Ref
Ti_3AlC_2	HF (50%, RT, 2 h)	—	[10]
Ti_3AlC_2	HF (48%, RT, 24 h)	TMAOH (24 h, RT)	[33]
Ti_3AlC_2	HF (40%, 45°C, 2 days)	TMAOH (3 days, 45°C)	[25]
Ti_3AlC_2, Ti_2AlC, Nb_2AlC	HCl + LiF (40°C, 45 h)	—	[34]
Ti_3AlC_2	HCl + LiF (50°C, 24 h)	—	[35]
	HCl + NH_4F(30°C, 24 h)	—	
	HCl + KF/NaF (40°C, 48 h)	—	
Ti_2AlC	HCl + LiF (48 h, 50°C)	—	
	HCl + NH_4F/KF(48 h, 40°C)	—	
	NaF+HCl (24 h, 60°C)	DMSO, (2 h) $NH_3.H_2O$ or (60°C, 24 h) urea	
Ti_3AlC_2	NH_4HF_2 (RT, >3 h)	—	[36]
Ti_3AlC_2	NaOH (270°C, 0.5 day)	—	[37]
Ti_4AlN_3	LiF/KF/NaF (550°C, 0.5 h)	TBAOH (5 min)	[38]
Ti_3AlC_2	—	DMSO (RT, 18 h)	[39]
Nb_2AlC	HF (50%, 55°C, 48 h)	Isopropylamine (RT,18 h)	[40]
V_2AlC	HF (48%, RT, 92 h)	TBAOH (RT, 4 h)	[41]
Ti_3AlCN	HF (30%, RT, 18 h)		
Ti_3AlC_2	HF (40%)	TPAOH (RT, 72 h)	[42]

(> 25%). However, the use of HF has led to hazards in humans [43]. Owing to the safety and environmental problems associated with the direct use of HF or even the in situ production of HF, MXenes cannot be used more extensively [43]. In addition to HF treatment, a fluorine-free etching process is more advantageous for producing MXene with a functional surface finish that can be controlled for a variety of biomedical applications. In a study by Li et al., multilayer TC-MXene was effectively produced at a high temperature of 270°C using an alkaline-assisted hydrothermal process with sodium hydroxide solution as the corrosive agent [37].

In addition to previous methods, molten salt methods have been used [38]. Using tetrabutylammonium hydroxide (TBAOH), the Ti_4N_3 layers were delaminated, yielding both multi- and single-layer MXenes. After etching, MXenes typically split their sheets using a delamination method to further examine the properties of their 2D structure [44]. Sonication and intercalants can be used to perform the delamination stage; the absence of intercalants results in smaller MXenes with greater flaws [24, 44]. The distance between two consecutive MXene sheets, or the c-lattice parameter, is frequently increased by cross-linking agents, making it simpler to separate or delaminate pure MXene sheets [44]. Dimethyl sulfoxide (DMSO), a polar organic solvent, and isopropyl amine are two examples of commonly used cross-linking agents [39, 40]. Large sheets from traditional top-down synthesized laminated MXenes can result in biosecurity problems and subpar therapeutic results. Large leaves of MXene are also inappropriate for intravenous (IV) delivery or for penetration and accumulation in malignant tissues [45]. Accordingly, corrosion optimization, etching time, and interstitial type are crucial factors to ensure the overall successful synthesis of biocompatible MXene nanosheets, according to global studies reported on top-down MXene synthesis.

8.2.2 Bottom-Up Synthesis Method

The bottom-up synthesis method using atomic-scale control is another less well-known technique for fabricating MXenes. Small organic/inorganic molecules or atoms are typically the starting point for bottom-up synthesis. Subsequently, crystals that can be assembled to form an ordered 2D sheet grow over time. This strategy's most popular technology is chemical vapor deposition (CVD), which can be used to create high-quality thin films on a variety of substrates (Figure 8.6a). An extensive range of films with internal sizes between 10 and 100 m was produced by optimizing the growth temperature and duration. The lack of surface functional groups may be indicated by the high crystallinity of the Mo_2C films and the lack of defects [46, 47]. Techniques other than CVD have also been investigated for the synthesis of MXene, such as the array approach and plasma-enhanced pulsed laser deposition (PELPD) [48]. PELPD was used to create the first ultrathin Mo_2C films by employing methane

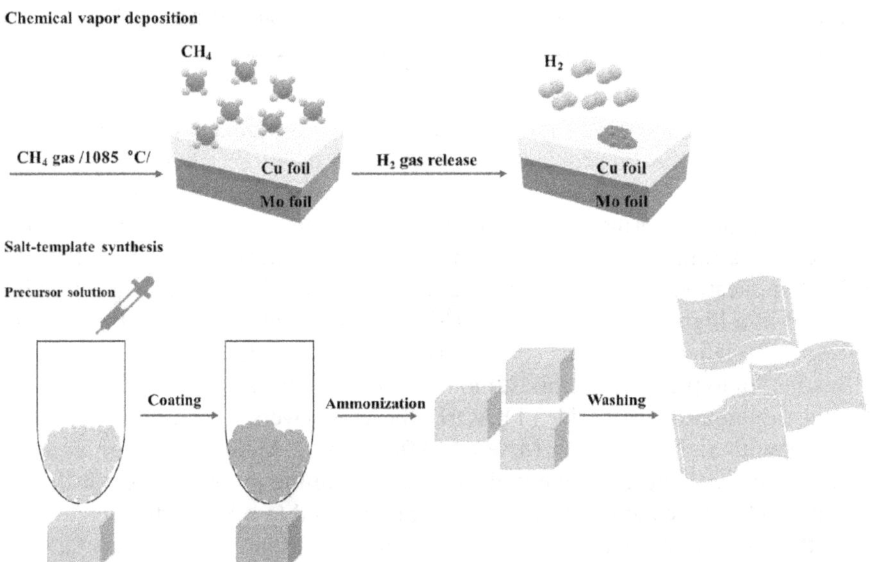

FIGURE 8.6 Systemized representation of bottom-up synthesis: (i) vapor deposition and (ii) salt-template synthesis [31].

plasma as a carbon source to combine with Mo vapor created by the pulsed laser. Other modifications, such as scalable salt-template synthesis, have also been used [45], as shown in Figure 8.6(b) [48, 49].

It has been demonstrated that the technique works on additional metal oxides, including V_2N and W_2N nanosheets. However, there is still room for improvement, as there is currently little information available on the bottom-up synthesis of biocompatible MXene [44]. Robust and secure approaches for large-scale synthesis are another crucial component of top-down and bottom-up synthesis that needs more research.

Table 8.3 provides a summary of the key distinctions between the top-down and bottom-up synthesis of MXene as well as the described material attributes.

The reported MXene synthesis methods are shown in Figure 8.7, along with the publication date of each method.

TABLE 8.3
Comparison of MXene Top-Down and Bottom-Up Synthesis Methods [26]

	Top-Down Approach	Bottom-Up Approach	Refs.
Antecedent	Beginning with the MAX phase of the 3D structure	Development of the MXene film starts from the atom up	[46]
Method	Uses etchants made of chemicals, such as HF acid	Use plasma-assisted deposition, salt modeling, or chemical vapor deposition	[47, 48]
Synthesis condition	Low-pressure synthesis	Controls gas flow as a source of carbon	[47]
Temperature	Needs to be at ambient temperature or below for synthesis	Needs synthesis at a high temperature of at least 1000°C	[46]
Morphology	Thin sheets with a thickness between 10 and 200 nm and large, irregular MXene sheets	Flawless multilayer thin films with lateral sizes between 10 and 100 μm	[33, 46]
Surface properties	After synthesis, -OH and -O were functionalized	Following synthesis, there were no functional groupings	[47]

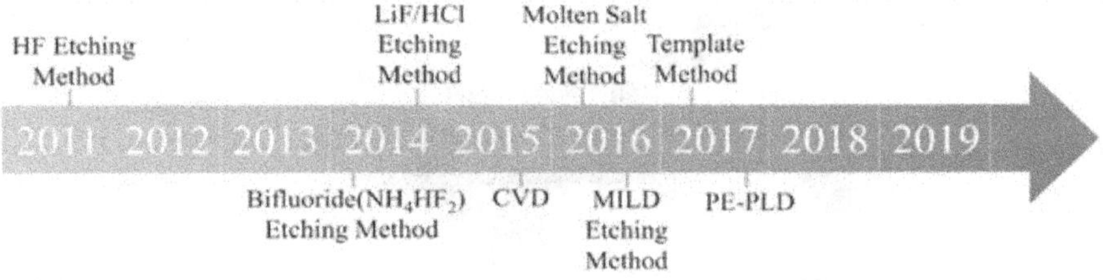

HF Etching Method
LiF/HCl Etching Method
Molten Salt Etching Method
Template Method
2011 2012 2013 2014 2015 2016 2017 2018 2019
Bifluoride(NH_4HF_2) Etching Method
CVD
MILD Etching Method
PE-PLD

FIGURE 8.7 Synthesis techniques for MXenes and the original paper's year, respectively.

8.3 STRUCTURES, PROPERTIES, AND SURFACE MODIFICATION OF MXENES

MXenes, a newly discovered two-dimensional substance, differ from other substances in that they exhibit superior electrical conductivity, excellent optics, extraordinary magnetic properties, superior hydrophilicity, and mechanical flexibility.

8.3.1 STRUCTURES OF MXENES

Currently, there are four different types of MXene structures: arranged double (M) metal transition MXenes, such as $(Cr_2V)C_2$ and $(Mo_2Ti_2)C_3$; arranged binary MXenes, such as $Mo_{1.33}C$ and $W_{1.33}C$ [50] single-transition metal MXenes, including Ti_2C and Nb_4C_3; and MXene solid solutions consisting of $(Ti, V)_3C_2$. The list of MAX and MXene nanocomposites with the elements used to build MAX and MXene is summarized in Figure 8.8.

There are 1D structures [Figure 8.9(b and c)], 3D structures [Figure 8.9(d)], and almost non-dimensional structures in addition to the typical two-dimensional structure of MXenes [Figure 8.9(a and e)] [51–58]. One-dimensional MXenes have been studied less than two-dimensional MXenes [51]. Although still in their infancy, they have many advantages, including a large number of active sites, superior dispersibility, tunable structure, and high hydrophilicity, [59]. Therefore, they can be applied to manufacture optoelectronic devices, catalysts, and energy storage and detection systems.

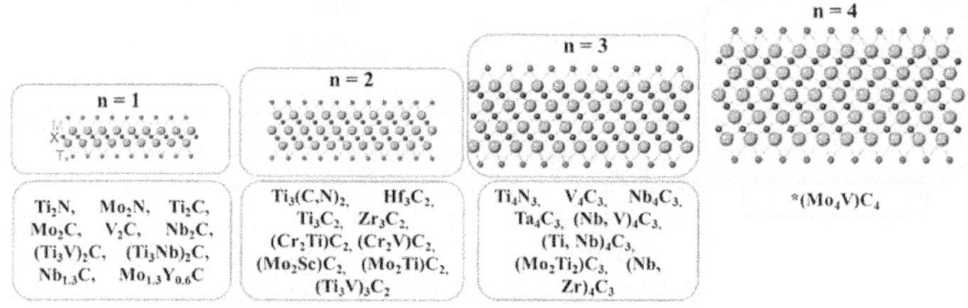

FIGURE 8.8 M_2X, M_3X_2, and M_4X_3 groups of MXenes.

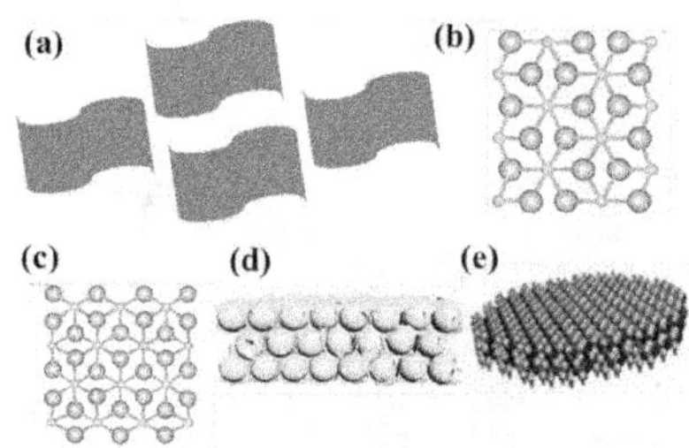

FIGURE 8.9 0D, 1D, 2D, and 3D structures from MXenes: (a) nano-sheet, (b) nano-ribbons, (c) zig-z module f nano-ribbons, (d) 3D film, and (e) QDs of MXenes [54].

8.3.2 Features of MXenes

The specific characteristics of 2D-structured MXene materials include hydrophilicity, biocompatibility, and electrical, magnetic, optical, and mechanical properties. Electronic properties have received the most research attention compared with other properties. Interestingly, all bare MXene monolayers are metals; however, the OH, F, and O terminals can lead to a metal-to-semiconductor transition.

The preparation method, surface finish, elemental composition, internal structure, and environmental variables such as humidity, pH, temperature, and other factors affect the electrical performance of MXenes [60]. Therefore, the following techniques can be used to modify the electrical properties: (i) change the "M" elements, which are mainly concerned with metallic MXenes; (ii) improve synthesis techniques that produce fewer defects, such as the use of fluorine-free synthesis; (iii) modify the surface groups to accommodate band gap gaps, for example, by removing the groups contained in OH or by adding groups contained in O; (iv) prolong the interlayer distance and increase the energy density; and (v) design new structures.

MXenes can be used in a variety of fields, including photocatalysis, sensing and biomedicine, photothermal therapy (PTT), photodynamic therapy (PDT), release-controlled drugs, and biosensors. Because MXenes exhibit high near-infrared (NIR) light absorption and photothermal conversion due to localized surface plasmon resonance, they are commonly used to eliminate malignancies. MXene has both ferromagnetic and antiferromagnetic properties, such as Cr_2C [61] and Ti_3C_2 [62, 63]. The composition of "M," surface terminals, and other elements affect the magnetic properties of MXenes, just as they affect their optical and electrical properties. The surface finishes of MXenes are responsible for their hydrophilicity. The surface of a cold-pressed MXene plate was measured using the contact angle with deionized water, and the results showed that MXenes are highly conductive and hydrophilic. Consequently, highly conductive MXenes interact more favorably with the polymer matrix, which favors their use in composite materials [64]. Hydrophilicity is often correlated with biocompatibility. Generally, higher hydrophilicity is due to improved biocompatibility. Ti_3AlC_2 is a typical Ti-based MXene that has been shown to be biocompatible, despite the fact that there are not many studies on the biocompatibility of MXenes [65, 66]. In addition, the mechanical properties have generated a lot of curiosity.

8.3.3 Surface Modification of MXenes

Bare MXene is unstable in physiological environments and is readily synthesized, despite possessing remarkable mechanical, optical, magnetic, hydrophilic, biocompatible, and electronic properties. Synthetic MXenes can be modified by electrostatic adsorption, hydrogen bonding, van der Waals forces, and other processes owing to their functional groups, high charge density, and large surface areas. Three different surface chemistry methods have been used to modify the surface of MXenes: polymer-based techniques, inorganic nanoparticle-based methods, and metal nanoparticle-based methods (Figure 8.10). Inorganic nanoparticles can be used to modify the outer surfaces of polymers [67–70]. A common example is the

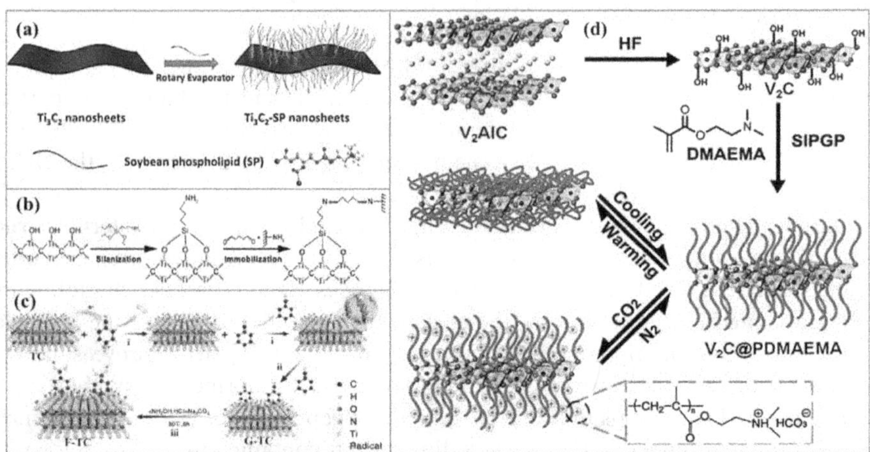

FIGURE 8.10 Polymer-based surface chemistry method of MXenes. (a) SP modifications [12], (b) lipase immobilization [73], (c) functionalization mechanism [74], (d) hybrid system V2C@ PDMAEMA. The Al atom is colored blue, the V atom is purple, and the C atom is gray [75].

traditional DDS drug delivery system that combines TC-MXene with suspended silica nanoparticles (MSNs). For the first time, Yang et al. created an engineered surface nanopore of 2D TC-MXene using a simple sol-gel technique. In alkaline media, tetraethyl orthosilicate (TEOS) was used as the silica precursor, and cetane trimethylammonium chloride (CTAC) was used as a mesopore orienteering agent. Although nanoparticles and inorganic compounds are commonly used for surface modification, precious-metal modifications can significantly improve the properties of MXenes and expand their range of use. This process usually involves changes to gold, silver, and platinum. Furthermore, the surfaces of MXenes can be modified with Co [71] and Bi [72] nanoparticles to improve their photocatalytic performance, charge transfer efficiency, and sensing performance.

8.4 USE OF MXENES IN BIOMEDICAL APPLICATION

MXenes exist as stacked 2D sheets connected by strong H-bonds and/or weak van der Waals (vdW) forces between the surface functional groups. Surface-based functional groups can be functionalized and chemically reactive. MXene surfaces can be customized for biomedical research investigations using a wide range of materials suitable for certain biomedical tests, as shown in Figure 8.11 [76–80].

8.4.1 Biosensors

Biosensors are used to selectively detect specific compounds in humans and mainly consist of (i) transducers that can convert biochemical signals into electrical and other signals such as optical signals, (ii) sensing elements such as immobilized biomolecules that can identify their respective analytes, and (iii) data interpretation units [81]. Owing to their large surface area, two-dimensional (2D) layered atomic

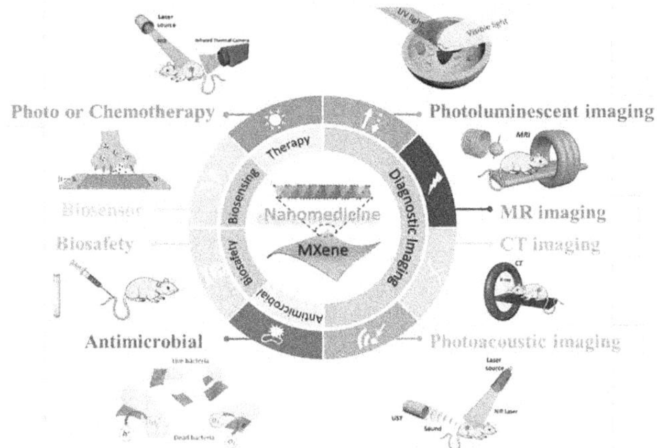

FIGURE 8.11 Overview of recently discovered 2d MXenes used in biomedical science, such as therapeutic uses, imaging, biosensors, antibiotic assessment, and biosafety assessment.

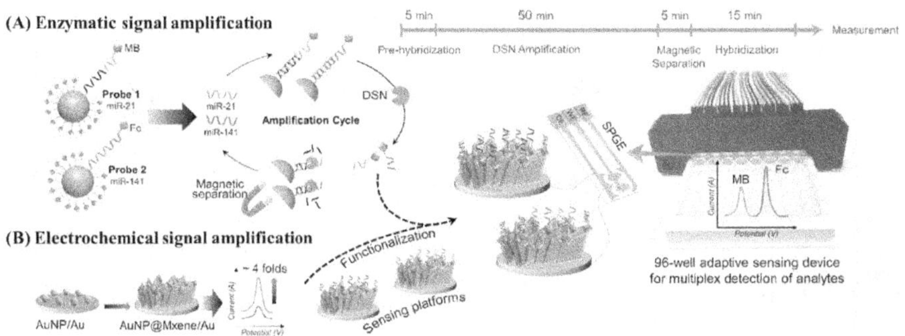

FIGURE 8.12 (a–b) MXenes for sensors in biomedical applications with multiple and simultaneous detection of miR-21 and miR-14 [82].

structure, excellent electrical properties, superior hydrophilicity, exceptional optical properties, and abundance of surface functional groups, MXenes are becoming common materials for fabricating biosensors. The term "electrical-signal-based biosensor" refers to a sensor used for the early diagnosis of disease that alters its surface electrical properties when combined with analytes, usually proteins, amino acids, RNA, H_2O_2, and others. However, because the redox center of the protein is tightly enclosed within the protein coat and enzymes are often easily inactivated, DET presents a challenge. MXene-based biosensors have been used to detect H_2O_2, nucleic acids [82–84], amino acids [85–86], and proteins [86] to aid in the identification and early diagnosis of cancer. The resulting biosensor has the following advantages: the detection limits of microRNA-21 and microRNA-141 were 204 and 138 aM, respectively, with a wide linear range of 500 to 50 nM [Figure 8.12(a and b)] and a short test period. duration of 80 min, multiplexing antifungal properties, sensitivity, and specificity (recognition of a single mutation) [82].

According to the illumination concept, optical signal-based biosensors can be broadly decomposed into electrochemical luminescence (ECL) and surface plasmon resonance (SPR) biosensors. With the following advantages, SPR biosensors have recently emerged as an important research tool in the life sciences and pharmaceutical industries. A staphylococcal Protein A (SPA) decorator was used to sequence and immobilize the anti-CEA monoclonal antibody (Abl) via its Fc region. A double-spliced version of the MWCNT-PDA-AgNPs-anti-CEA polyclonal antibody (MWPAg-Ab2) conjugate was used to provide a dynamic range of 2 10^{-16} to 2 10^{-8} M with limited detection of 0.07 fM to determine CEA. The biodetection method appears to be reproducible and has a good degree of specificity for CEA in real serum samples [Figure 8.13(a and b)] [87, 88]. It can be used to measure ACE levels in human serum for early cancer detection.

In addition to the two previously mentioned techniques, target analytes can be detected using chemiluminescence and near-infrared (NIR) chemiluminescence immunoassays. Cai et al. developed the NIR photo immunoassay for prostate-specific antigen (PSA) using liposomes encapsulated in Ti_3C_2 QDs with exceptional photothermal efficiency [89, 90]. In addition, Ti_3C_2 QDs strongly absorb NIR laser radiation and convert light energy into heat (Figure 8.14).

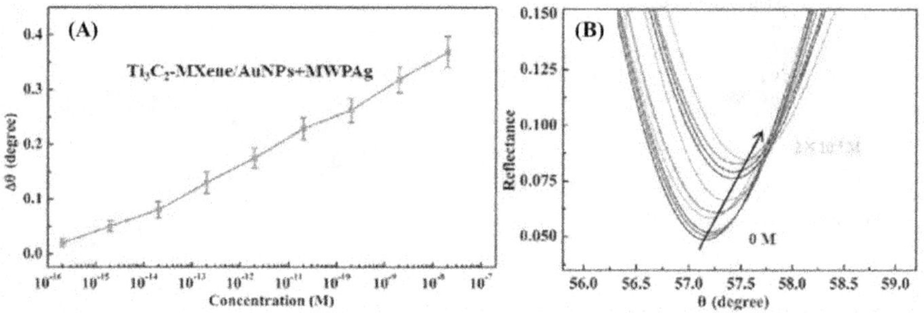

FIGURE 8.13 (a–b) The correlation between changes in SPR signal and CEA concentrations in the MWPAg-Ab2 signal enhancer [87, 88].

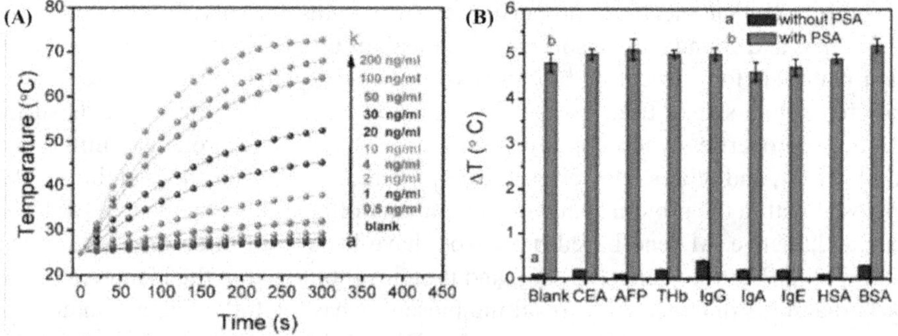

FIGURE 8.14 (a–b) The photothermal immunoassay's temperature responses are depicted (vs. time under 808 nm illumination) in the left image, and its selectivity is shown in the right image [89, 90].

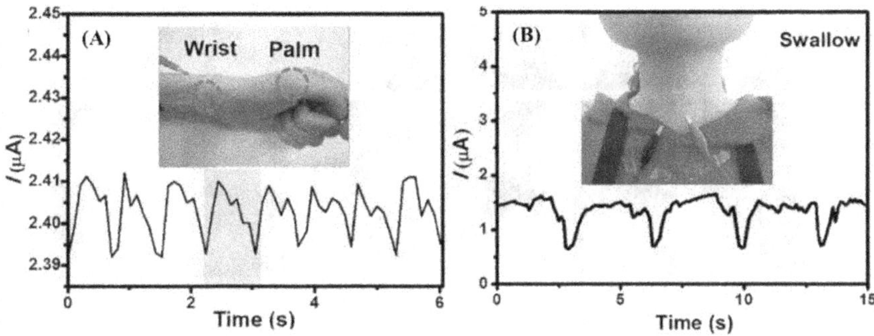

FIGURE 8.15 Swallowing and feeling of wrist pulse (A-B) [94].

In addition to certain biomolecules that can alter the electrical signal of MXene-based sensors, pressure and voltage can be used because of the superior mechanical and electronic properties of MXenes. To detect human activity and physiological signals in real time, sensors can be designed to be worn in various ways [91–94], as shown in Figure 8.15(a and b). Feasible worn sensors have been developed in practical application. This portable pressure sensor has a detection limit in the pressure range of 30 kPa–10.2 Pa and exhibits high performance after more than 10,000 cycles, which confirms good repeatability [94].

8.4.2 Bioimaging

Two-dimensional nanoscale materials are increasingly being used in diagnostic imaging, and the performance of diagnostic imaging can be enhanced by directly manipulating their physicochemical properties [95]. Diagnostic imaging devices such as PAI, magnetic resonance imaging (MRI), and CT can be used to determine tumor and PTT direction, and multimodal imaging is sometimes used for phototherapy [96–99]. MXenes exhibit high absorption and conversion efficiencies for a wide spectrum of light. Figure 8.16 illustrates the efficient tumor accumulation of Nb_2C MSN-SNO with strong tumor contrast of PAI in vivo after intravenous injection of synthetic Nb_2C-MSN-SNO nanosheets generated in mice bearing 4T1 tumors at different time intervals. Furthermore, the PA signal gradually increased

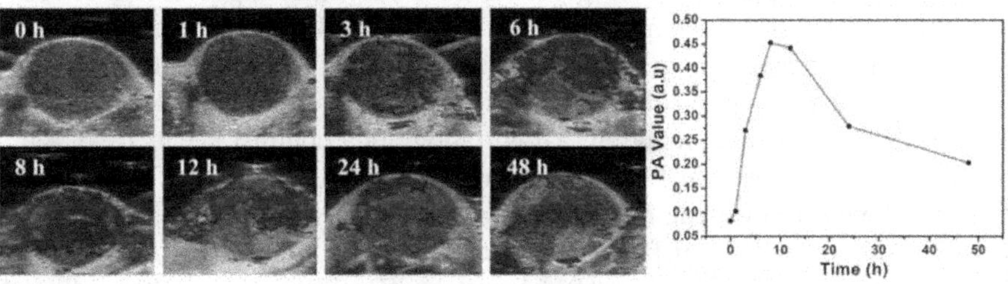

FIGURE 8.16 The intravenous administration of Nb_2C-MSN-SNO; PA pictures in vivo and related tumor signal intensities are displayed [100].

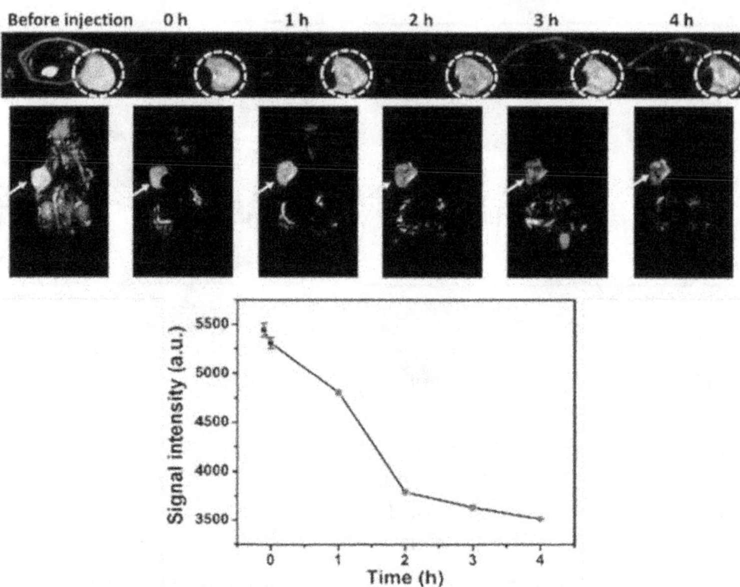

FIGURE 8.17 Transverse, coronal section, and corresponding signal intensities of the T2-weighted MRI of 4T1 tumor-bearing mice shown before and after intravenously injecting Ti_3C_2-IONP-SPs [102].

over time and peaked 8 h after injection owing to the accumulation of Nb_2C-MSN-SNO at the tumor site. The technology used in MRI has no ionizing radiation, high soft tissue contrast, and good spatial resolution. Accordingly, interest in the use of this technique for bioimaging and other contrast agents has increased to further increase sensitivity and produce information-rich images [100, 101]. By generating superparamagnetic Fe_3O_4 nanocrystals in situ on the surface of TC-MXenes, Liu et al. generated two-dimensional superparamagnetic TC-MXenes that appeared to be excellent cancer treatments. Therapeutic guidance potential has been demonstrated with the grown TC-IONP MXene composite, which shows a high T-2 relaxation capacity of 394.2 $mM^{-1}s^{-1}$ and appears to be a method of generating effective contrast-enhanced magnetic resonance imaging for malignancies (Figure 8.17). In addition, the developed MXene composites have a significant photothermal conversion efficiency (48.6%) for removing tumor tissues and killing cancer cells.

Three-dimensional tomography of anatomical structures is based on the difference in the X-ray absorption of lesions and tissues; therefore, X-ray computed tomography is used for medical imaging. X-ray tomography contrast agents (CTCAs) are nanomaterials composed mostly of (Z) elements [95]. As shown in Figure 8.18, Liu et al. developed a new type of superparamagnetic MXene nanotherapeutic based on Ta_4C_3 MXene and its surface superparamagnetic iron oxide-modified synthetic MXenes (Ta_4C_3-IONP-SPs) to treat breast cancer effectively. The results confirm that Ta_4C_3-IONP-SPs can produce CT images directly, depending on the atomic number and coefficient of the Ta component.

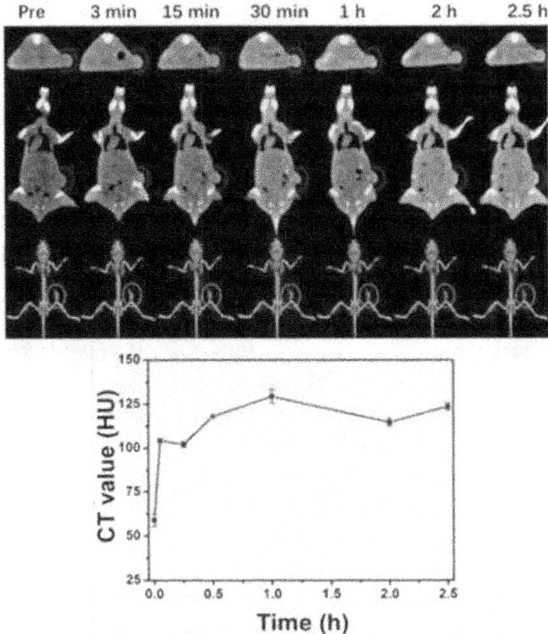

FIGURE 8.18 The right image exhibits time-dependent CT values following intravenous injection of Ta_4C_3-IONP-SPs, while the left image displays transverse, coronal, and 3D-rendering CT scans of a mouse with a 4T1 tumor [14].

The unique characteristics of MXene-derived QDs include chemical and photochemical stability, high dispersion, and luminescence, which can be controlled by varying the size, shape, and composition of the QDs. In 2017, Xue et al. fabricated monolayer Ti_3C_2 multi-quantum dots (MQDs). These MQDs could be easily dissolved in water via a simple hydrothermal process. The average size of the MQD could be adjusted by varying the reaction temperature. The strong quantum confinement of the generated MQDs allowed the visualization of excitation-dependent PL spectra with a quantum efficiency of approximately 10%. MQDs may be candidates for biocompatible multicolor cell imaging probes, according to RAW 264.7 cell labeling (Figure 8.19). Furthermore, the prepared MQDs were strongly selective for Zn^{2+} ions [103].

8.4.3 Tissue Engineering

MXenes are less commonly used in tissue engineering than in other biomedical applications. Cardiac [104, 105], skin [15], neural [106, 107], and bone tissue engineering [108, 109] are the major applications of MXene in tissue engineering. To fabricate durable and biocompatible TC-MXene-reinforced polylactic acid (PLA) nanocomposite nanocomposites, Chen et al. developed a reliable procedure using n-octyltriethoxysilane (OTES) as a mediator. Owing to the strong contact between

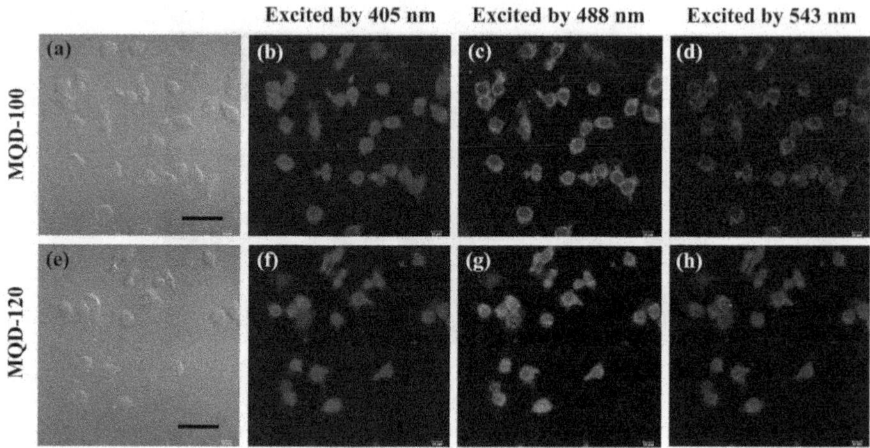

FIGURE 8.19 RAW264 imaging in bright-field and confocal mode. Seven cells are visible [103].

OTES-TC-MXene and PLA, the ultimate tensile strength (UTS) of the OTES-TC-MXene and PLA nanocomposite films was significantly increased. Furthermore, the inclusion of TC-MXene nanosheets confers exceptional biocompatibility, leading to increased cell adhesion, proliferation, and osteogenic differentiation, all of which are beneficial for bone resorption [Figure 8.20 (a–e)] [109]. Then Zhang et al. created TC-MXene multilayer films to test the biocompatibility, osteogenic potential, and guided bone remodeling of TC-MXenes in vitro and in vivo. The results of this investigation demonstrated a high degree of cytocompatibility of the TC-MXene membrane and its ability to promote osteogenic differentiation in vitro. MXene membranes exhibit good biocompatibility, osteogenic potential, and in vivo osteogenic activity when implanted in the subcutaneous and craniofacial regions of mice [110].

Ye et al. combined conductive MXene Ti_2C with mussel-inspired hydrophilicity and dopamine biocompatibility to create a biofunctional heart patch using poly(ethylene glycol) diacrylate (PEGDA)-Gel (ECP). The manufacturing process of the product is as follows: First, using an ultrasonic bath in an easy water bath, MXene Ti_2C was evenly dispersed by etching into a dopamine-N-formulated preparation, N'-methylene-bisacrylamide, methacrylate-gelatin, and polyethylene glycol. In vitro tests were performed on the ECP with conductivity before implantation in the mouse MI model [Figure 8.21 (i)]. In vitro, mouse aortic endothelial cells injected with Ti_2C-8 cryogel showed 3D vascular-like structures. After 7 days of co-culture of the Ti_2C cryogel and cardiomyocytes (CMs), the sarcomeres were well organized, and the initial insertion disc between the mature CMs was produced. In addition, there was obvious vasoconstriction, especially in newly created small arteries [Figure 8.21 (j and k)] [105].

8.4.4 Antibacterial Applications

The antibacterial activity of MXenes is influenced by their hydrophilicity, high electrical conductivity, surface oxygen-containing functional groups, atomic layer

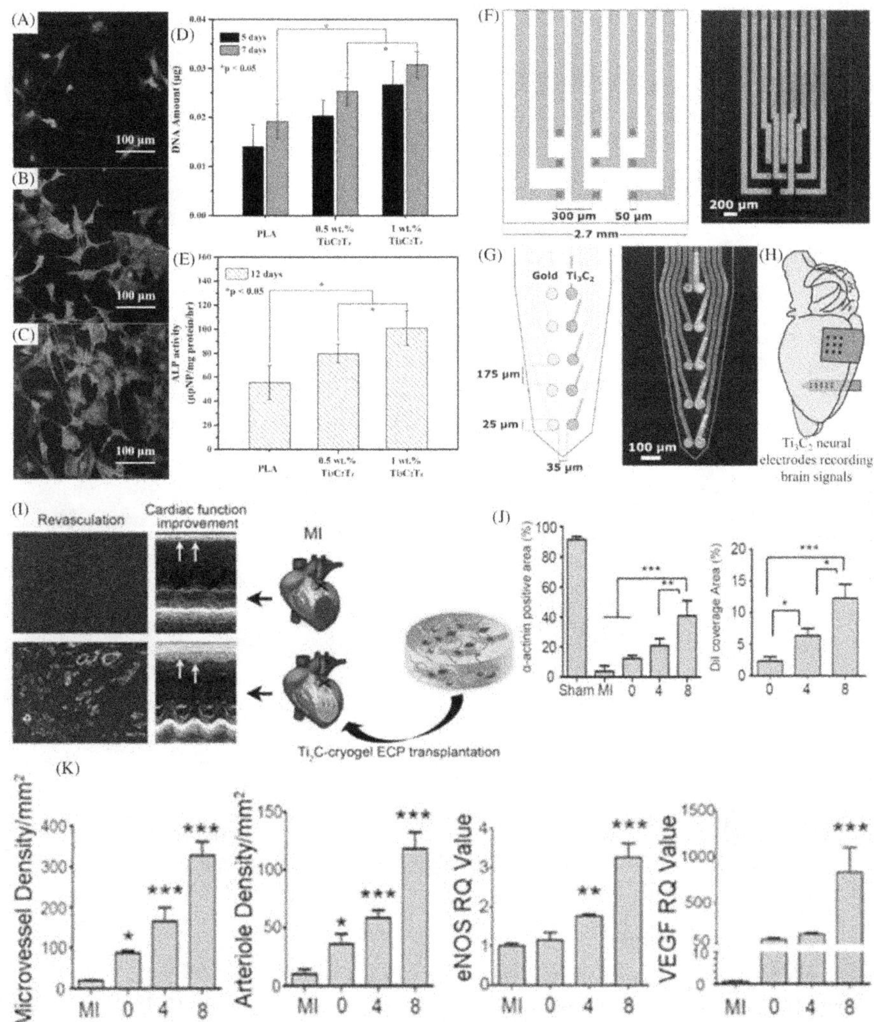

FIGURE 8.20 Engineering of tissue with MXenes. (a–c) CLSM images, (d) DNA test that assesses cell viability [109], (e–g) scheme of Ti_3C_2/Au intracortical electrode array with bright-field microscopy, (h) application of TC-MXene arrays in rat brain [111], (i) MI model, (j) polarization of unrepaired M2 macrophage in the MI site, and (k) the infarcted heart [105].

thickness, and optical characteristics, particularly the localized surface plasmon resonance (LSPR) effect. Kashif et al. discovered that TC-MXene was more effective than GO in inhibiting the growth of Gram-negative *E. coli* and Gram-positive *Bacillus subtilis* in a colloidal aqueous solution. The direct interaction between TC-MXene and the cell membrane, which results in membrane breakdown and consequent cell death, was determined to be caused by an antibacterial mechanism, according to the findings of several experiments. TC-MXene nanosheets were used

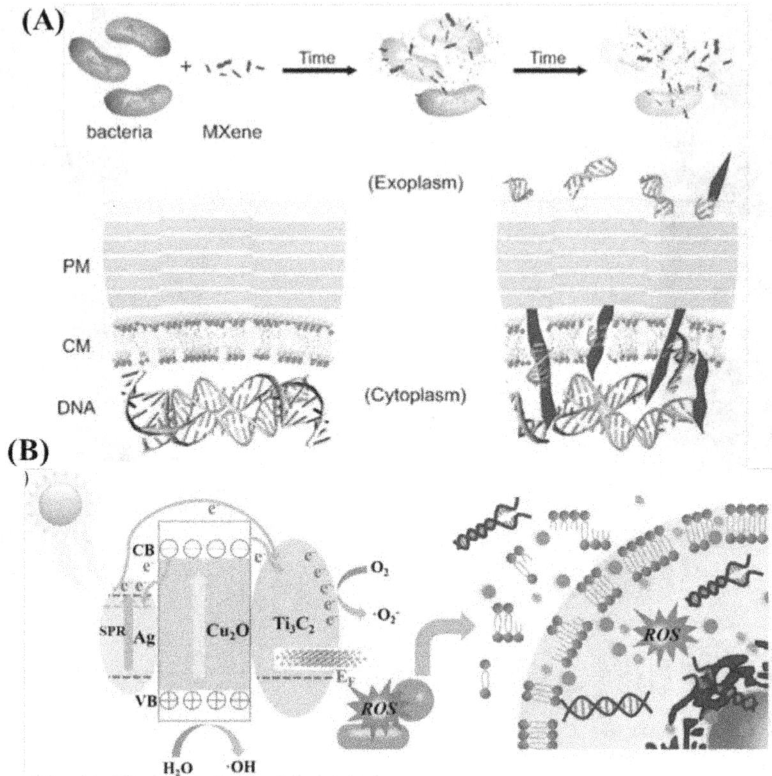

FIGURE 8.21 MXenes as an antibiotic: (a) Schematic representation and (b) band diagram with the antibacterial action of Ag@Ti_3C_2@Cu_2O (PM: peptidoglycan mesh; CM: cytoplasmic membrane) [113–114].

in this study. According to quantitative analyses using flow cytometry, complementary technologies, and fluorescence imaging, the smallest nanosheets exhibited the highest antibacterial activity against Gram-negative and Gram-positive bacteria. According to Figure 8.21 (b) [112–114], the charge channel produced by the heterostructure between the nanomaterials considerably improves the efficiency of charge transfer and separation. Additionally, the lifetime of active electrons, as well as electron creation, have primacy.

8.4.5 Therapeutics

Cancer is one of the biggest risks to human life and health that the planet is facing. The human body suffers significant damage from traditional surgical procedures, which are prone to infection and recurrence. Photothermal therapy (PTT) is a new, less invasive treatment strategy for killing tumor cells by absorbing NIR light and transforming it into heat. MXenes have been extensively investigated for PTT of malignant tumors because of their increased surface area, high photothermal conversion efficiency, hydrophilicity, and abundance of functional groups [115]. Lin et al. demonstrated for the first time the use of atom-thick MXene nanosheets for the photothermal ablation of cancer, as shown in Figure 8.22 (a–c). PTT can be used to make multidrug-resistant bacteria more susceptible to common antibiotics, contributing to the antibacterial therapeutic goal in addition to tumor resection. In summary, cooperative therapy appears to be biocompatible and effective for treating MRSA infections in vivo [Figure 8.22 (d–g)] [116].

PDT is a non-invasive treatment that is as effective as photothermal therapy. Currently, there are fewer studies on MXenes in PDT compared to PTT, and most MXene-based PDTs are combined with PTT for adjuvant tumor treatment. In Figure 8.23, ROS generation in cells exposed to the 808-nm laser was detected using a 2′,7′-dichlorofluorescein diacetate (DCFH-DA) probe (0.8 W cm^{-2}, 10 min). When hyaluronidase (HAase) solubilized the HA coating to expose TC-MXene-DOX, bright green DCF fluorescence in the cells indicated that TC-MXene-DOX induced

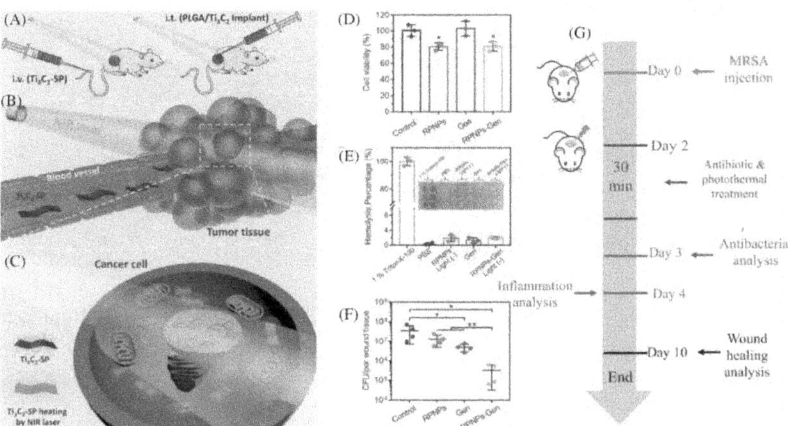

FIGURE 8.22 MXenes used for medical purposes, (a) A phase-changeable PLGA/Ti_3C_2 implant and Ti_3C_2-SP nanosheets were administered intravenously and into the tumor, (b) EPR effect, (c) NIR-laser (d) cancer cells are photothermally destroyed. (e) NIH-3T3 cell viability in the presence of light, (f) the samples' hemolysis % and the associated photos, and (g) MRSA CFU in each tissue of the wound. Combinatorial therapy protocol schematic [116].

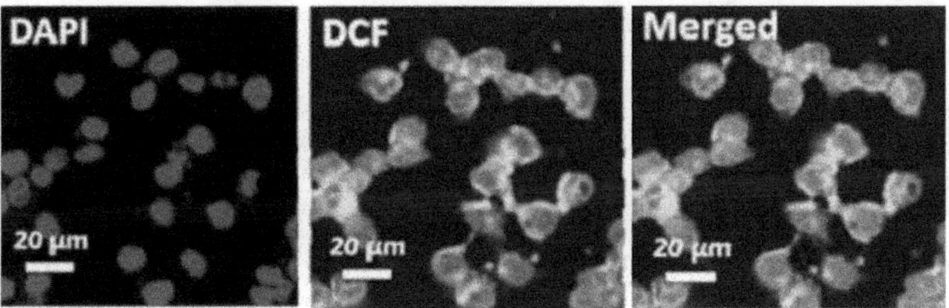

FIGURE 8.23 ROS in cells are visible in a fluorescence picture [118].

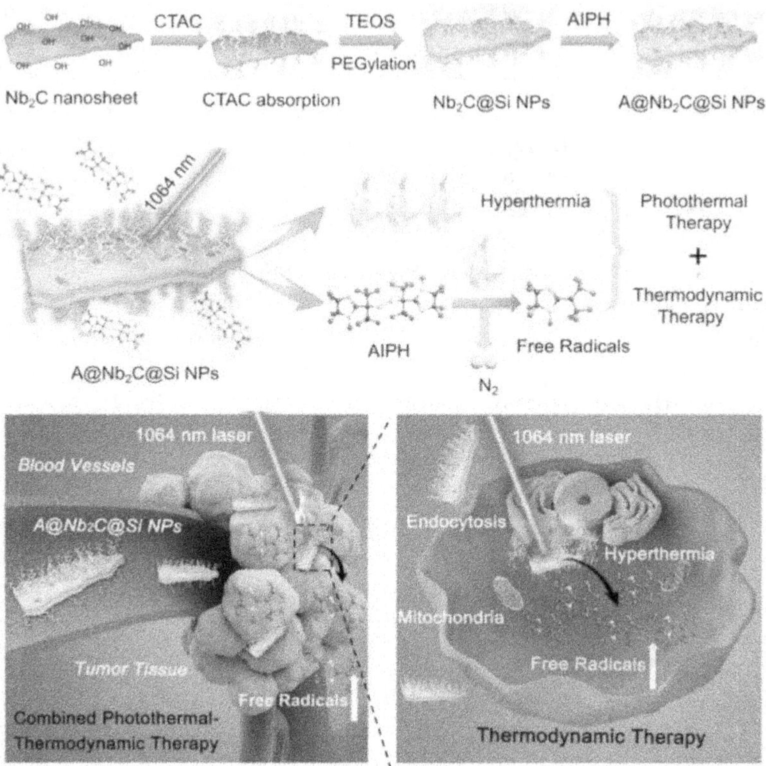

FIGURE 8.24 Photothermal therapy of AIPH at Nb_2C at $mSiO_2$ NP under light was combined with thermodynamic therapy brought on by free radicals produced by the thermal decomposition of AIPH [119].

a large amount of ROS. In addition, previous studies have shown that the LSPR effect of metals (especially gold nanoparticles) is a determining factor in their ability to generate ROS upon exposure to visible light; therefore, the LSPR effect of transition metal-containing MXenes is also a factor, while the large surface area of MXene nanosheets likely contributes to the enhancement of LSPR [117, 118].

Because the tumor microenvironment is hypoxic, some oxygen-dependent therapies may not be effective for treating tumors. For example, an oxygen-deficient atmosphere prevents ROS synthesis of reactive oxygen species. According to pharmacokinetic testing, longer cycle lengths were highly consistent with the biodistribution results. According to the results of in vivo toxicity experiments, the developed AIPH@Nb_2C@$mSiO_2$ nanoplatforms exhibited good biocompatibility and biosafety (Figure 8.24) [119].

8.4.6 Drug Delivery Systems

Chemotherapy, surgery, and radiation with cancer drugs, such as cisplatin, methotrexate, cytarabine, paclitaxel, and doxorubicin, are common cancer treatments.

However, most cancer drugs are cytotoxic; therefore, they affect non-cancerous cells in the human body, in addition to cancer cells. Therefore, the gradual and controlled release of drugs is an important focus of cancer treatment research. A new class of 2D-MXenes has attracted much interest because of their potential as drug carriers. Alternatively, near-infrared radiation stimulation can be used to induce controlled drug release. To interact with DOX, a cationic model drug, Liu et al. created a nanoplatform based on TC-MXene-DOX nanosheets via layer-by-layer adsorption using TC-MXene. With a loading capacity of up to 84.2%, TC-MXene can transport cationic anticancer medications. Because of the HA coating on the nanosheets, this strategy aggressively targets tumor cells that over-express CD44 and successfully inhibits DOX permeability when the environment is neutral. TC-MXene-DOX is typically extremely sensitive to light, heat, pH, and enzymes (Figure 8.25) [120]. Han et al. utilized an adaptable sol-gel chemistry technique to create a particular layer of "healing interstitial holes" on the surface of 2D Nb_2C MXene. However, this approach is not ideal because of its drawbacks and lack of surface chemistry [120].

8.4.7 Antimicrobial Activity

Different 2D nanomaterials have been examined for their antibacterial activity using various techniques. For instance, adding Zn-Ti double hydroxide (LDH) to bacterial cultures under light dramatically inhibits the development of microorganisms such as *S. cerevisiae*, *S. aureus*, and *Escherichia coli*. This impact appears to be caused by the size and LDH generation effects. To thoroughly explore the environmental and health implications of these new 2D MXenes, we tested the antibacterial activity of monolayer and multilayer TC-MXenes shared by two bacterial models, *E. coli* and *Bacillus subtilis*, and compared them with those of other carbon-based nanomaterials (Figure 8.26 a). *Escherichia coli* and *Bacillus subtilis* were used to test the antibacterial effects of TC-MXenes. TC-MXene treatment of both bacterial cell types in colloid solution for 4 h (200 g ml^{-1}) revealed concentration-dependent antibacterial activity and survived >98% loss of capacity (Figure 8.26 c). To assess the cytotoxic capability of TC-MXenes, LDH released from bacterial cells exposed to various concentrations of TC-MXene nanosheets for 4 h was utilized (Figure 8.26 d). The antibacterial mechanism of TC-MXenes was theorized as follows: First, TC-MXenes can successfully cling to the surface of microorganisms owing to their sharp edges. Second, the membrane can be damaged by microbes exposed to the jagged edges of MXenes. Third, TC-MXenes disturb the cell architecture and cause microbial mortality by interacting with biomolecules found in the cytoplasm and cell wall of microbes [Figure 8.26 (d)] [121].

8.4.8 Evaluations of MXenes' Cytotoxicity and Compatibility

The potential cytotoxicity of the first MXenes relevant for clinical application has not been adequately studied. Nasrallah et al. investigated the biocompatibility of $Ti_3C_2T_x$ using a zebrafish embryo model to ascertain the toxicity of the compound in vivo [Figure 8.27 (a–d)]. Titanium content in zebrafish embryos was measured using

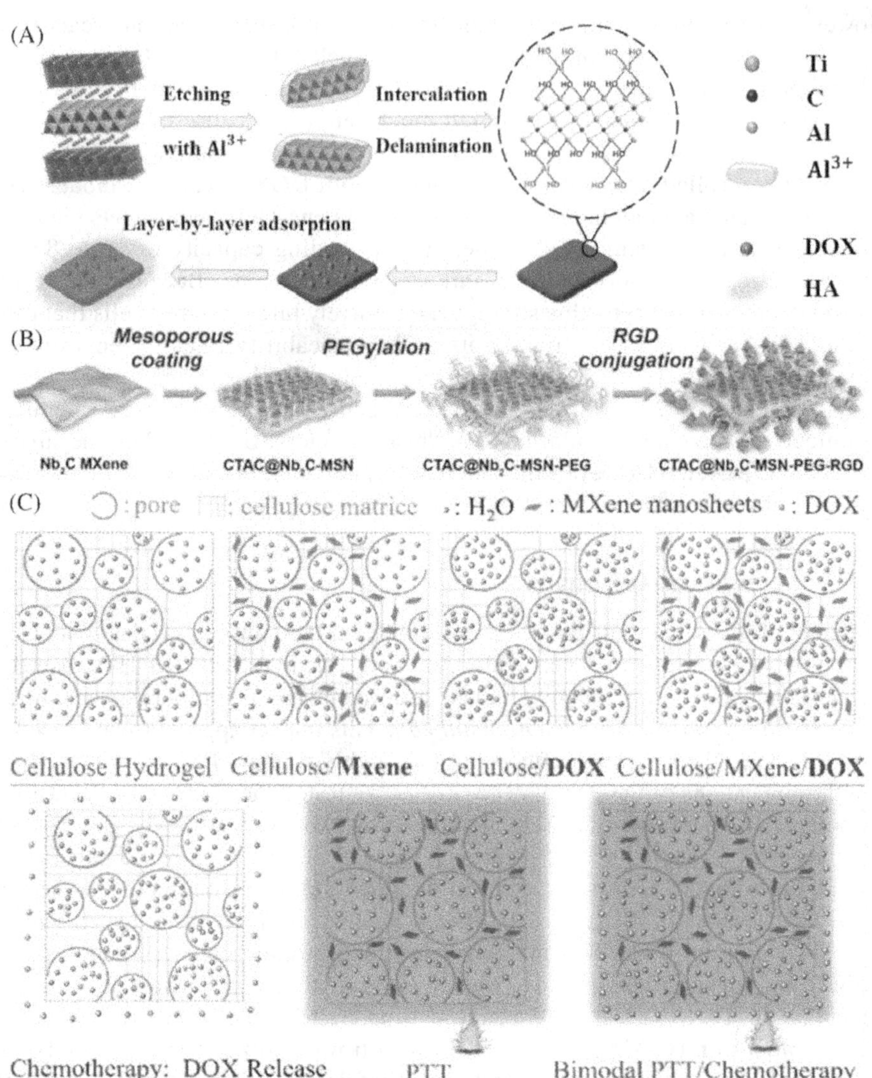

FIGURE 8.25 MXenes for medication administration. (a) Layer-by-layer absorption manufacturing scheme [118]. (b) CTAC surface modification and synthesis at Nb_2C-MSN [13]. (c) Release of hydrogel under various conditions and four different cellulose-based hydrogel types [120].

inductively coupled plasma mass spectrometry (ICP-MS), and titanium content was found to be dose-dependent. $Ti_3C_2T_x$ was bound or absorbed in proportion to titanium concentration. When the nanosheet concentration in the acute toxicity test exceeded 25 M, the nanosheets started to gather around the embryo, showing that 100 g mL^{-1} $Ti_3C_2T_x$ exhibited very low acute toxicity. Additionally, studies have been conducted

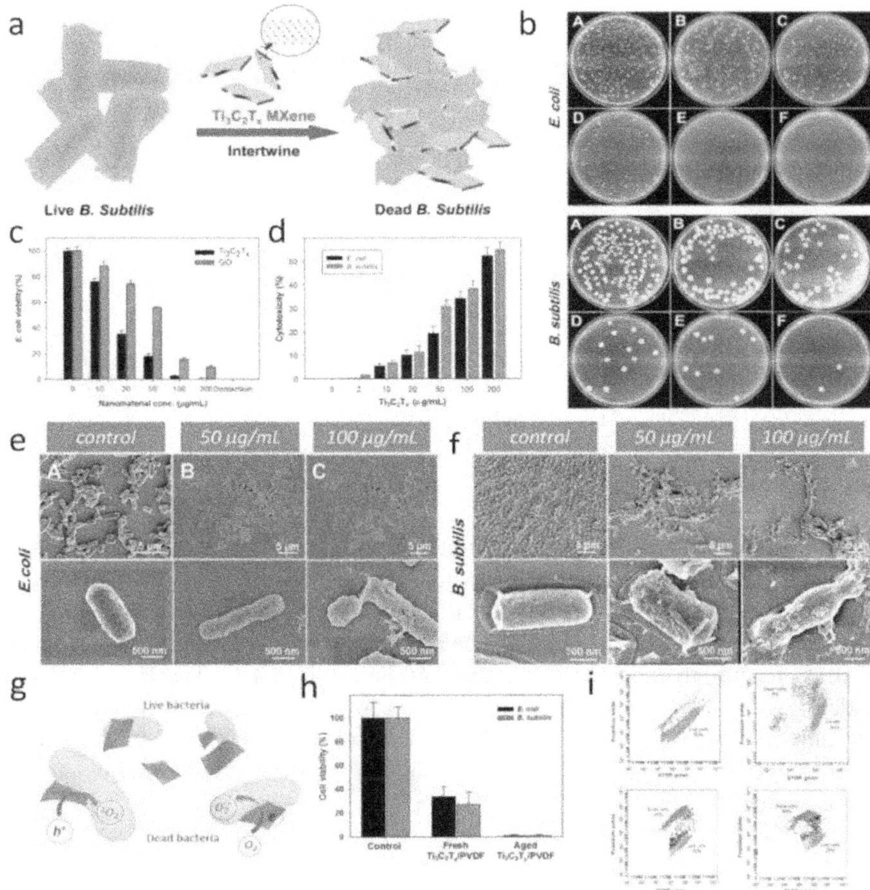

FIGURE 8.26 (a) Schematic representation and (b) agar plates on which *E. coli* (top) and *B. subtilis* (bottom) bacterial cells, (c) analysis of cell viability, (d) TC-MXene cytotoxicity research using bacterial cell-released LDH, (e) low- and high-magnification SEM pictures, (g) antibacterial mechanism diagram, (h) research on the cell viability of *B. subtilis* and *E. coli* grown on TC-MXene membranes supported by both new and used PVDF, and (i) *E. coli* and *B. subtilis* bacterial cell flow cytometry analysis [122].

on its impact on the neurological system and motor behavior. Neurotoxicity and exercise performance evaluations of $Ti_3C_2T_x$-treated embryos showed normal locomotor function [123].

A research team led by Jastrzebska et al. examined the bioactivity of 2D nano-TC-MXene-sheets using two cancer cell lines (A549 and A375) and two normal cell lines (MRC-5 and HaCaT). Calcein-AM and MTT assays revealed that both normal and cancer cells were viable. Tumor cell viability declined more quickly than normal cell viability, even when MXenes were reduced with increasing concentrations. Additionally, it was discovered that the survival probability of normal cells

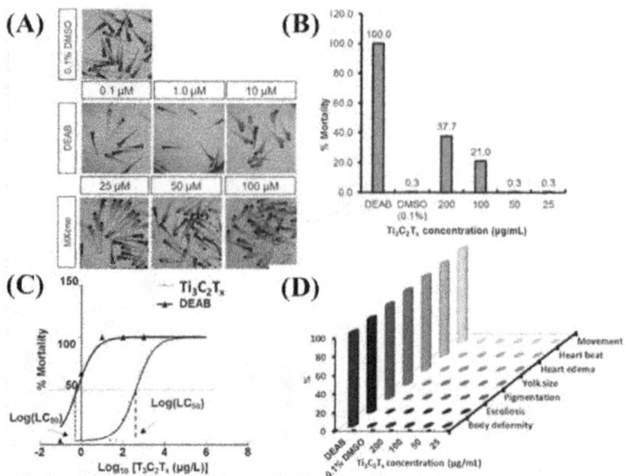

FIGURE 8.27 Evaluations of the cytotoxicity and cytocompatibility of MXenes [123].

might approach 70% when the concentration of MXenes is not less than 125 mg l^{-1}. It is possible that MXenes increase ROS generation [Figure 8.2 (a and b)] [124] as a mechanism of toxicity. In addition to TiC, Ti_2AlC, and Ti_3AlC_2, Scheibe et al. also produced $Ti_3C_2T_x$ at the end of 2019. The results showed that oxidative and mechanical stress from high concentrations (about 400 g mL^{-1}) of TiC, Ti_2AlC, and Ti_3AlC_2 particles with a diameter of 44 μm could damage cells, and among all $Ti_3C_2T_x$ variants, only $Ti_3C_2T_x$ at the highest concentration was less cytotoxic to MSU1.1 cells. In addition, cytotoxicity was cell type-dependent, with fewer cells of malignant origin than viable cells [Figure 8.28 (c and d)] [124].

8.5 CONCLUSION

MXenes are new 2D materials that have attracted much attention because of their unique properties, such as large surface area, electrochemical and LSPR effects, mechanical and magnetic properties, raw layer thicknesses, mortality, and biocompatibility. This chapter provides an overview of the structure, synthesis, surface functionalization, biological applications, and cytotoxicity of MXenes. Although significant advances have been made in areas such as tissue engineering, bioimaging, photothermal therapy, sensors for biomedical applications, and antimicrobial therapy, some aspects remain unclear. Among them, firstly, the biocompatibility data of MXene is sparse, most related to Ti-based MXene, and the research literature highlights the great potential of novel 2D structural materials in biomedicine and other related fields. However, the environmental impact of the material must also be considered because the mentioned MXene materials are synthesized by a top-down method using reagents that can be harmful to nature, the environment, and people. In addition, there are environmentally friendly top-down methods, but there is still little data on their use in biomedical research. Therefore, there is a need to expand the use of MXenes in biomedical research by developing new advanced synthesis methods, resulting in the large-scale processing and amplification of different members

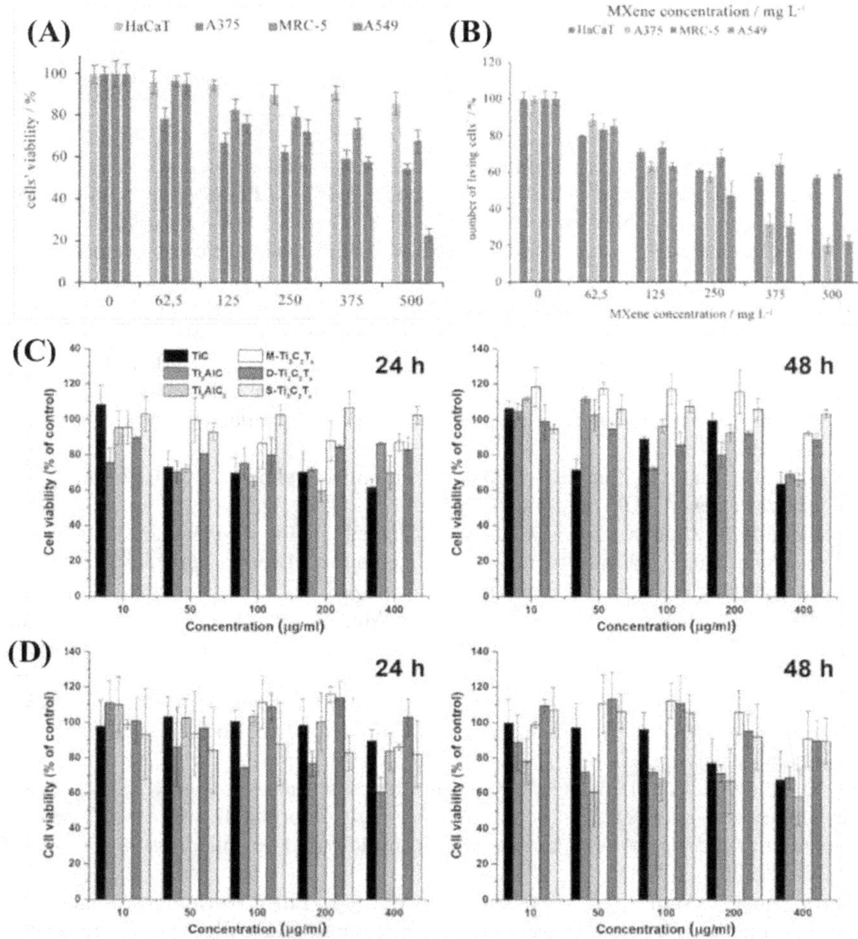

FIGURE 8.28 (a) The MTT assay findings. Results of the calcein-AM assay (b). (c, d) With reference to untreated control cells, the percentage of tetrazolium salt converted to formazan was determined [124].

of the MXene family. To date, research on the performance of MXenes has mainly focused on their electrical and optical properties, and little attention has been paid to their magnetic properties. Compared with the first two properties, magnetic properties have fewer applications, and biomedical applications use even fewer magnetic properties than MRI imaging. In addition, many of the ways in which MXenes work, including how they promote ROS generation, remain unclear. A better understanding of this mechanism will help optimize materials, improve their properties, and provide more efficient applications. Finally, other biomedical applications of MXenes, including dynamic ultrasound therapy, should be explored. Note that composites created by combining two materials sometimes improve overall performance.

REFERENCES

[1] Y. Xie, H. Zhang, H. Huang, Z. Wang, Z. Xu, H. Zhao, Y. Wang, N. Chen, W. Yang, High-voltage asymmetric MXene-based on-chip micro-supercapacitors, *Nano Energy.* 2020, 74, 104928.
[2] H. Huang, X. Chu, H. Su, H. Zhang, Y. Xie, W. Deng, N. Chen, F. Liu, H. Zhang, B. Gu, W. Deng, W. Yang, Massively manufactured paper-based all-solid-state flexible micro-supercapacitors with sprayable MXene conductive inks, *J Power Sources.* 2019, 415, 1.
[3] M. Ghidiu, M.R. Lukatskaya, M.-Q. Zhao, Y. Gogotsi, M.W. Barsoum, Conductive two-dimensional titanium carbide 'clay' with high volumetric capacitance, *Nature.* 2014, 516, 78.
[4] M.R. Lukatskaya, O. Mashtalir, C.E. Ren, Y. Dall'Agnese, P. Rozier, P.L. Taberna, M. Naguib, P. Simon, M.W. Barsoum, Y. Gogotsi, Cation intercalation and high volumetric capacitance of two-dimensional titanium carbide, *Science.* 2013, 341, 1502.
[5] S. Cao, B. Shen, T. Tong, J. Fu, J. Yu, 2D/2D Heterojunction of ultrathin MXene/Bi_2WO_6 nanosheets for improved photocatalytic CO_2 reduction, *Adv Funct Mater.* 2018, 28, 1800136.
[6] Z. Guo, J. Zhou, L. Zhu, Z. Sun, MXene: a promising photocatalyst for water splitting, *J Mater Chem A.* 2016, 4, 11446.
[7] J. Zhang, Y. Zhao, X. Guo, C. Chen, C.-L. Dong, R.-S. Liu, C.-P. Han, Y. Li, Y. Gogotsi, G. Wang, Single platinum atoms immobilized on an MXene as an efficient catalyst for the hydrogen evolution reaction, *Nat Catal.* 2018, 1, 985.
[8] W.T. Cao, C. Ma, D.-S. Mao, J. Zhang, M.-G. Ma, F. Chen, MXene-reinforced cellulose nanofibril inks for 3D-printed smart fibres and textiles, *Adv Funct Mater.* 2019, 29, 1905898.
[9] W.T. Cao, H. Ouyang, W. Xin, S. Chao, C. Ma, Z. Li, F. Chen, M.-G. Ma, A stretchable high-output triboelectric nanogenerator improved by MXene liquid electrode with high electronegativity, *Adv Funct Mater.* 2020, 2004181.
[10] K. Wang, Z. Lou, L. Wang, L. Zhao, S. Zhao, D. Wang, W. Han, K. Jiang, G. Shen, Bioinspired interlocked structure-induced high deformability for two-dimensional titanium carbide (MXene)/natural microcapsule-based flexible pressure sensors, *ACS Nano.* 2019, 13, 9139.
[11] D. Wang, L. Wang, Z. Lou, Y. Zheng, K. Wang, L. Zhao, W. Han, K. Jiang, G. Shen, Biomimetic, biocompatible and robust silk Fibroin-MXene film with stable 3D cross-link structure for flexible pressure sensors, *Nano Energy.* 2020, 78, 105252.
[12] H. Lin, X. Wang, L. Yu, Y. Chen, J. Shi, Two-dimensional ultrathin MXene ceramic nanosheets for photothermal conversion, *Nano Lett.* 2017, 17, 384.
[13] X. Han, X. Jing, D. Yang, H. Lin, Z. Wang, H. Ran, P. Li, Y. Chen, Therapeutic mesopore construction on 2D Nb_2C MXenes for targeted and enhanced chemo-photothermal cancer therapy in NIR-II biowindow, *Theranostics.* 2018, 8, 4491.
[14] Z. Liu, H. Lin, M. Zhao, C. Dai, S. Zhang, W. Peng, Y. Chen, 2D superparamagnetic tantalum carbide composite MXenes for efficient breast-cancer theranostics, *Theranostics.* 2018, 8, 1648.
[15] L. Mao, S. Hu, Y. Gao, L. Wang, W. Zhao, L. Fu, H. Cheng, L. Xia, S. Xie, W. Ye, Z. Shi, G. Yang, Biodegradable and electroactive regenerated bacterial cellulose/MXene ($Ti_3C_2T_x$) composite hydrogel as wound dressing for accelerating skin wound healing under electrical stimulation, *Adv Healthc Mater.* 2020, 9, 2000872.
[16] P.V. Shinde, M.K. Singh, Synthesis, characterization, and properties of graphene analogs of 2D material, *Fundam Sens Appl 2D Mater.* 2019, 5, 91–143.
[17] Z. Wang, W. Zhu, Y. Qiu, X. Yi, A.V.D. Busshe, A. Kane, H. Gao, K. Koshi, E. Hurt, Biological and environmental interactions of emerging two-dimensional nanomaterials, *Chem Soc Rev.* 2016, 45, 1750–1780.

[18] Z. Huang, X. Cui, S. Li, J. We, P. Li, Y. Wang, C.S. Lee, Integrated proteogenomic characterization of HBV-related hepatocellular carcinoma, *Nanophotonics.* 2020, 9, 2233–2249.

[19] L. Bai, W. Yi, T. Sun, Y. Tian, P. Zhang, J. Si, X. Hou, J. Hou, Surface modification engineering of two-dimensional titanium carbide for efficient synergistic multitherapy of breast cancer, *J Mater Chem B.* 2020, 8, 6402–6417.

[20] M. Naguib, O. Mashtalir, J. Carle, V. Presser, J. Lu, L. Hultman, Y. Gogotsi, M.W. Barsoum, Two-dimensional transition metal carbides, *ACS Nano.* 2012, 6, 1322.

[21] M.H. Kang, D. Lee, J. Sung, J. Kim, B.H. Kim, J. Park, Structure and chemistry of 2D materials. *Comprehens Nanosci Nanotechnol.* 2019, 1–5, 55–90.

[22] A. Sundaram, J.S. Ponraj, C. Wan, W.K. Peng, R.K. Manavalan, S.C. Dhanabalan, H. Zhang, J. Gaspar, Engineering of 2D transition metal carbides and nitrides MXenes for cancer therapeutics and diagnostics, *J Mater Chem B.* 2020, 8, 4990–5013.

[23] H. Huang, R. Jiang, Y. Feng, H. Ouyng, N. Zhou, X. Zhang, Y. Wei, Recent development and prospects of surface modification and biomedical applications of MXenes, *Nanoscale.* 2020, 12, 1325–1338.

[24] Y. Wang, W. Feng, Y. Chen, Chemistry of two-dimensional MXene nanosheets in theranostic nanomedicine, *Chin Chem Lett.* 2020, 31, 937–946.

[25] Y. Ibrahim, A. Mohamed, A.M. Abdelgawad, K. Eid, A.M. Abdullah, A. Elzatahry, Unveiling fabrication and environmental remediation of MXene-Based nanoarchitectures in toxic metals removal from wastewater: strategy and mechanism, *Nanomaterials.* 2020, 10, 10.

[26] A. Zamhuri, G.P. Lim, N.L. Ma, K.S. Tee, C.F. Soon, MXene in the lens of biomedical engineering: synthesis, applications and future outlook, *BioMed Eng Online*, 2021, 20, 33.

[27] L. Bai, W. Yi, T. Sun, Y. Tian, P. Zhang, J. Si, X. Hou, J. Hou, Surface modification engineering of two-dimensional titanium carbide for efficient synergistic multitherapy of breast cancer, *J Mater Chem B.* 2020, 8, 6402–6417.

[28] B. Shao, Z. Liu, G. Zeng, H. Wang, Q. Liang, Q. He, M. Cheng, C. Zhou, L. Jiang, B. Song, Two-dimensional transition metal carbide and nitride (MXene) derived quantum dots (QDs): synthesis, properties, applications and prospects, *J Mater Chem A.* 2020, 8, 7508–7535.

[29] A. Zhou, *Cognitive-behavioral perspectives on body image*, New York: Elsevier; 2012, 21–46.

[30] S. Venkateshalu, A.N. Grace, MXenes—A new class of 2D layered materials: synthesis, properties, applications as supercapacitor electrode and beyond, *Appl Mater Today.* 2020, 18, 100509.

[31] G.P. Lim, C.F. Soon, M. Morsin, M.K. Ahmad, N. Nayan, K.S. Tee, Synthesis, characterization and antifungal property of $Ti_3C_2T_x$ MXene nanosheets. *Ceram Int.* 2020, 46(12), 20306–20312.

[32] J. Pang, R.G. Mendes, A. Bachmatiuk, L. Zhao, H.Q. Ta, T. Gemming, H. Liu, Z. Liu, M.H. Rummeli, Applications of 2D MXenes in energy conversion and storage systems, *Chem Soc Rev.* 2019, 48, 72–133.

[33] A.W. Rozmysłowska, J. Mitrzak, A. Szuplewska, M. Chudy, J. Woźniak, M. Petrus, T. Wojciechowski, A.S. Vasilchenko, A.M. Jastrzebska, A simple, low-cost and green method for controlling the cytotoxicity of MXenes, *Materials.* 2020, 13, 1–18.

[34] M. Ghidiu, M.R. Lukatskaya, M.Q. Zhao, Y. Gogotsi, M.W. Barsoum, Conductive two-dimensional titanium carbide 'clay' with high volumetric capacitance, *Nature.* 2015, 516, 78–81.

[35] F. Liu, A. Zhou, J. Chen, J. Jia, W. Zhou, L. Wang, Q. Hu, Preparation of Ti_3C_2 and Ti_2C MXenes by fluoride salts etching and methane adsorptive properties, *Appl Surf Sci.* 2017, 416, 781–789.

[36] J. Halim, M.R. Lukatskaya, K.M. Cook, J. Lu, C.R. Smith, L.A. Näslund, S.J. May, L. Hultman, Y. Gogotsi, P. Eklund, M.W. Barsoum, Transparent conductive two-dimensional titanium carbide epitaxial thin films, *Chem Mater.* 2014, 26, 2374–2381.
[37] T. Li, L. Yao, Q. Liu, J. Gu, R. Luo, J. Li, X. Yan, W. Wang, P. Liu, B. Chen, Fluorine-free synthesis of high-purity $Ti_3C_2T_x$ (T=OH, O) via alkali treatment, *Angew Chem Int Ed.* 2018, 57, 6115–6119.
[38] P. Urbankowski, B. Anasori, T. Makaryan, D. Er, S, Kota, P.L. Walsh, M. Zhao, V.B. Shenoy, M.W. Barsoum, Y. Gogotsi, Synthesis of two-dimensional titanium nitride Ti_4N3 (MXene), *Nanoscale.* 2016, 8, 11385–11391.
[39] O. Mashtalir, M. Naguib, V.N. Mochalin, Y. Dall'Agnese, M. Heon, M.W. Barsoum, Y. Gogotsi, Intercalation and delamination of layered carbides and carbonitrides, *Nat Commun.* 2013, 4, 1–7.
[40] O. Mashtalir, M.R. Lukatskya, M.Q. Zhao, M.W. Zhao, M.W. Barsoum, Y Gogotsi, Amine-assisted delamination of Nb_2C MXene for Li-Ion energy storage devices, *Adv Mater.* 2015, 27, 3501–3506.
[41] M. Naguib, R.R. Unocic, B.L. Armstrong, J. Nanda, Large-scale delamination of multi-layers transition metal carbides and carbonitrides "MXenes", *Dalton Trans.* 2015, 44, 9353–9358.
[42] H. Lin, X. Wang, L. Yu, Y. Chen, J. Shi, Two-dimensional ultrathin MXene ceramic nanosheets for photothermal conversion, *Nano Lett.* 2017, 17, 384–391.
[43] S.Y. Pang, Y.T. Wong, S. Yuan, Y. Liu, M.K. Tsang, Z. Yang, H. Huang, W.T. Wong, J. Hao, Universal strategy for HF-free facile and rapid synthesis of two-dimensional MXenes as multifunctional energy materials, *J Am Chem Soc.* 2019, 141, 9610–9616.
[44] M. Soleymaniha, M.A. Shahbazi, A.R. Rafieerad, A. Maleki, A. Amiri, Sweet-MXene hydrogel with mixed-dimensional components for biomedical applications, *Adv Healthcare Mater.* 2019, 8, 1–26.
[45] B. Yang, Y. Chen, J. Shi, Material chemistry of two-dimensional inorganic nanosheets in cancer theranostics, *Chem.* 2018, 4, 1284–1313.
[46] L. Verger, C. Xu, V. Natu, H.M. Cheng, W. Ren, M.W. Barsoum, Overview of the synthesis of MXenes and other ultrathin 2D transition metal carbides and nitrides, *Curr Opin Solid State Mater Sci.* 2019, 23, 149–163.
[47] C. Xu, L. Wang, Z. Liu, L. Chen, J. Guo, N. Kang, X.L. Ma, H.M. Cheng, W. Ren, Large-area high-quality 2D ultrathin Mo_2C superconducting crystals, *Nat Mater.* 2015, 14, 1135–1141.
[48] X. Xiao, H. YU, H. Jin, M. Wu, Y. Fang, J. Sun, Z. Hu, T. Li, J. Wu, L. Huang, Salt-templated synthesis of 2D metallic MoN and other nitrides, *ACS Nano.* 2017, 11, 2180–2186.
[49] Y. Gogotsi, B. Anasori, The rise of MXenes, *ACS Nano.* 2019, 13, 8491.
[50] S. Zhao, W. Kang, J. Xue, MXene nanoribbons, *J Mater Chem C.* 2015, 3, 879.
[51] S.Y. Pang, W.F. Io, L.W. Wong, J. Zhao, J. Hao, Efficient energy conversion and storage based on robust fluoride-free self-assembled 1D niobium carbide in 3D nanowire network, *Adv Sci.* 2020, 7, 1903680.
[52] K. Li, M. Liang, H. Wang, X. Wang, Y. Huang, J. Coelho, S. Pinilla, Y. Zhang, F. Qi, V. Nicolosi, Y. Xu, 3D MXene architectures for efficient energy storage and conversion, *Adv Funct Mater.* 2020, 2000842.
[53] K. Li, X. Wang, S. Li, P. Urbankowski, J. Li, Y. Xu, Y. Gogotsi, An ultrafast conducting polymer@ MXene positive electrode with high volumetric capacitance for advanced asymmetric supercapacitors, *Small.* 2020, 16, 1906851.
[54] S.A. Shah, T. Habib, H. Gao, P. Gao, W. Sun, M.J. Green, M. Radovic, Template-free 3D titanium carbide ($Ti_3C_2T_x$) MXene particles crumpled by capillary forces, *Chem Commun.* 2017, 53, 400.

[55] L. Wang, H. Liu, X. Lv, G. Cui, G. Gu, Facile synthesis 3D porous MXene $Ti_3C_2T_x$@RGO composite aerogel with excellent dielectric loss and electromagnetic wave absorption, *J Alloy Compd.* 2020, 828, 154251.

[56] H. Shi, M. Yue, C.J. Zhang, Y. Dong, P. Lu, S. Zheng, H. Huang, J. Chen, P. Wen, Z. Xu, Q. Zheng, X. Li, Y. Yu, Z.-S. Wu, 3D flexible, conductive, and recyclable Ti_3C_2Tx MXene-melamine foam for high-areal-capacity and long-lifetime alkali-metal anode, *ACS Nano.* 2020, 14, 8678.

[57] J. Qin, L. Hao, X. Wang, Y. Jiang, X. Xie, R. Yang, M. Cao, Toward understanding the enhanced pseudocapacitive storage in 3D SnS/MXene architectures enabled by engineered surface reactions, *Chem-Eur J.* 2020, 26, 11231.

[58] B. Shao, Z. Liu, G. Zeng, H. Wang, Q. Liang, Q. He, M. Cheng, C. Zhou, L. Jiang, B. Song, Two-dimensional transition metal carbide and nitride (MXene) derived quantum dots (QDs): synthesis, properties, applications and prospects, *J Mater Chem A.* 2020, 8, 7508.

[59] Z. Guo, X. Zhu, S. Wang, C. Lei, Y. Huang, Z. Nie, S. Yao, Fluorescent Ti_3C_2 MXene quantum dots for an alkaline phosphatase assay and embryonic stem cell identification based on the inner filter effect, *Nanoscale.* 2018, 10, 19579.

[60] X. Li, F. Liu, D. Huang, N. Xue, Y. Dang, M. Zhang, L. Zhang, B. Li, D. Liu, L. Wang, H. Liu, X. Tao, Nonoxidized MXene quantum dots prepared by microexplosion method for cancer catalytic therapy, *Adv Funct Mater.* 2020, 30, 2000308.

[61] F. Ran, T. Wang, S. Chen, Y. Liu, L. Shao, Constructing expanded ion transport channels in flexible MXene film for pseudocapacitive energy storage, *Appl Surf Sci.* 2020, 511, 145627.

[62] X. Jiang, A.V. Kuklin, A. Baev, Y. Ge, H. Ågren, H. Zhang, P.N. Prasad, Two-dimensional MXenes: from morphological to optical, electric, and magnetic properties and applications, *Phys Rep.* 2020, 848, 1.

[63] H. Hadipour, Y. Yekta, Ab initio study of the effective Coulomb interactions and Stoner ferromagnetism in M_2C and $M_2\ CO_2$ MX– enes (M= Sc, Ti, V, Cr, Fe, Zr, Nb, Mo, Hf, Ta), *Phys Rev B.* 2019, 100, 195118.

[64] Z. Jing, H. Wang, X. Feng, B. Xiao, Y. Ding, K. Wu, Y. Cheng, Superior thermoelectric performance of ordered double transition metal MXenes: $Cr_2TiC_2T_2$ (T = –OH or –F), *J Phys Chem Lett.* 2019, 10, 5721.

[65] M. Iqbal, J. Fatheema, Q. Noor, M. Rani, M. Mumtaz, R.K. Zheng, S.A. Khan, S. Rizwan, Co-existence of magnetic phases in two-dimensional MXene, *Mater Today Chem.* 2020, 16, 100271.

[66] P. Zhang, L. Wang, Z. Huang, J. Yu, Z. Li, H. Deng, T. Yin, L. Yuan, J.K. Gibson, L. Mei, L. Zheng, H. Wang, Z. Chai, W. Shi, Aryl diazonium-assisted amidoximation of MXene for boosting water stability and uranyl sequestration via electrochemical sorption, *ACS Appl Mater Inter.* 2020, 12, 15579.

[67] F. Wang, C. Yang, M. Duan, Y. Tang, J. Zhu, TiO_2 nanoparticle modified organ-like Ti_3C_2 MXene nanocomposite encapsulating hemoglobin for a mediator-free biosensor with excellent performances, *Biosens Bioelectron.* 2015, 74, 1022.

[68] R.P. Pandey, K. Rasool, V.E. Madhavan, B. Aissa, Y. Gogotsi, K.A. Mahmoud, Ultrahigh-flux and fouling-resistant membranes based on layered silver/MXene ($Ti_3C_2T_x$) nanosheets, *J Mater Chem A.* 2018, 6, 3522.

[69] J. Zheng, B. Wang, A. Ding, B. Weng, J. Chen, Synthesis of MXene/DNA/Pd/Pt nanocomposite for sensitive detection of dopamine, *J Electroanal Chem.* 2018, 816, 189.

[70] R. Tang, S. Zhou, C. Li, R. Chen, L. Zhang, Z. Zhang, L. Yin, Janus-Structured Co-Ti_3C_2 MXene quantum dots as a Schottky catalyst for high-performance photoelectrochemical water oxidation, *Adv Funct Mater.* 2020, 30, 2000637.

[71] Y. He, L. Ma, L. Zhou, G. Liu, Y. Jiang, J. Gao, Preparation and application of bismuth/MXene nano-composite as electrochemical sensor for heavy metal ions detection, *Nanomaterials.* 2020, 10, 866.

[72] H. Lin, X. Wang, L. Yu, Y. Chen, J. Shi, Insights into 2D MXenes for versatile biomedical applications: current advances and challenges ahead, *Nano Lett.* 2017, 17, 384.
[73] C. Ding, J. Liang, Z. Zhou, Y. Li, W. Peng, G. Zhang, F. Zhang, X. Fan, Photothermal enhanced enzymatic activity of lipase covalently immobilized on functionalized $Ti_3C_2T_X$ nanosheets, *Chem Eng J.* 2019, 378, 122205.
[74] J. Chen, K. Chen, D. Tong, Y. Huang, J. Zhang, J. Xue, Q. Huang, T. Chen, CO_2 and temperature dual responsive "smart" MXene phases, *Chem Commun.* 2015, 51, 314.
[75] Q. Yang, H. Yin, T. Xu, D. Zhu, J. Yin, Y. Chen, X. Yu, J. Gao, C. Zhang, Y. Chen, Y. Gao, Engineering 2D mesoporous Silica@ MXene-integrated 3D-printing scaffolds for combinatory osteosarcoma therapy and NO-augmented bone regeneration, *Small.* 2020, 16, 1906814.
[76] H. Lin, Y. Wang, S. Gao, Y. Chen, J. Shi, Theragnostic 2D Tantalum Carbide (MXene), *Adv Mater.* 2018, 30, 1703284.
[77] H. Liu, C. Duan, C. Yang, W. Shen, F. Wang, Z. Zhu, A novel nitrite biosensor based on the direct electrochemistry of hemoglobin immobilized on MXene-Ti_3C_2, *Sens Actuators B.* 2015, 218, 60.
[78] A. Jastrzębska, A. Szuplewska, T. Wojciechowski, M. Chudy, W. Ziemkowska, L. Chlubny, A. Rozmysłowska, A. Olszyna, In vitro studies on cytotoxicity of delaminated Ti_3C_2 MXene, *J Hazard Mater.* 2017, 339, 1.
[79] K. Rasool, K.A. Mahmoud, D.J. Johnson, M. Helal, G.R. Berdiyorov, Y. Gogotsi, Efficient antibacterial membrane based on two-dimensional $Ti_3C_2T_x$ (MXene) nanosheets, *Sci Rep.* 2017, 7, 1598.
[80] S.M. George, B. Kandasubramanian, Advancements in MXene-Polymer composites for various biomedical applications, *Ceram Int.* 2020, 46, 8522.
[81] M. Mohammadniaei, A. Koyappayil, Y. Sun, J. Min, M.H. Lee, Gold nanoparticle/MXene for multiple and sensitive detection of oncomiRs based on synergetic signal amplification, *Biosens Bioelectron.* 2020, 159, 112208.
[82] F. Duan, C. Guo, M. Hu, Y. Song, M. Wang, L. He, Z. Zhang, R. Pettinari, L. Zhou, Construction of the 0D/2D heterojunction of $Ti_3C_2T_x$ MXene nanosheets and iron phthalocyanine quantum dots for the impedimetric aptasensing of microRNA-155, *Actuat B-Chem.* 2020, 310, 127844.
[83] L. Liu, Y. Wei, S. Jiao, S. Zhu, X. Liu, A novel label-free strategy for the ultrasensitive miRNA-182 detection based on MoS2/Ti3C2 nanohybrids, *Biosens Bioelectron.* 2019, 137, 45.
[84] S. Hroncekova, T. Bertok, M. Hires, E. Jane, L. Lorencova, A. Vikartovska, A. Tanvir, P. Kasak, J. Tkac, Ultrasensitive $Ti_3C_2T_X$ MXene/chitosan nanocomposite-based amperometric biosensor for detection of potential prostate cancer marker in urine samples, *Processes.* 2020, 8, 580.
[85] H. Wang, J. Sun, L. Lu, X. Yang, J. Xia, F. Zhang, Z. Wang, Competitive electrochemical aptasensor based on a cDNA-ferrocene/MXene probe for detection of breast cancer marker Mucin1, *Anal Chim Acta.* 2020, 1094, 18.
[86] H. Zhang, Z. Wang, Q. Zhang, F. Wang, Y. Ti_3C_2 MXenes nanosheets catalyzed highly efficient electrogenerated chemiluminescence biosensor for the detection of exosomes, *Bioelectron.* 2019, 124, 184.
[87] Q. Wu, N. Li, Y. Wang, Y. Liu, Y. Xu, S. Wei, J. Wu, G. Jia, X. Fang, F. Chen, X. Cui, A 2D transition metal carbide MXene-based SPR biosensor for ultrasensitive carcinoembryonic antigen detection, *Biosens Bioelectron.* 2019, 144, 111697.
[88] Y. Li, Z. Kang, L. Kong, H. Shi, Y. Zhang, M. Cui, D. Yang, MXene-Ti_3C_2/CuS nanocomposites: enhanced peroxidase-like activity and sensitive colorimetric cholesterol detection, *Mat Sci Eng C.*2019, 104, 110000.
[89] G. Cai, Z. Yu, P. Tong, D. Tang, Ti_3C_2 MXene quantum dot-encapsulated liposomes for photothermal immunoassays using a portable near-infrared imaging camera on a smartphone, *Nanoscale.* 2019, 11, 15659.

[90] Y. Ma, Y. Yue, H. Zhang, F. Cheng, W. Zhao, J. Rao, S. Luo, J. Wang, X. Jiang, Z. Liu, N. Liu, Y. Gao, 3D synergistical MXene/reduced graphene oxide aerogel for a piezoresistive sensor, *Acs Nano*. 2018, 12, 3209.

[91] Y. Ma, N. Liu, L. Li, X. Hu, Z. Zou, J. Wang, S. Luo, Y. Gao, A highly flexible and sensitive piezoresistive sensor based on MXene with greatly changed interlayer distances, *Nat Commun*. 2017, 8, 1207.

[92] Y. Cai, J. Shen, G. Ge, Y. Zhang, W. Jin, W. Huang, J. Shao, J. Yang, X. Dong, Stretchable $Ti_3C_2T_x$ MXene/carbon nanotube composite based strain sensor with ultrahigh sensitivity and tunable sensing range, *Acs Nano*. 2018, 12, 56.

[93] Y. Guo, M. Zhong, Z. Fang, P. Wan, G. Yu, A wearable transient pressure sensor made with MXene nanosheets for sensitive broad-range human–machine interfacing, *Nano Lett*. 2019, 19, 1143.

[94] H. Lin, Y. Chen, J. Shi, Insights into 2D MXenes for versatile biomedical applications: current advances and challenges ahead, *Adv Sci*. 2018, 5, 1800518.

[95] Z. Xie, S. Chen, Y. Duo, Y. Zhu, T. Fan, Q. Zou, M. Qu, Z. Lin, J. Zhao, Y. Li, L. Liu, S. Bao, H. Chen, D. Fan, H. Zhang, Biocompatible two-dimensional titanium nanosheets for multimodal imaging-guided cancer theranostics, *ACS Appl Mater Inter*. 2019, 11, 22129.

[96] D.-Y. Zhang, H. Xu, T. He, M.R. Younis, L. Zeng, H. Liu, C. Jiang, J. Lin, P. Huang, Cobalt carbide-based theranostic agents for in vivo multimodal imaging guided photothermal therapy, *Nanoscale*. 2020, 12, 7174.

[97] Y. Cheng, F. Yang, G. Xiang, K. Zhang, Y. Cao, D. Wang, H. Dong, X. Zhang, Ultrathin tellurium oxide/ammonium tungsten bronze nanoribbon for multimodality imaging and second near-infrared region photothermal therapy, *Nano Lett*. 2019, 19, 1179.

[98] C. Dai, Y. Chen, X. Jing, L. Xiang, D. Yang, H. Lin, Z. Liu, X. Han, R. Wu, Two-dimensional tantalum carbide (MXenes) composite nanosheets for multiple imaging-guided photothermal tumor ablation, *ACS Nano*. 2017, 11, 12696.

[99] H. Yin, X. Guan, H. Lin, Y. Pu, Y. Fang, W. Yue, B. Zhou, Q. Wang, Y. Chen, H. Xu, Nanomedicine-enabled photonic thermogaseous cancer therapy, *Adv Sci*. 2020, 7, 1901954.

[100] M. Soleymaniha, M.A. Shahbazi, A.R. Rafieerad, A. Maleki, A. Amiri, Promoting role of MXene nanosheets in biomedical sciences: therapeutic and biosensing innovations, *Adv Healthc Mater*. 2019, 8, 1801137.

[101] Z. Liu, M. Zhao, H. Lin, C. Dai, C. Ren, S. Zhang, W. Peng, Y. Chen, 2D magnetic titanium carbide MXene for cancer theranostics, *J Mater Chem B*.2018, 6, 3451

[102] Q. Xue, H. Zhang, M. Zhu, Z. Pei, H. Li, Z. Wang, Y. Huang, Y. Huang, Q. Deng, J. Zhou, S. Du, Q. Huang, C. Zhi, Photoluminescent Ti_3C_2 MXene quantum dots for multicolor cellular imaging, *Adv Mater*. 2017, 29, 1604847.

[103] G.P. Awasthi, B. Maharjan, S. Shrestha, D.P. Bhattarai, D. Yoon, C.H. Park, C.S. Kim, Synthesis, characterizations, and biocompatibility evaluation of polycaprolactone—MXene electrospun fibers, *Colloids Surf A*.2020, 586, 124282.

[104] G. Ye, Z. Wen, F. Wen, X. Song, L. Wang, C. Li, Y. He, S. Prakash, X. Qiu, Mussel-inspired conductive Ti_2C-cryogel promotes functional maturation of cardiomyocytes and enhances repair of myocardial infarction, *Theranostics*. 2020, 10, 2047.

[105] J. Zhang, Y. Fu, A. Mo, Multilayered titanium carbide MXene film for guided bone regeneration, *Int J Nanomedicine*. 2019, 14, 10091.

[106] H. Rastin, B. Zhang, A. Mazinani, K. Hassan, J. Bi, T.T. Tung, D. Losic, 3D bioprinting of cell-laden electroconductive MXene nanocomposite bioinks, *Nanoscale*. 2020, 12, 16069.

[107] A. Rafieerad, W. Yan, G.L. Sequiera, N. Sareen, E. Abu-El-Rub, M. Moudgil, S. Dhingra, Quantum dots: application of Ti_3C_2 MXene quantum dots for immunomodulation and regenerative medicine, *Adv. Healthc. Mater*. 2019, 8, 1900569.

[108] K. Chen, Y. Chen, Q. Deng, S.-H. Jeong, T.-S. Jang, S. Du, H.-E. Kim, Q. Huang, C.-M. Han, Strong and biocompatible poly (lactic acid) membrane enhanced by $Ti_3C_2T_z$ (MXene) nanosheets for Guided bone regeneration, *Mater Lett*. 2018, 229, 114.

[109] N. Driscoll, K. Maleski, A.G. Richardson, B. Murphy, B. Anasori, T.H. Lucas, Y. Gogotsi, F. Vitale, Fabrication of Ti_3C_2 MXene microelectrode arrays for in vivo neural recording, *J Vis Exp*. 2020, 156, e60741.

[110] N. Driscoll, A.G. Richardson, K. Maleski, B. Anasori, O. Adewole, P. Lelyukh, L. Escobedo, D.K. Cullen, T.H. Lucas, Y. Gogotsi, F. Vitale, Two-dimensional Ti_3C_2 MXene for high-resolution neural interfaces, *ACS Nano*. 2018, 12, 10419.

[111] K. Rasool, M. Helal, A. Ali, C.E. Ren, Y. Gogotsi, K.A. Mahmoud, Antibacterial activity of $Ti_3C_2T_x$ MXene, *ACS Nano*. 2016, 10, 3674.

[112] A.A. Shamsabadi, M.S. Gh, B. Anasori, M. Soroush, Antimicrobial mode-of-action of colloidal $Ti_3C_2T_x$ MXene nanosheets, *ACS Sustainable Chem Eng*. 2018, 6, 16586–16596.

[113] H. Feng, W. Wang, M. Zhang, S. Zhu, Q. Wang, J. Liu, S. Chen, 2D titanium carbide-based nanocomposites for photocatalytic bacteriostatic applications, *Appl Catal B-Environ*. 2020, 266, 118609.

[114] S. Zada, W. Dai, Z. Kai, H. Lu, X. Meng, Y. Zhang, Y. Cheng, F. Yan, P. Fu, X. Zhang, H. Dong, Algae extraction controllable delamination of vanadium carbide nanosheets with enhanced near-infrared photothermal performance, *Angew Chem Int Edit*. 2020, 132, 6663.

[115] L. Tan, Z. Zhou, X. Liu, J. Li, Y. Zheng, Z. Cui, X. Yang, Y. Liang, Z. Li, X. Feng, S. Zhu, K.W.K. Yeung, C. Yang, X. Wang, S. Wu, Overcoming multidrug-resistant MRSA using conventional aminoglycoside antibiotics, *Adv Sci*. 2020, 7, 1902070.

[116] K. Huang, Z. Li, J. Lin, G. Han, P. Huang, Two-dimensional transition metal carbides and nitrides (MXenes) for biomedical applications, *Chem Soc Rev*. 2018, 47, 5109.

[117] G. Liu, J. Zou, Q. Tang, X. Yang, Y. Zhang, Q. Zhang, W. Huang, P. Chen, J. Shao, X. Dong, Surface modified Ti_3C_2 MXene nanosheets for tumor targeting photothermal/photodynamic/chemo synergistic therapy, *ACS Appl Mater Inter*. 2017, 9, 40077.

[118] H. Xiang, H. Lin, L. Yu, Y. Chen, Hypoxia-irrelevant photonic thermodynamic cancer nanomedicine, *ACS Nano*. 2019, 13, 2223.

[119] C. Xing, S. Chen, X. Liang, Q. Liu, M. Qu, Q. Zou, J. Li, H. Tan, L. Liu, D. Fan, H. Zhang, Two-dimensional MXene (Ti_3C_2)-Integrated cellulose hydrogels: toward smart three-dimensional network nanoplatforms exhibiting light-induced swelling and bimodal photothermal/chemotherapy anticancer activity, *ACS Appl Mater Inter*. 2018, 10, 27631.

[120] K. Rasool, M. Helal, A. Ali, C.E. Ren, Y. Gogotsi, K.A. Mahmoud, Antibacterial activity of $Ti_3C_2T_x$ MXene, *ACS Nano*. 2016, 10, 3674.

[121] K. Rasool, K.A. Mahmoud, D.J. Johnson, M. Helal, G.R. Berdiyorov, Y. Gogotsi, Efficient antibacterial membrane based on two-dimensional $Ti_3C_2T_x$ (MXene) nanosheets, *Sci Rep*. 2017, 7, 1598.

[122] G.K. Nasrallah, M. Al-Asmakh, K. Rasool, K.A. Mahmoud, Ecotoxicological assessment of $Ti_3C_2T_x$ (MXene) using a zebrafish embryo model, *Environ Sci Nano*. 2018, 5, 1002.

[123] A.M. Jastrzebska, A. Szuplewska, T. Wojciechowski, M. Chudy, W. Ziemkowska, L. Chlubny, A. Rozmyslowska, A. Olszyna, In vitro studies on cytotoxicity of delaminated Ti_3C_2 MXene, *J Hazard Mater*. 2017, 339, 1.

[124] B. Scheibe, J.K. Wychowaniec, M. Scheibe, B. Peplinska, M. Jarek, G. Nowaczyk, L. Przysiecka, Cytotoxicity assessment of Ti–Al–C based MAX phases and $Ti_3C_2T_x$ MXenes on human fibroblasts and cervical cancer cells, *ACS Biomater Sci Eng*. 2019, 5, 6557.

9 Nanobiomaterials of Human Bones for Anatomy and Physiology

Siwaluk Srikrajang and Won-Chun Oh

9.1 INTRODUCTION

Musculoskeletal disorder is a major concern in the public health system, affecting approximately 17.71 billion people worldwide in various age ranges of the general population (1). Abnormality of the bones, joints, and skeletal supporting tissues can be a source of orthopedic disease, leading to pain and functional and activity disturbance (2–4). Due to the various factors that can cause bones and supporting tissues to have defects, such as congenital abnormalities, infections, trauma, neoplasms, or degenerative changes caused by aging (5), recent medical approaches have recognized that bionanomaterials are an important technology for clinical implications in the treatment of bone defects as well as injuries to supportive tissues (6), such as bone implants, bone grafting, and bone cell and tissue growth engineering. Therefore, this chapter aims to summarize an updated review of bionanomaterials' utility in the treatment of skeletal system-related disease. The normal anatomy and physiology of the bones and supportive structure are described for understanding bone normal structure, bone supportive tissues, bone microstructure, bone growth and remodeling processes, and bone mineralization. Finally, concepts of biomaterials for tissue replacement, nanotechnology for tissue replacement, and biomaterials for bone grafting and joint replacement are summarized in order to establish the currently understood implications of bionanomaterials in clinical settings using background knowledge of bone anatomy and physiology, both in general and in in-depth explanations.

9.2 ANATOMY OF THE BONE

The skeletal system provides three main functions for the human body: support, protection, and movement. Human bones are varied in shape and size, depending on specific function. For example, long bones in the arms and legs are appropriate to support weight and generate movement, while flat bones, such as the sternum, ribs and skull, provide the main function of protecting cardinal internal organs (brain, lungs, and heart). The structure of bones includes several important components, including the periosteum (a tough, fibrous membrane that covers the bone), the

DOI: 10.1201/9781003425427-12

compact bone (a dense and hard outermost layer of bone tissue), the spongy bone (a porous and honeycomb-like structure of an inner layer of bone tissue), the bone marrow (soft tissue that fills the inner cavities of long bones and produces red blood cells [erythrocytes], white blood cells [granulocytes, monocytes, lymphocytes], and platelets) (7), and various joints and connective tissues that allow for movement and flexibility. Overall, the bone structure is a complex and essential component of the human body. The following sections explain the normal anatomical structure and physiological functions of the human bone in the skeletal system.

9.2.1 The Human Skeleton System

The human skeleton system is composed of bones, joints, muscles, and connective tissues (i.e., fascia, ligaments, tendons, and cartilages). The complex skeletal structure in the human body is made up of 206 bones, which are divided into two sections: the axial skeleton (80 bones) and appendicular skeleton (126 bones). The axial skeleton is defined as "the bones of the long body axis," including the skull, spine (vertebral column, spinal column), sternum (chest bone), and ribs (thoracic cage). The appendicular skeleton includes the arms and leg bones, as well as the bones that connect the limbs to the axial skeleton.

The upper appendicular skeletal system includes the "shoulder girdle" and the arm bones. The shoulder girdle is a complex structure that connects the arms to the trunk and allows for their stability and motions. The clavicle (collar bone), the scapula (shoulder blade), and the head of the upper arm bone, called the humerus, form together as the shoulder girdle or the shoulder complex. Arms in anatomical definition are classified into three sections: the arm (upper arm), forearm (lower arm), and hand. The arms are also called the upper extremities or upper limbs in anatomical terms. Similarly, the lower appendicular skeleton consists of the "pelvic girdle" and the leg bones. Leg bones are also divided into three sections: leg (upper leg/thigh), foreleg (lower leg), and foot. In the same way, legs are called the lower extremities or lower limbs in anatomical terms. The pelvic girdle is composed of the coxal bones (hip) and sacrum bone, connecting the vertebral column and leg bones on both sides. Descriptions and illustrations of the human bones are presented in Table 9.1.

9.2.2 The Supportive Skeleton Structures

9.2.2.1 Soft Tissue

9.2.2.1.1 Muscles

Muscle is a contractile soft tissue that accounts for at least 33–40% of body mass (8). There are approximately 600 muscles in the human body from head to toe. According to the skeleton supportive structure, muscles perform four important functions: generating movement in manipulation and locomotion, maintaining body posture, joint stabilization, and body temperature preservation with heat generation. Human muscles can be divided into three types depending on characteristic, locations, and functions: skeletal muscle, smooth muscle, and cardiac muscle (9). Skeletal

TABLE 9.1
Bones of the Axial and Appendicular Skeleton

Body Part	Bones (Both Sides)	Illustration	Structural Function
Axial skeleton			
Skull	— Frontal (1) — Parietal (2) — Temporal (2) — Occipital (1) — Ethmoid (1) — Sphenoid (1) — Facial bones (14)	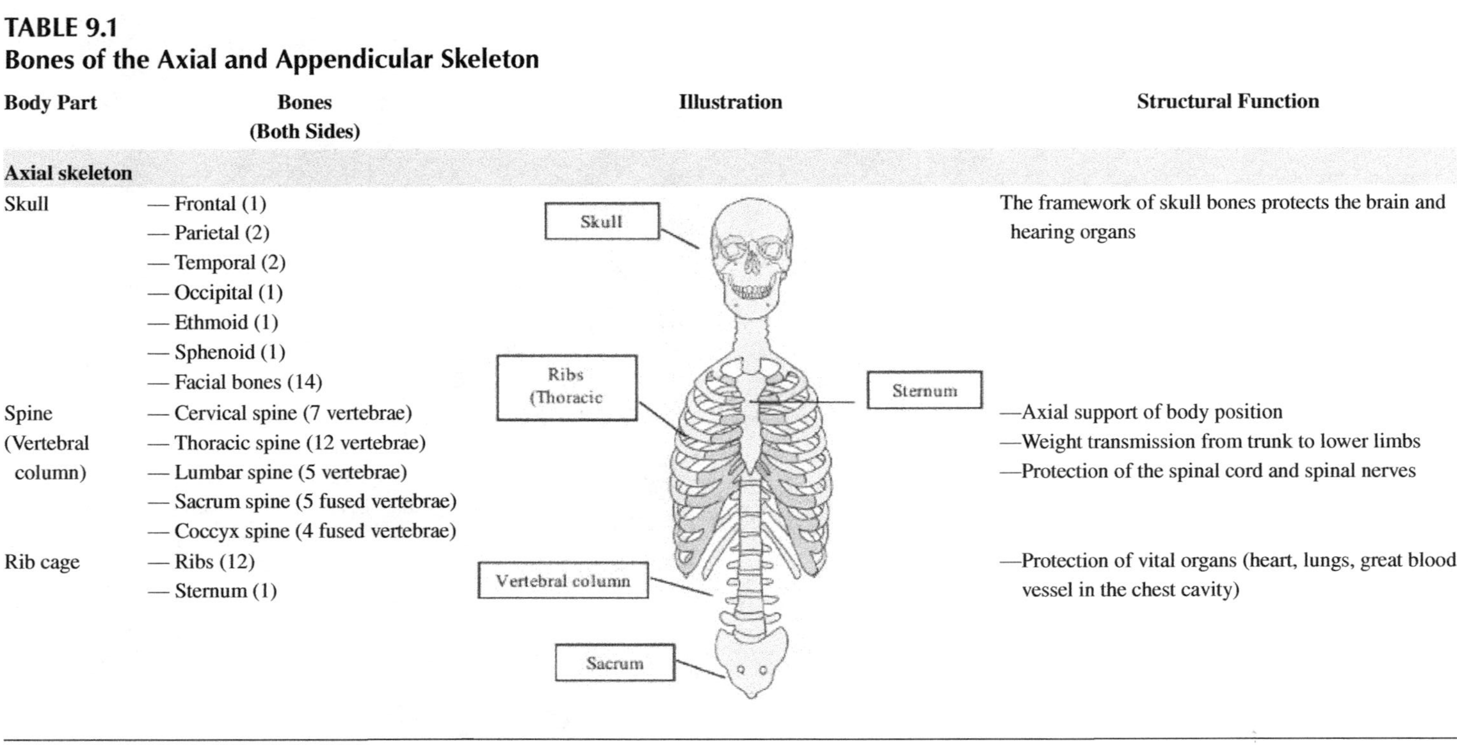	The framework of skull bones protects the brain and hearing organs
Spine (Vertebral column)	— Cervical spine (7 vertebrae) — Thoracic spine (12 vertebrae) — Lumbar spine (5 vertebrae) — Sacrum spine (5 fused vertebrae) — Coccyx spine (4 fused vertebrae)		—Axial support of body position —Weight transmission from trunk to lower limbs —Protection of the spinal cord and spinal nerves
Rib cage	— Ribs (12) — Sternum (1)		—Protection of vital organs (heart, lungs, great blood vessel in the chest cavity)

(Continued)

TABLE 9.1 (Continued)
Bones of the Axial and Appendicular Skeleton

Body Part	Bones (Both Sides)	Illustration	Structural Function
Appendicular skeleton (Upper part)			
Shoulder girdle	— Clavicle (2) or collar bones — Scapula (2) or shoulder blade	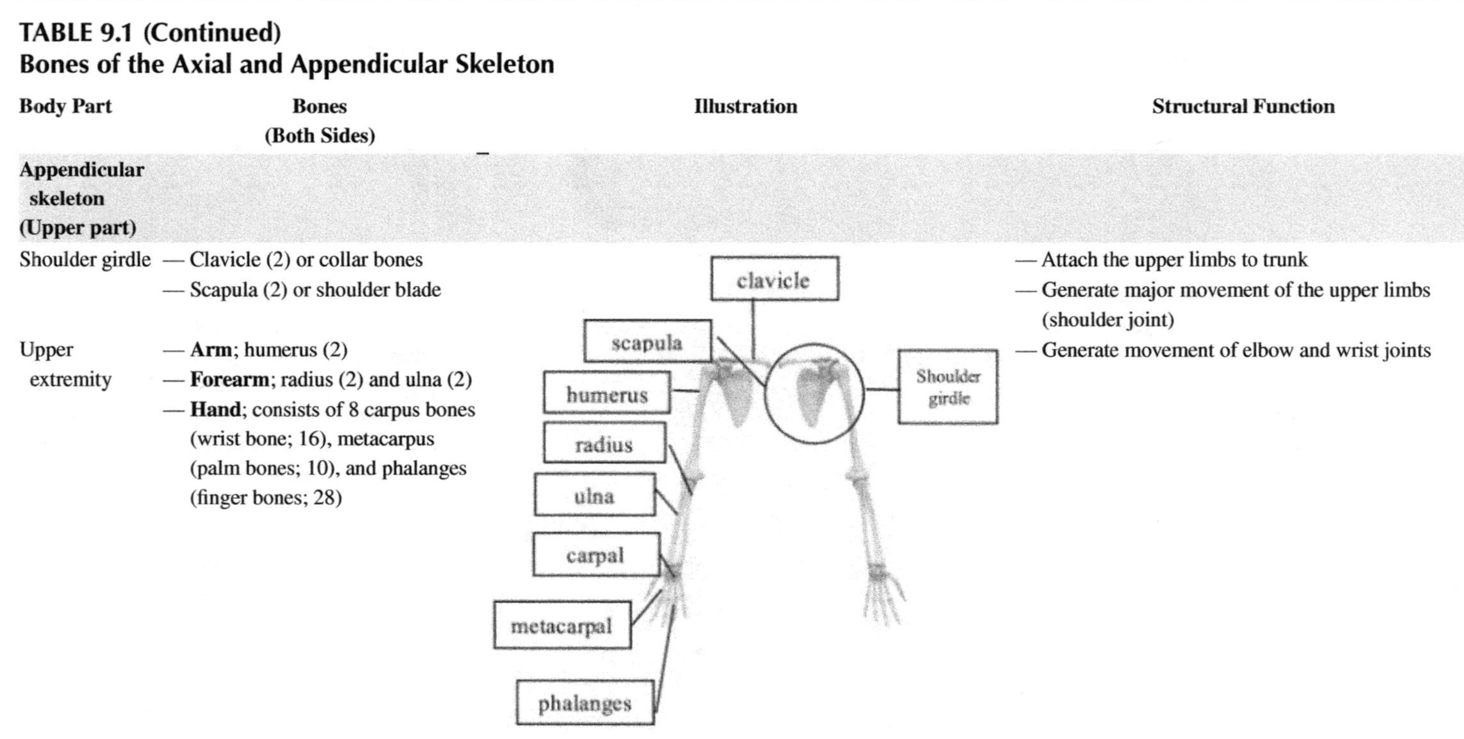	— Attach the upper limbs to trunk — Generate major movement of the upper limbs (shoulder joint)
Upper extremity	— **Arm**; humerus (2) — **Forearm**; radius (2) and ulna (2) — **Hand**; consists of 8 carpus bones (wrist bone; 16), metacarpus (palm bones; 10), and phalanges (finger bones; 28)		— Generate movement of elbow and wrist joints

Appendicular skeleton (Lower part)			
Pelvic (hip) girdle (coxal and sacral bones)	— **Coxal bone**; consists of ilium (1) — Ischium (2) — Pubis (1)	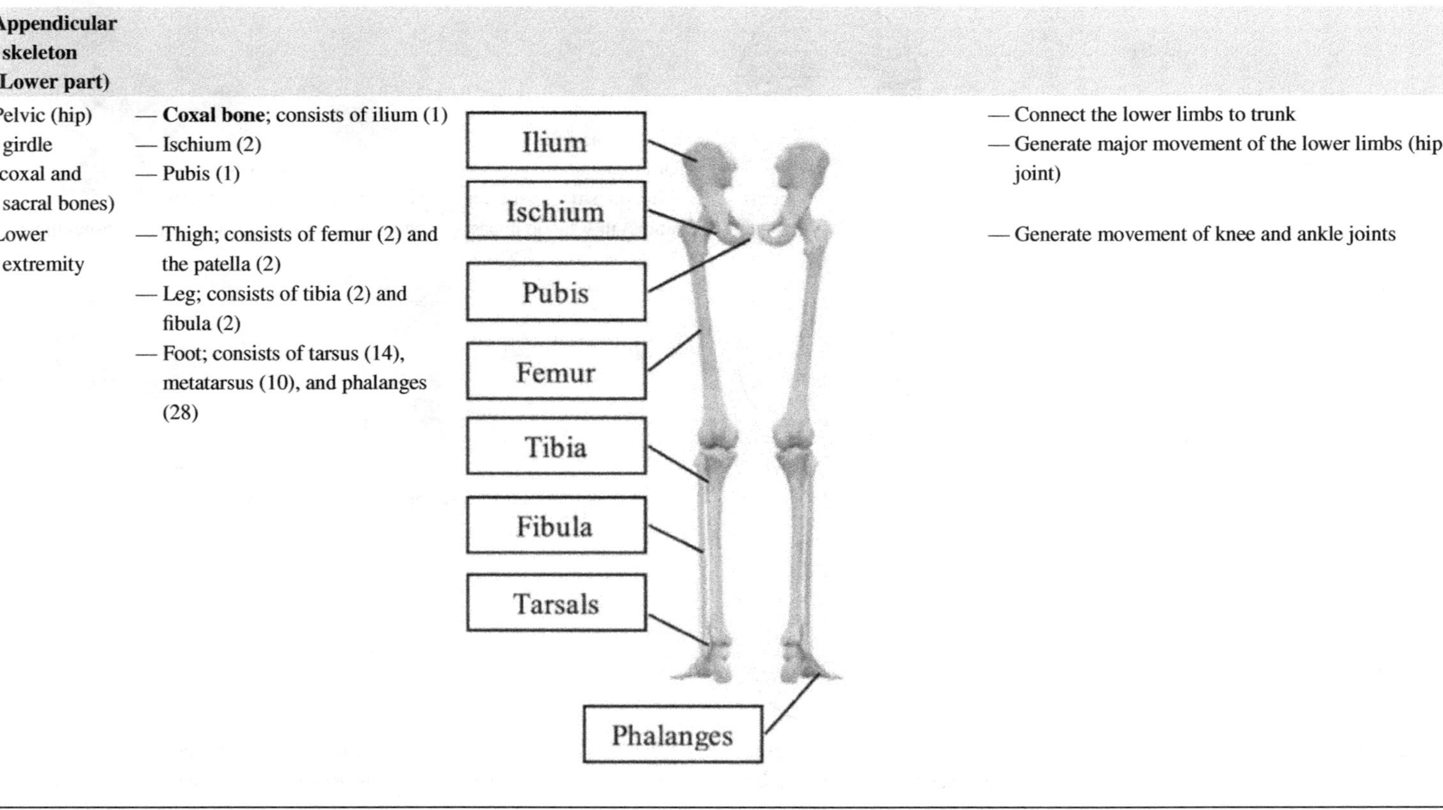	— Connect the lower limbs to trunk — Generate major movement of the lower limbs (hip joint)
Lower extremity	— Thigh; consists of femur (2) and the patella (2) — Leg; consists of tibia (2) and fibula (2) — Foot; consists of tarsus (14), metatarsus (10), and phalanges (28)		— Generate movement of knee and ankle joints

TABLE 9.2
Types of Human Muscle Tissues

Muscle Type	Nervous Control	Location	Characteristics	Function
Skeletal muscle	Voluntary muscle (somatic nervous system control)	Directly attach to the bone	Very long fiber, cylindrical, multinucleate with obviously strips called *striations*	— Slow to fast muscle contraction — Generate conscious movement — Maintain position
Cardiac muscle	Involuntary muscle (autonomic nervous system and hormonal control)	Heart (most in the heart wall)	Long fiber with branching, uni-binucleate and striation 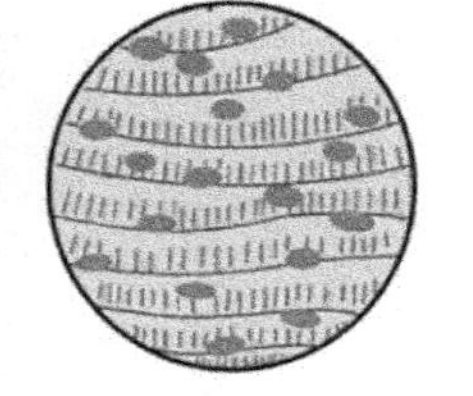	— Slow muscle contraction — Generate heartbeat for blood circulation
Smooth muscle	Involuntary muscle (autonomic nervous system and local chemical control)	Visceral organ wall (stomach, intestine, uterus, urinary bladder, esophagus, bronchi, blood vessels etc.)	Spindle-shapes cell, single nucleus, no striation 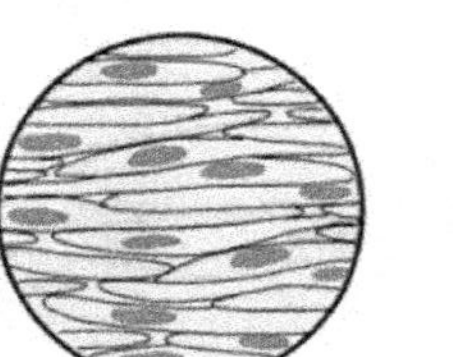	— Very slow muscle contraction — Controlling specific function of internal organs

muscle is primarily responsible for controlling movement and posture. Cardiac muscle regulates heart function. Smooth muscle is present throughout the internal organs. Skeletal muscle is the only one of the three types of human muscle that can be controlled consciously (voluntary control). A description of each muscle type is presented in Table 9.2.

Muscle organization differs between muscle types. For the skeletal muscles, each muscle fiber (or myofiber) is surrounded by connective tissue, which includes epimysium, perimysium, and endomysium. The epimysium is the dense connective tissue that envelopes the entire muscle in the outermost layer. It serves as a defensive barrier and attaches the muscle to tendon and bone. The perimysium surrounds groups of muscle fibers called fascicles; it is categorized as the middle layer. It helps to separate and protect individual muscle fibers within a fascicle. The endomysium, the connective tissue, lies in the innermost layer for surrounding individual muscle fibers (muscle cells). The endomysium has blood arteries and nerves, which transport nutrients and neurological signals to the muscle fibers, are its distinctive characteristics (10). These connective tissue layers work together to hold the muscle fibers in place, provide support and protection, and allow for the transfer of force between muscle fibers and tendons (Figure 9.1) (9).

The microscopic anatomy of a skeletal muscle consists of muscle fibers, which are elongated cells containing myofibrils, and the connective tissue that surrounds and supports the muscle fibers. Myofibrils are composed of actin and myosin proteins, which act as the contractile unit that can generate muscle contraction. The myofibrils, organized into repeating units called sarcomeres, contain a series of thick filaments (myosin) and thin filaments (actin) that slide past each other, or "cross bridge," during muscle contraction. In addition to the myofibrils, muscle fibers also contain other important organelles such as mitochondria, which produce energy for muscle contraction; the sarcoplasmic reticulum, which stores and releases calcium ions for the muscle contraction process; and the sarcolemma, or muscle cell membrane (11).

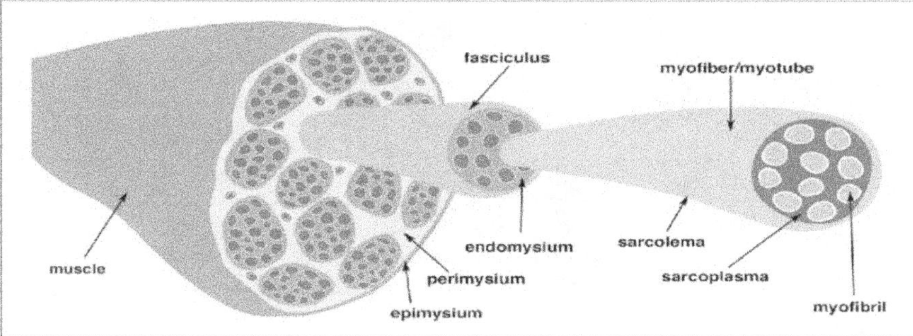

FIGURE 9.1 Muscle microscopic anatomy; epimysium covers the entire muscle. The perimysium covers the fasciculus (bundles of muscle fibers). Endomysium surrounds each muscle fiber. Sarcolemma envelopes individual myofibers (12).

9.2.2.1.2 Synovial Tissue

In the human body, a joint is formed by the articulation of two or more bones with the joint cavity, which is filled with synovial fluid. This type of joint is called a synovial joint. Most of the joints in the human body are synovial joints. The important characteristic of this joint type is freedom of movement due to the lubricant property of synovial fluid. The components of synovial joints are divided into five parts: articular cartilage, articular capsule, joint cavity, synovial membrane, and synovial fluid (13, 14).

Articular cartilage is a smooth type of hyaline cartilage that covers the articulating bone surface of diarthroses and acts as the load absorber of the joint. Articular cartilage is discussed in Section 9.2.2.2. A key feature of synovial joints is the joint cavity, which is located between the diarthrosis bones and filled with synovial fluid for lubrication and load absorption (Figure 9.2). The internal surface of a joint that is not covered by articular cartilage is covered by the synovial membrane, a loose connective tissue that runs perpendicular to the fibrous capsule. Synovial fluid is a viscous liquid that fills the joint cavity. This fluid is filtrated from blood flow in capillaries in the synovial membrane. The major composition of synovial fluid is lubricant elements, including hyaluronic acid, surface active phospholipids, and proteoglycan 4, which are released by synoviocytes (synovial membrane cells) and chondrocytes (articular cartilage cells) (15). The synovial fluid acts as a weight-bearing film to reduce friction between the cartilages. A weeping lubricating mechanism can explain the functioning of synovial fluid. Depending on whether the joint is compressed or relieved during normal joint movement, the fluid is forced into and out of the cartilage. The fluid can lubricate the cartilage surface and bring nourishment to the cartilage area (16).

9.2.2.1.3 Ligaments and Tendons

Ligaments and tendons are both types of connective tissues in the body, but they have slightly different functions and properties. Ligaments are tough, fibrous tissues

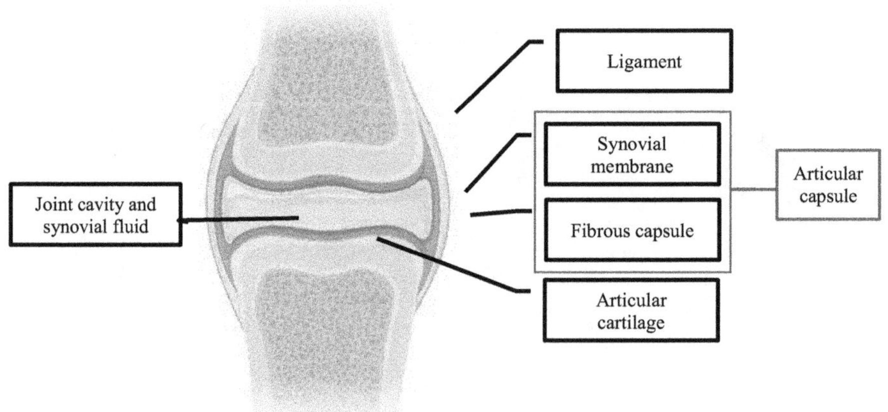

FIGURE 9.2 The synovial joint structure (17).

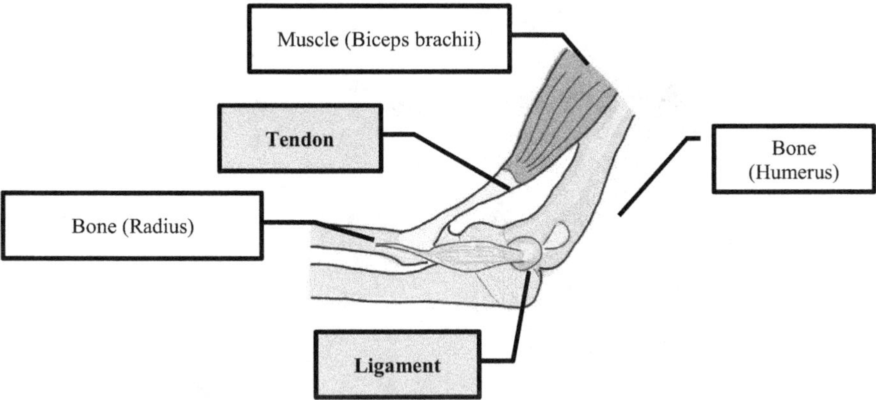

FIGURE 9.3 Ligament (bone-to-bone connection) and tendon (muscle-to-bone connection).

that connect bones to other bones, providing stability for joints. They are made up of collagen fibers and are designed to be strong and relatively inflexible. In order to protect joints from injury, ligaments contribute to limiting excessive joint movement, such as the anterior cruciate ligament of the knee joint, which protects anterior hyper-translation of the knee joint during weight bearing movement. Tendons, on the other hand, are also made of collagen fibers, but they connect muscles to bones, such as the biceps tendon, which attaches the biceps brachii muscle to the radius bone and transfers the force of biceps contraction to move the forearm bone (Figure 9.3). Their main function is to transmit the force generated by the muscle to the bone, enabling movement. Tendons are designed to be flexible and elastic, allowing them to stretch and recoil with the movement of the muscle.

Overall, both ligaments and tendons play important roles in the musculoskeletal system and support and enable movement. However, because of their diverse functions, they have slightly different properties and types of connective tissue.

9.2.2.2 Hard Tissue

9.2.2.2.1 Articular Cartilage

Cartilage is a hydrated and resilient connective tissue that is classified into three types based on component and function: hyaline cartilage, fibrocartilage, and elastic cartilage. In diarthrodial joints, such as the elbows, knees, and ankles, articular cartilage, one of the hyaline cartilages, covers the articular surface of the bones (Figure 9.4). Articular cartilage is an essential structure, providing a lubricated, smooth surface for joint articulation and enabling the transmission of loads with less friction between the bone surface during joint movement. This structure has a 2–4-mm thickness (18) and does not contain vasculature, lymphatic, or neural supply (19). Cartilage is composed of specialized cells called "chondrocytes," which are coated by extracellular matrix (18). Depending on the anatomical area of the articular cartilage, chondrocytes differ in size, shape, and number after developing from mesenchymal stem

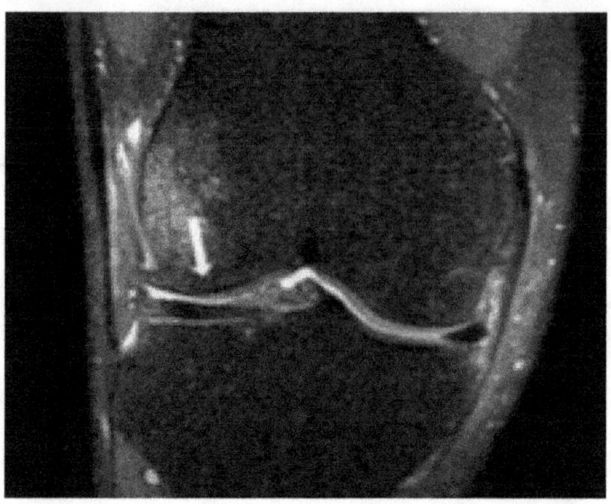

FIGURE 9.4 Magnetic resonance imaging shows normal knee joint articular cartilage (white arrow) (22).

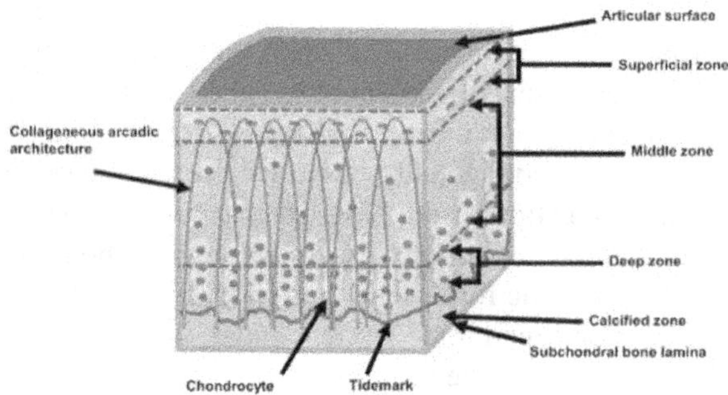

FIGURE 9.5 Zones of articular cartilage (26).

cells (18, 20). Because of the chondrocyte need an optimal microenvironment and chemical components, each chondrocyte creates a specific environment and localizes its own area of extracellular matrix (21).

Articular cartilage is divided into three zones according to collagen ultrastructure: superficial zone, middle zone, and deep zone (23) (Figure 9.5). Below the deep zone, the calcific layer and subchondral cartilage are located for securing the cartilage to bone. The characteristics and function of each zone of articular cartilage are presented in Table 9.3.

Combined solid and liquid materials make up articular cartilage; water and dissolved substances contain about 60–85% of the total wet weight of the cartilage (27), whereas the main solid component of the cartilage is collagen type II (60% of its dry weight), large aggregating proteoglycans (25–30% of its dry weight), specific non-collagenous proteins, and a small amount of proteoglycans (15–20% of its dry weight) (28). The extracellular matrix contains a significant amount of collagen type II (90–95% of all collagen type), which forms fibers that are entangled with

TABLE 9.3
Zones of Articular Cartilage (23–25)

Zone	Depth	Composition	Importance
Superficial zone (tangential zone)	10%–20% of cartilage thickness	—Parallel aligning of collagen types II, IV, and IX to the articular surface —Flattened-shaped chondrocytes —Low proteoglycan content —High water concentration	—The protection and stabilization of deeper layers —Tensile strength (resist tension, sheer, and compressive loads)
Middle zone (transitional zone)	40%–60% of cartilage thickness	—Proteoglycans —Obliquely arranged thickened collagen fibers —Spherical-shaped chondrocytes	Tensile strength (resistance to compressive load)
Deep zone	30%–40% of cartilage thickness	—Proteoglycans —Collagen vertical arranged to the articular surface —Column-shaped chondrocytes —High proteoglycan content —Low water concentration	Tensile strength (greatest resistance to compressive loads)

proteoglycan aggregation. Several collagen subtypes are also investigated as a minor part of cartilage, such as types I, IV, V, VI, IX, X, XI, XII, and XIV (18, 19, 29). The collagen-related functions in the cartilage are tensile properties and cartilage differentiation. In addition, the development of the cartilage growth plate can be found in children and adolescent with prevention of multiple epiphyseal dysplasia (an abnormal end to long bone development), and repair of damaged articular cartilage. Table 9.4 includes descriptions of articular cartilage components.

According to the specific characteristics of the cartilage, which are avascular, aneural, and alymphatic. These properties lead to limited tissue repair after injury (23). In general, chondrocytes maintain and restore articular cartilage; however, as older people, cartilage degrades and the chondrocytes lose their ability to restore and heal the cartilage, resulting in osteoarthritis. Osteoarthritis is a common musculoskeletal disorder, especially in older adults, who account for 10–13% of the population worldwide (31). Osteoarthritis or degenerative joint disease is affected by progressive destruction of articular cartilage followed by abnormal recovery processes such as subchondral bone sclerosis, subchondral bone cysts, hypertrophic spurs, and marginal osteophytes (32) (Figures 9.6 and 9.7). This abnormality of cartilage structure resulting in joint pain, joint movement limitation, joint effusion, and inability to perform effective weight-bearing activities such as prolong standing, walking, and running. Osteoarthritis is a major health concern that can lead to lower physical function and a decreased

TABLE 9.4
Major Components of Articular Cartilage (23, 30)

Component	Characteristics	Related-Function
Water	—Filling in interfibrillar space of collagen fiber —Filling in intracellular space of the matrix	—Transport and distribute nutrients to chondrocytes —Solution of minerals (sodium, calcium, chloride, and potassium) —Provide pressure and frictional resistance to the cartilage
Collagen	Polypeptide chains with triple helix and hydroxyproline providing stability attached via hydrogen bonds	—Form fibril network of extracellular matrix —Provide shear and tensile properties of the extracellular matrix
Proteoglycans (aggrecan, versican, fibromodulin, lumican, biglycan (DS-PGI), decorin (DS-PGII), epiphycan (DS-PGIII),)	A core protein attached with glycosaminoglycan chains via covalent bonds	—Provides osmotic properties to resist compressive load on the articular cartilage — Significant role in collagen fibrillogenesis process
Non-collagenous protein and glycoprotein	Example; cartilage oligomeric matrix protein, fibronectin, elastin, fibrillin, chondroadherin, and so on	—Organization and maintenance of the extracellular matrix macromolecular structure

* DS-PG (I, II, III) = Dermatan sulfate proteoglycan (I, II, III)

quality of life. Especially in severe cases, it can cause muscle wasting and joint deformity (33). Therefore, the engineering articular cartilage is needed for bone surface and cartilage replacement.

9.2.2.2.2 Intervertebral Discs

The intervertebral disc is a specific structure located between the vertebral body of the spinal column (or spine). The function of the intervertebral disc is to enable spinal movement, weight bearing, and flexibility of the spine with the composition of different layers of flexible connective tissues, including the annulus fibrosus, the nucleus pulposus, and the end plates (Figure 9.8) (11).

The annulus fibrosus is the outermost layer of intervertebral disc and is made from fibrocartilage (type I collagen fiber). This layer supports disc stability; limits the expansion of nucleus pulposus during spinal movement; and resists the impact load on the vertebrae, especially in the directions of flexion (bending forward), extension (bending backward), and lateral flexion (lateral bending). In addition, it holds together the vertebrae and distributes the load of the spinal column during body movement. The annulus fibrosus is a vascularized tissue of the intervertebral disc

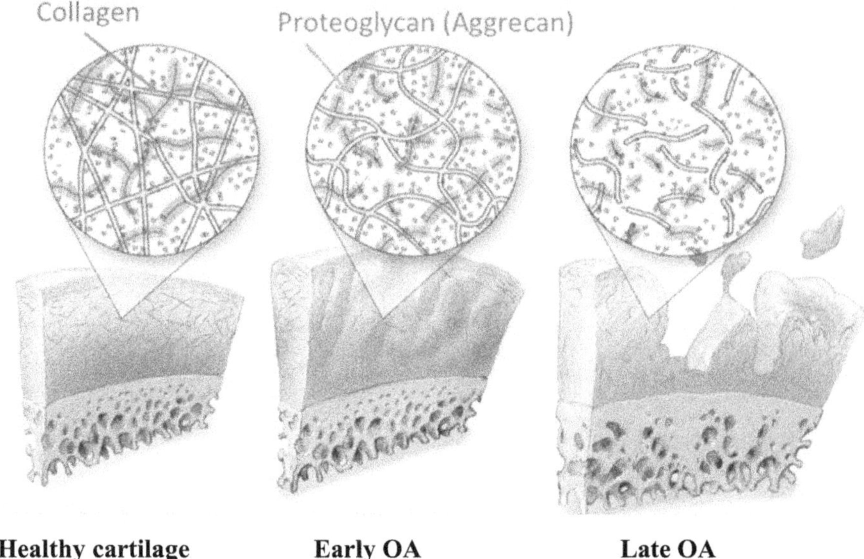

FIGURE 9.6 Articular cartilage in case of osteoarthritis (OA). The degenerative changes in early stages to late stage of OA are progressive dehydration, loss of proteoglycan, and thinning and disruption of collagen. Severe cartilage thinning and eventual detachment of the subchondral bone are indications of late OA (34).

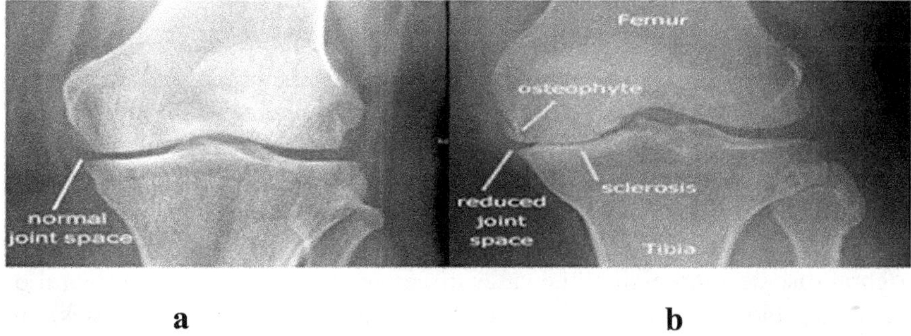

FIGURE 9.7 Demonstration of normal knee joint (a) and osteoarthritis knee joint (b) presenting osteophyte (bone spur), sclerosis, and reduced joint space (35).

that supports nutrient and waste transport. However, the localization of blood vessels in annulus fibrosus varies with age and tissue-damaging characteristics, with vascular ingrowths within the inner layers of the annulus fibrosus being the most common reported in a previous study (36).

The nucleus pulposus is the innermost layer of the intervertebral disc. It is composed of a semifluid substance (hydrated gelatinous tissue) that provides the spine with elasticity and compressibility, allowing for flexibility and movement. The nucleus pulposus components are mainly water and negatively charged proteoglycan,

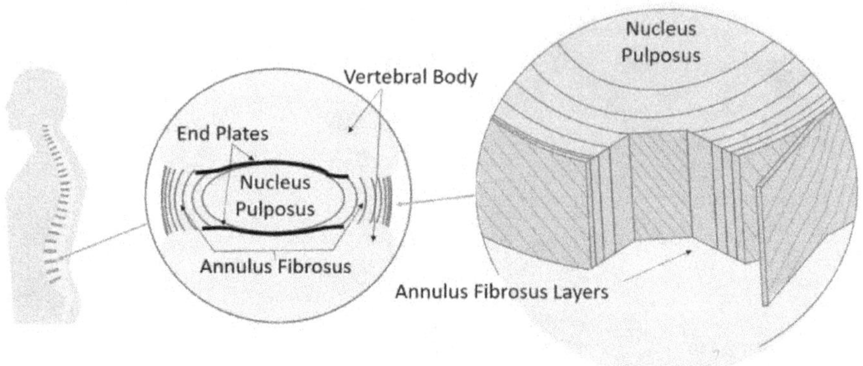

FIGURE 9.8 Intervertebral disc components (39).

with collagen and non-collagenous protein for providing tensile strength of the disc. In the nuclease pulposus, Aggrecan is the large proteoglycan that is commonly found (5–15% of wet tissue weight), as well as small proteoglycans such as decorin, biglycan, fibromodulin, lubrican, and versican (37, 38).

End plates are thin hyaline cartilage layers that separate the intervertebral disc from the adjacent vertebral bodies. They contribute to a smooth surface for disc movement and load transfer between the intervertebral disc and the vertebrae (37).

The intervertebral disc is a major cause of spinal pain from degeneration, microtrauma, and herniation (or displacement) because the intervertebral discs are located between the spinal bodies, which is the location that spinal nerves run through. Anatomical changes in the intervertebral discs can cause nerve root compression and the following symptoms in patients: radiating pain (pain radiating down to the arms in cervical disc pathology and legs in lumbar disc pathology), numbness, and weakness. Indeed, several biomaterial engineering approaches have been developed to restore the hydration of the intervertebral disc using glycosaminoglycan species. Furthermore, according to the surgical technique of intervertebral disc decompression known as discectomy (removing intervertebral disc fragment), a bioengineering strategy may be required for annulus fibrosus repair after discectomy (39).

9.2.2.3 Biomaterials for Tissue Replacement

Artificial implant devices for tissue replacement consist of various materials such as metallic alloys, polymers, ceramics, hydrogels, or composites. As materials are nanosized, nanocomposites are also emerging as major materials in recent years. They are required to have a large number of permissions and widely different properties. Special effects such as resistance of chemical and biochemical degradation and minimal learning of structural components or additions with the exception of drug delivery systems, sutures, and other degradable biomaterials systems are required. Polymer-based implants for specific purposes may not have much mechanical strength of the implants due to surface changes as well as pathological calcification.

TABLE 9.5
Examples of Uses of Tissue Related to Biomaterials

Organ/Tissue	Examples
Heart	Pacemaker, artificial valve, artificial heart
Eye	Contact lens, intraocular lens
Ear	Artificial stapes, cochlea implant
Bone	Bone plate, intramedullary rod, joint prosthesis, bone cement, bone defect repair
Kidney	Dialysis machine
Bladder	Catheter and stent
Muscle	Sutures, muscle stimulator
Circulation	Artificial blood vessels
Skin	Burn dressings, artificial skin
Endocrine	Encapsulated pancreatic islet cells

TABLE 9.6
Disadvantages of Biomaterials Used in the Body

Biomaterial	Disadvantages
Metals	Corrosion
	Density
	Processing
Ceramics	Brittle—fractures easily in tension
Polymers	Not strong
	Not rigid
	Time-dependent deformation (creep; stress relaxation)
	Degradation

On the other hand, released substances are often biologically affected by surrounding tissue or other factors. Due to the heterogeneous biological characteristics of biomaterials, there is a possibility of systemic toxicity and hypersensitivity reactions. Examples of uses of tissue related to biomaterials are listed in Table 9.5.

In long-term aspects, the primary required characteristics of biomaterials for implants should be hemocompatibility (blood compatibility), nontoxicity, durabity, non-irritant to biological tissue, resistant to platelet and thrombosis deposition, non-degradability in physiological settings, non-absorption of blood elements, and not releasing foreign substances. Human-friendly biomaterials used in tissue are largely divided into metals, ceramics, polymers, and natural polymers. Metal materials include stainless steel, cobalt alloys, and titanium alloys, while ceramic materials include aluminum oxide, zirconia, and calcium phosphate. In addition, representative examples of polymer materials include silicon, poly(ethylene), polyvinyl chloride, polyurethanes, and polylactides. Collagen, gelatin, elastin, silk, and polysaccharides

TABLE 9.7
Materials and Applications to Replace Biological Tissues

	Example
Metals	Joint replacement
	Dental roots
	Orthopedic fixation
	Stents
Ceramics	Dental implants
	Orthopedic implants (some)
Polymers	Sutures
	Blood vessels (e.g., vascular grafts)
	Joint socket (knee, shoulder)
	Ear, nose
	Soft tissues in general

are widely used as natural polymer materials. Disadvantages of these materials are listed in Table 9.6.

Materials and applications to replace biological tissues are shown in Table 9.7.

Tissue regeneration using materials using animal biology is a promising field for the future. Typical animal biomaterials are fibrin and collagen. Fibrin consists of nanofibrous structures and fibrils, and fibrin obtained from patients is identified as a personalized approach to treatment. Mader et al. (40) used fibrin sealant implants impregnated with antibiotics to treat bone bacterial infection as an incurable disease. In the rabbit model, antibiotic-containing fibrin facilitated bone regeneration and reconstruction and provided an easy and fast suitable method for antibiotic delivery for orthopedic infections. Antibiotics are distributed in fibrous nanofiber matrices, and the spread of these antibiotics makes it easy for them to be delivered locally and deeply to the diseased bones. The rate of release of antibiotics depends on the type of antibiotic used, but release has been consistent for several days at concentration levels above the minimum concentration required to remove all cases and most common orthopedic pathogens.

As a protein mass, the nanofiber structure is used, and collagen, a super complex, can be found in various protein structures and other species in the ECM. Heterogeneous or homogeneous collagen implants induce immune responses and are useful in some patients, but these collagens are easily available in a relatively pure form. In addition, the use of collagen for implants has the advantage of biomimicry, biodegradability, and being easy to handle (Table 9.8). In addition, it is a biomaterial widely used in biology and medicine because it is possible to cultivate tissues in 3D shapes through their encapsulation. Collagen implants promote cell bonding and adhesion to different biomolecules between homogeneous and heterogeneous cells. Atala et al. (41) used a biodegradable collagen nanofiber skeleton for human bladder design. A human bladder was designed for patients using urinary tract biology and muscle cells obtained from culture-expanded bladder biopsy with culture by encapsulation

TABLE 9.8
Some Natural Resorbable Polymers

Natural Polymers	Applications
Collagen	Artificial skin, coatings to improve cellular adhesion, drug delivery, guided tissue regeneration in dental applications, orthopedic applications, soft tissue augmentation, tissue engineering, scaffold for reconstruction of blood vessels, wound closure
Fibrinogen and fibrin	Tissue sealant
Gelatin	Capsule coating for oral drug delivery, hemorrhage arrester
Cellulose	Adhesion barrier, hemostat
Various polysaccharides such as chitosan, alginate	Drug delivery, encapsulation of cells, sutures, wound dressings
Starch and amylose	Drug delivery

of collagen nanofibers. The advantage of transplantation by binding collagen and autologous cells does not require immunosuppression. Thus, these medical technologies are excellent examples of the medical application of nanotechnology. In the example of the human bladder design process, a new organ must be prepared for transplantation after an incubation period of eight weeks. Through these processes, the transplanted bladder was bound to the stump of the native bladder. In the process of studying such a new organ, it was confirmed that the kidney function of the new organ was restored to normal and metabolic complications did not occur. In addition, one of the important factors, urinary stones, were not formed. Protocol biopsy showed a three-layer structure consisting of submucosal and muscle, that is, lumen of urea cells surrounded by all expected components of normal bladder tissue. As another example, Vacenti and colleagues used collagen with built-in titanium wire instead of PLGA in the first experiment (42). Here, a method for reconstructing ear tissue was developed by combining the advantages of biologically and medically induced collagen nanofibers with the mechanical properties of bio-inert titanium wires. The initial partial expansion and size reduction occurred after two weeks of in vitro incubation, probably related to the onset of ECM formation. No further reduction was observed in the in vivo tissue for the next six weeks. In humans, it will depend on the treatment conditions, but in this method, it is expected that stronger shrinkage will be exerted by the skin and surrounding tissues depending on the condition.

9.2.3 Nanobiomaterials for Bone Cells and Tissue

9.2.3.1 Osteoblasts

An osteoblast is a type of bone cell that functions to synthesize new bone matrix, support the bone structure, and repair damaged or aged bone surfaces. Osteoblasts accounted for 4–6% of the total bone cells (43), which are developed from osteochondral progenitor cells in the periosteum (a fibrous membrane that covered the outer

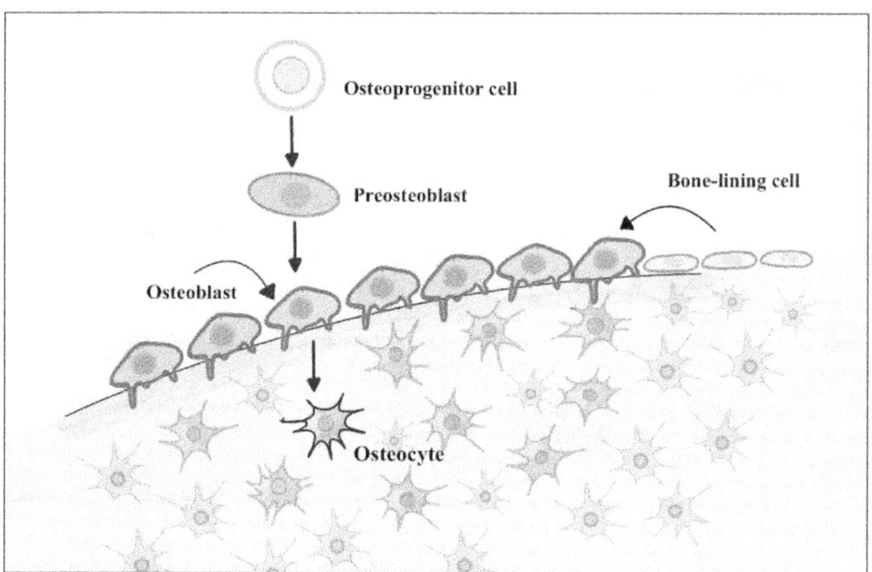

FIGURE 9.9 The development of osteoblasts to osteocytes and bone lining cells.

bone surface) and the bone marrow that forms the bone in a process called bone ossification. Preosteoblasts are formed in bone marrow by mesenchymal stem cells, which then proliferate to become osteoblasts (Figure 9.9). Osteoblasts are divided into active and inactive cells. Active osteoblasts function by creating a bone matrix that covers the surface of the aged bone, followed by the development of bone new layers. The osteoblast cells lie on the bony surface, which is later encircled by the bone matrix once they have undergone maturation to become osteocytes. In the process of bone ossification, osteoblasts produce the osteoid (unmineralized, recently formed bone tissue). Subsequently, the osteoid increases on the matrix, and the cells differentiate into osteocytes. Some osteoblasts survive and differentiate into bone lining cells (or inactive osteoblasts) on the new bone surface, which are programmed to undergo apoptosis (self-programmed cell death). As a result, osteoblasts are typically found in high-metabolism areas where bone formation occurs. The cell organs of osteoblasts have the same characteristics as any other protein-synthesized cell: a large nucleus, secretory vesicles, the Golgi tendon organ, and an extensive endoplasmic reticulum. Several genetic markers and proteins are generated by osteoblasts toward the bone formation surface, such as type I collagen, glycoprotein, bone sialoprotein, alkaline phosphates, osteocalcin, osteopontin, osteonectin, macrophage colony-stimulating factor (M-CSF), and runt-related transcription factor 2 (RUNX2 gene). These diverse genetic markers produced by each osteoblast can explain the differences in microarchitecture of various anatomic skeletal sites. In addition, the different osteoblasts have different responses to mechanical and chemical stimulators. This information is used to indicate osteoblast function to respond to treatment agents in bone disease (43).

9.2.3.2 Osteocytes

An osteocyte is a type of bone cell that is the differentiated form of osteoblasts and becomes embedded in the bone matrix. They are considered the mature form of bone cells that conserve the bone matrix. The regulation of bone remodeling and mineral metabolism is significantly influenced by osteoclasts. The cell body of osteocytes is located in pores called lacunae and extends its filipodium through a network of small canals called canaliculi for connection to other cells and the bone surface. Nutrients, signaling molecules, and waste materials can be exchanged through canaliculi (Figure 9.10) (44). Osteocytes generate several substances to regulate the exchange process of mineral transmission in the lacuno-canalicular network, with the main objective being intercellular adhesion, such as galectin 3, osteocalcin, the cell adhesion receptor for hyaluronate, and bone matrix proteins (43, 45). Osteocytes need to communicate with other cells through the lacuno-canalicular network via the gap junction between cells. The protein connexin is the cellular protein that maintains the gap junction, which allows osteoblasts to be chemically, electrically, and metabolically linked to each other, which is an important cell activity for survival (46).

Osteocytes also play a role in sensing and responding to mechanical loads such as shear force, bending, and stretching of bone by converting the mechanical signal to a biological signal. During transitional periods of external load on the bone, fluid movement in canaliculi induces various responses in osteocytes. This mechanotransduction process needs specific substances, including prostaglandin E2, cyclooxygenase 2, runt-related transcription factor 2, kinase compound, and nitrous oxide. Mechanotransduction is considered a mechanism of bone strength preservation and preventing bone injury. Osteocyte cell death (apoptosis) or dysfunction can lead to

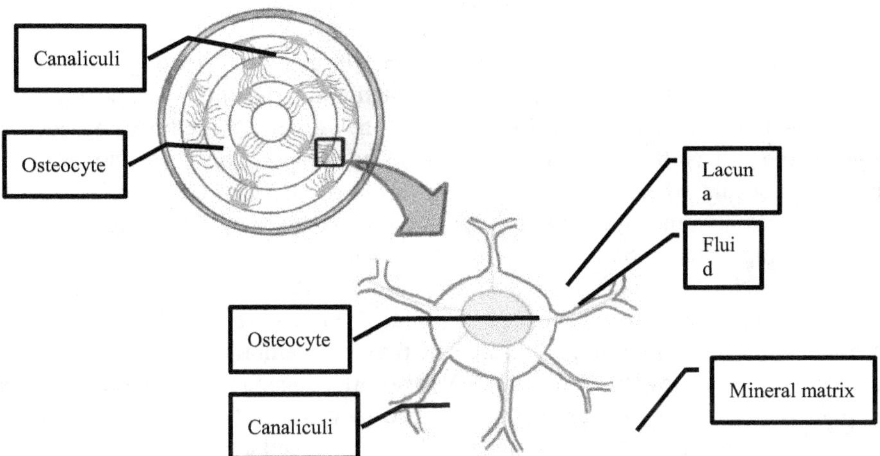

FIGURE 9.10 Sketch of osteocyte with lacuna and canaliculi surrounded by mineral matrix.

bone loss and contribute to bone degenerative diseases such as osteoporosis and osteoarthritis (47). Osteocyte apoptosis can occur from disruption of the gap junction between cells due to age-related degeneration, estrogen hormone deficiency, and glucocorticoid consumption, all of which increase bone resorption. In addition, genetic mutations have been reported to be linked with bone substance alterations and osteocyte malformations in humans, such as in osteogenesis imperfecta (OI) types I and V, X-linked hypophosphatemia, and the sphingomyelin synthase 2 mutation. Also, the therapeutic strategy in osteoporosis may focus on preventing osteoblast apoptosis (43, 48).

9.2.3.3 Osteoclasts and Bone Resorption

An osteoclast is the only type of bone cell that is capable of the resorption or breakdown of bone tissue in both normal and pathologic conditions. The important process of this cell type is osteclastogenesis, which refers to the formation of multinucleated osteoclast cells from the fusion of mononuclear precursors of the monocyte-macrophage lineage found in bone marrow (49). In the process of osteoclast formation, two cytokines that are important for this process are receptor activator of NF-B ligand (RANKL) and macrophage-colony stimulating factor (M-CSF or CSF-1). These two cytokines are derived from marrow stromal cells and osteoblastic precursors in bone marrow. M-CSF/CSF-1 binds to the specific receptor, colony-stimulating factor-1 receptor (c-FMR), and activates signaling through mitogen-activated protein kinases (MAP kinases) and a family of protein-serine/threonine kinases (ERKs) during the early phase of osteoclastogenesis. This activated cytokine is necessary for osteoclast precursor differentiation and proliferation and for cytoskeletal rearrangement during bone resorption. The RANKL binds to the receptor activator of NF-B (RANK), which is its specific receptor, on the formation site. The activating signaling through NF-B, the c-Fos protein, the nuclear factor of activated T cells c1 (NFATc1), and phospholipase C (PLC) occurs to induce differentiation of osteoclast precursors into osteoclasts (43).

Osteoclasts can dissolve the mineralized matrix of bone, releasing calcium and other minerals into the bloodstream and destructing the organic bone matrix, a process known as bone resorption (Figure 9.11). They also regulate bone remodeling, which is the process by which old bone is removed and new bone is developed. Bone resorption occurs when osteoclasts secrete H+ ions and the cathepsin K enzyme, which can result in the digestion of the proteinaceous matrix of the collagen fiber. At the beginning, osteoclasts firmly bind to the bone surface using a specialized actin ring (actin-rich podosomes), cytoskeleton rearrangement, and cellular polarization (50). After that, osteoclasts generate complicated, villus-like membranes termed as "ruffled borders" to maximize the area of the cell membrane that contacting the resorption lacuna (Howship's lacuna) (51). Resorption lacunae are little pits, grooves, or depressions in bone that are being resorbed by osteoclasts. At the ruffled membranes, hydrochloric acid, mediating acidification, as well as various forms of vesicles that contain lysosomal cathepsin, the phosphatase TRAP (tartrate-resistant acid phosphatase), and proteolytic MMPs (matrix metalloproteinases) are released form osteoclasts to change the cell environment to an acidic status, which leads to the dissolution of minerals in the bone matrix (43). Digestion of crystalline hydroxyapatite

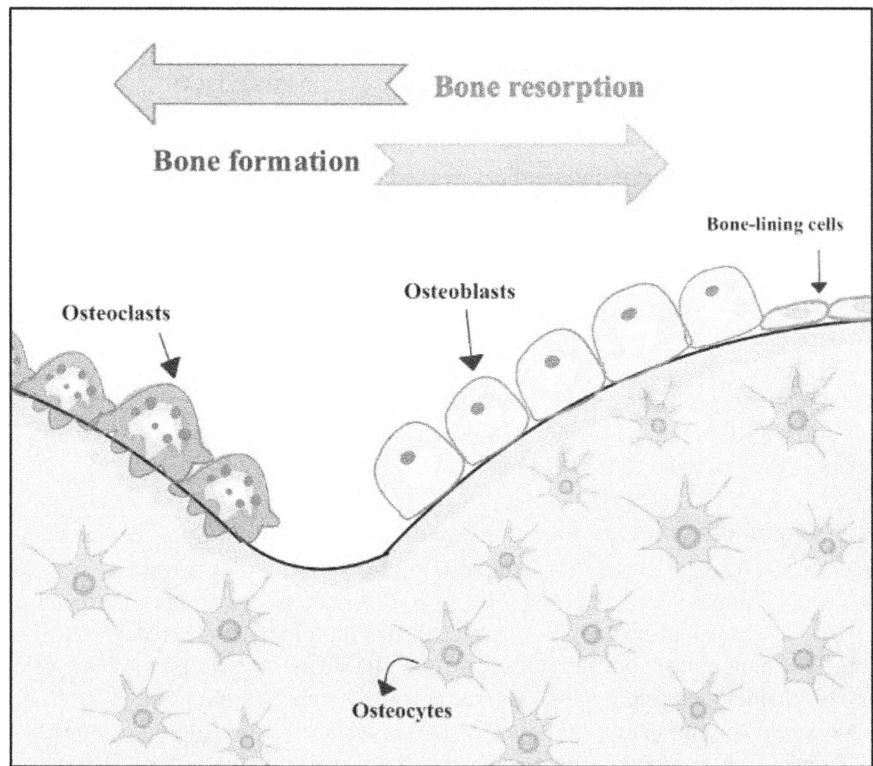

FIGURE 9.11 Demonstration of osteoblasts and osteoclasts functions.

and organic matrix compounds in the local area of the adhesion site can cause the release of great amounts of soluble calcium, phosphate, and bicarbonate, which can be degradable in bone tissue (52).

Osteoclasts are essential for maintaining healthy bones throughout life. However, excessive osteoclast activity can result in bone degenerative diseases and osteoporosis. Osteoclasts are also involved in the regulation of bone density and the bone repair after fractures. In addition, chronic inflammation and hormone deficiency may induce T and B lymphocytes to express RANKL and increase osteoclast formation in a pathologic state such as rheumatoid arthritis, estrogen hormone deficiency, and inflammatory joint disease (53).

9.2.3.4 Bone Matrix

The majority of the bone tissue is composed of the solid material known as the bone matrix. It is made up of both organic and inorganic substances, which provide strength and proper structure to the bones. The organic component of the bone matrix is primarily made up of collagen fibers (30% of the dry weight) to provide tensile strength and elasticity. The inorganic component of the bone matrix is primarily composed of minerals, mostly hydroxyapatite ($Ca_{10}(PO_4)_6(OH)_2$), which is a crystalline compound made up of calcium, phosphorus, and oxygen. Hydroxyapatite makes up about 65% of the bone's dry weight and can be found in teeth and other hard tissues (54). Although the chemical composition of hydroxyapatite is very similar to that of the mineral apatite, which is also found in bones and teeth, hydroxyapatite is distinguished by the presence of hydroxyl (OH) groups in its chemical structure. Aside from its role in dental and bone formation and maintenance, hydroxyapatite has a variety of industrial and medical applications. It's used in dental and orthopedic implants, bone cements, and wound dressings as a filler or coating material. It's also used to treat osteoporosis and bone deficiencies.

Non-collagenous proteins such as osteocalcin and osteopontin are found in the bone matrix and play a role in the regulation of bone formation and remodeling. Osteopontin is a phosphorylated glycoprotein produced by osteoblasts, osteocytes, and other cell types. It regulates bone formation and remodeling by promoting the attachment and proliferation of osteoblasts and inhibiting osteoclast-mediated bone resorption. Osteopontin is also involved in the formation of bone matrix and the mineralization process (55).

9.2.3.5 Nanotechnology for Bone Tissue Engineering

Recently, tissue engineering has been looking forward to many advances in materials, cellular tissue modification, and physiologically active substances (Figure 9.12). In particular, it shows progressive characteristics in the fields of cartilage, eyes, and skin. New nanomaterials are required in various fields such as gums, bone, blood vessels, bladders, muscle tendons, pancreas, cardiac muscle, and nerves.

As one of the important treatments, inner open-wedge high tibial osteoarthritis (MOWHTO) is used for the treatment of inner monocortical knee osteoarthritis (56). In order to treat osteoarthritis, the alignment of the lower extremities can be changed through osteotomy and professional plate fixation. However, MOWHTO has many drawbacks. Examples of such shortcomings include delayed healing of fracture areas, loss of correction lines, hinge fractures, and abnormalities in implant areas (57). Other shortcomings are that they cannot provide sufficient mechanical support to meet patients' needs for initial maximum weight maintenance (Figure 9.13) (58). Recent studies have shown that structural transplants at fracture sites can increase bone healing as a complete condition for early weight maintenance (59–62). Another important problem is economic and time-related problems in which excessive transplantation of artificial bone grafts into the fractured area increases medical costs, surgical trauma, and surgical time. In addition, it is important to determine the optimal location and optimize the structure of the bone graft, as many artificial bone grafts may not fully absorb the fractured area (63), so it is recommended to examine the optimal location of the bone graft and the suitability of the load distribution at the fractured area after examination. The purpose of this measure is to determine the optimal location of the bone graft using finite-element analysis, analyze the load distribution of the bone graft at the optimal location, and present the optimal conditions.

One of the biggest problems is the lack of tissue and organs provided by donors. They are designed to mimic the shape and size of damaged tissue in certain patients. Consisting of bio-affinity materials that can be designed to mimic the shape and size of damaged tissue, scaffolds are one of the potential marketable areas for personalized

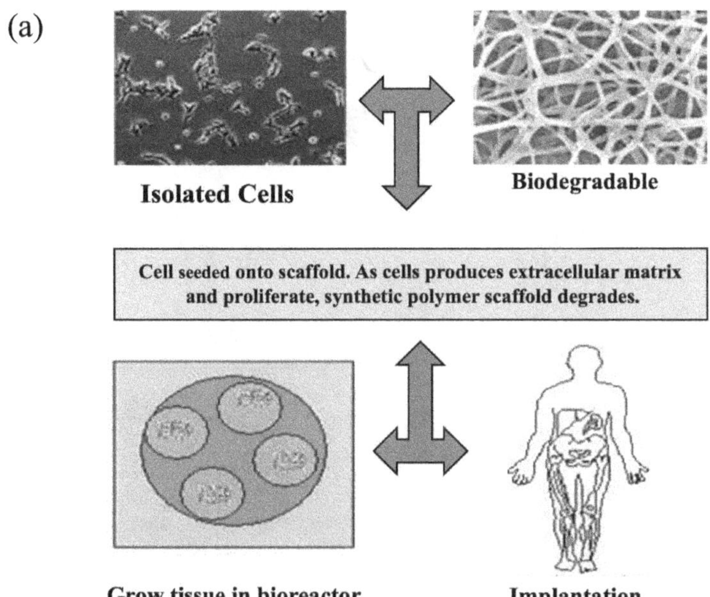

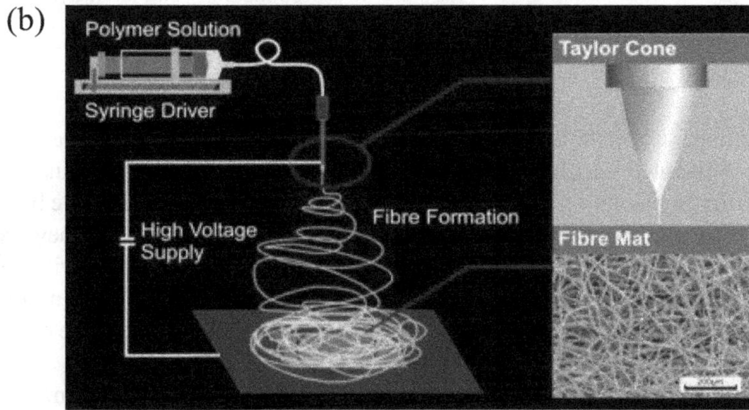

FIGURE 9.12 (a) Schematic procedure of tissue engineering. (b) Electrospinning and self-assembly are two promising techniques under investigation to fabricate nanodimensional fibers for tissue engineering.

nanomedical product applications (Figure 9.14). First, it is essential to design mainly geometries in order to design skeletons. As a next step, it is important to determine the type of biochemical cues and the physicochemical and mechanical properties of the tissue when attached to the material to be bonded. The most famous case in recent years is the transplant of artificial ear tissue into the back of mice (42). A template is made using a polymer base material, and polyglycolic acid is immersed in a 1% polylactic acid solution to form a molded body to shape a human earlobe. Alternatively, in the polyglycolic acid-polylactic acid template, cartilage cells separated from small

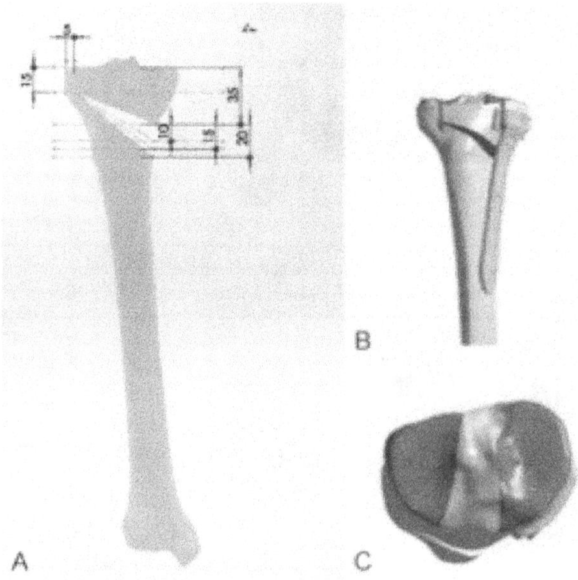

FIGURE 9.13 (A) The coronal view of the high tibial osteotomy displays its dimensions and position. (B) Tibial proximal axial loads were measured medially and laterally. (C) The two force loads range from the medial and lateral plateau (64).

joint cartilage are sown and transplanted into subcutaneous bags on the dorsal side of mice. Separated cartilage cells have customized biocompatibility, and cells that survive after implantation inside a biodegradable polymer mass regenerate body parts. Recently, Vacenti's group proposed an improved new method, and as a new example, an internal titanium wire skeleton embedded in a porous collagen matrix was used. The titanium wire was bent to simulate the elevation of the human ear, and a PLGA lump was used to create conditions that could maintain the size and shape of the structure for a long time (65).

There are several methods to be used in trial replacements using biomaterials. As shown in the Figure 9.15, there are bioresorbable viscous grafts, biodegradable neural guidance channels, skin grafts, and bone replacement methods.

Types of tissues include ephemeral tasks, connective tasks, micelle tasks, and near tasks. Epithelial tissue covers the body surface and forms the lining for most internal cavities. Skin is an organ made up of epithelial tissue. Its main function is protection. Connective tissues are bone, cartilage, dense fibrous tissue, loose connective tissue and fat. Their main functions are support, protection, and binding tissues together. Muscle tissue is used in the main parts of the skeleton, voluntary contraction of skeleton parts, and attached to bone via tendons. Its main roles are used in the major parts of voluntary movement, involuntary movement of the internal organs, and wall of blood vessels.

Nerve tissue is composed of neural cells or neurons located in brain, spinal cord, and nerves. It produces nervous impulses and must act on all parts of the body. Soft tissues serve as connectors or supports or surround other structures and organs of the

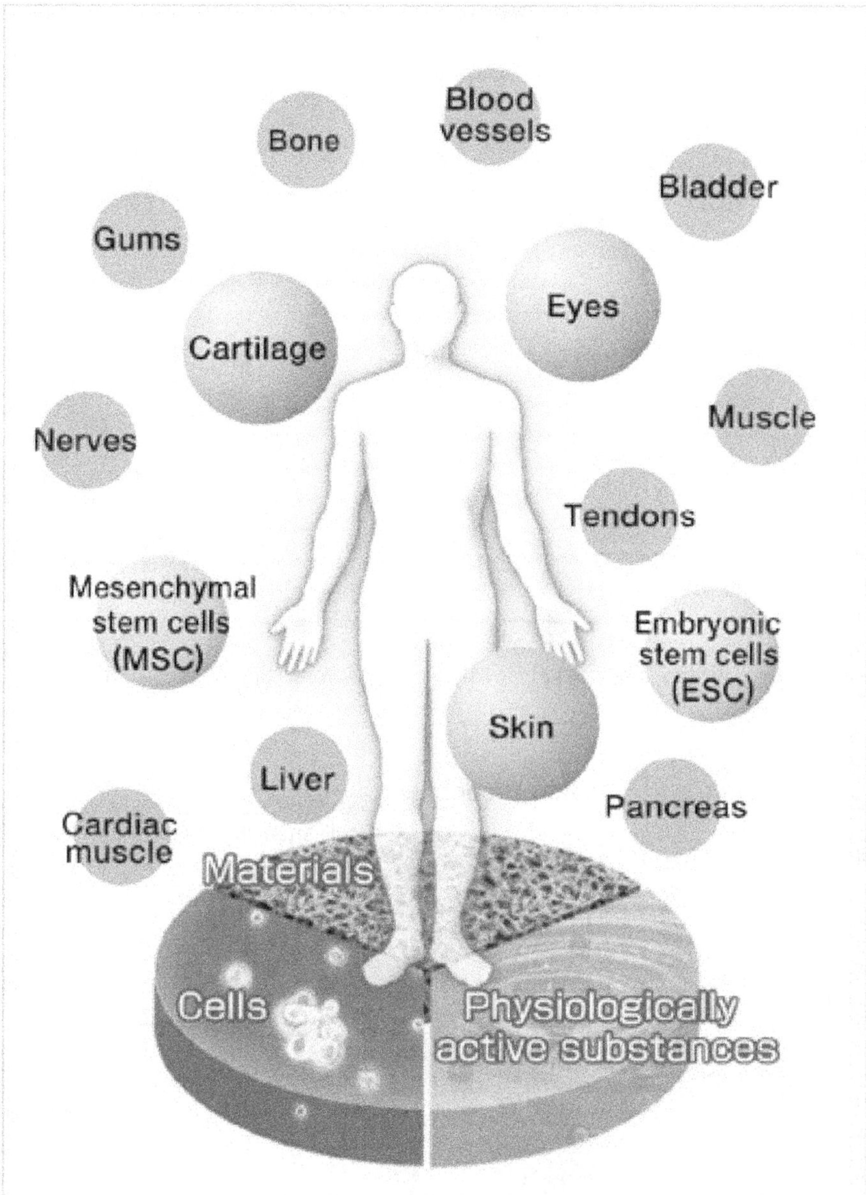

FIGURE 9.14 Tissue engineering.

body. For example, muscles support and move bone. Tendons play the role of connecting muscles to bones. Ligaments connect bone to bone. Joint capsule is sealed by the synovial fluid. Fascia is a sheet or band of fibrous connective tissue. Its main roles are developing, separating, and binding together muscles, organs, and other soft structures of the body and so on. Hard tissues are cartilage and bone. The mechanical properties of tissues are presented in Table 9.9.

Replacement Methods	Diagram
Bioresorbable vascular graft	
Biodegradable nerve guidance channel	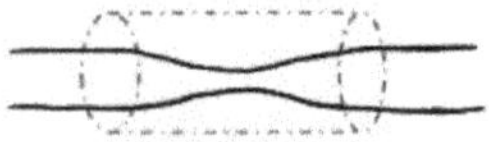
Skin graft	
Bone replacement	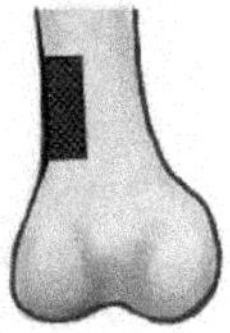

FIGURE 9.15 Several methods to be used in trial replacements using biomaterials.

TABLE 9.9
Biological Tissues: Mechanical Properties

Tissue	Modulus (MPa)	Tensile Strength (MPa)	Strain at Break (%)
Soft Tissues			
Smooth muscle, relaxed	0.006	–	300
Smooth muscle, contracted	0.01	–	300
Carotid artery	0.084 ± 0.22	–	–
Cerebral artery	15.69	4.34	50
Cerebral vein	6.85	2.82	83
Pericardium	20.4 ± 1.9	–	34.9 ± 1.1
Patellar tendon (29–50 yrs. old)	660 ± 266	64.7 ± 15	14 ± 6
ACL ligament (21–30 yrs. old)	345 ± 22.4	36.4 ± 2.5	15 ± 0.8
Hard Tissues			
	Modulus (GPa)	**Tensile Strength (MPa)**	**Strain at Break (%)**
Cortical bone	17–24	90–130	1–3
Cancellous bone	0.1–4.5	10–20	5–7
Cartilage	0.001–0.01	10–40	15–20

9.3 PHYSIOLOGY OF BONE

9.3.1 Functions of the Bone

Human bone is a significant structure that forms the human body shape and stabilize the erected position during human motion. Bones can perform several important functions, as follows.

1. Movement: Bone acts as the lever for skeletal muscle attachment with the ligaments and tendons that support the movement mechanism. Movement occurs when muscle contractions, whether voluntary or involuntary, move the bone in the direction that the muscle operates. As a result, functional activity and locomotion in daily living can be performed, such as walking, running, grasping, throwing, swimming, or driving.
2. Support and protection: Bones provide a hard framework of the body shape and protect the visceral organs, especially in the crucial organs of human body, such as the brain, spinal cord, and heart.
3. Mineral storage: Bone is a deposit of minerals, mainly calcium and phosphate, and these minerals can be released to the bloodstream if any part of the body requires the minerals in the form of soluble chemicals in the vascular system.
4. Blood cell formation: Red blood cells, white blood cells, and platelets are produced within the red bone marrow via the process of hematopoiesis (11).

9.3.2 Bone Growth and Modeling

9.3.2.1 Normal Bone-Growth Process

The bone growth process, also known as ossification or osteogenesis, is the process of bone formation that is developed and maintained throughout life. The bone growth process leads to skeletal development during the first 6–7 weeks of the embryonic stage and continues until the age of 25 or adulthood. Bones no longer grow in size, but their thickness increases throughout life. Initially, bones consist of fibrous membrane and hyaline cartilage. Then, when the bone is formed and replaced, the process is referred to in two ways: when the bone is formed from a fibrous membrane, it is referred to as intramembranous ossification, and when the bone is formed by replacing hyaline cartilage, it is referred to as endochondral ossification. In summary, normal bone growth processes are classified into two types: intramembranous ossification and endochondral ossification.

1. Intramembranous ossification: The formation of bone from a fibrous connective tissue precursor, such as skull bones and clavicles (flat bones). The fibrous membrane is derived from mesenchymal stem cells, and the process involves three steps. First, the formation of the bone matrix, starting with the differentiation of mesenchymal stem cells to osteoblasts; the osteoblasts

then secrete osteoid. After the mineralization process, osteoid becomes bone matrix, and osteoblasts in lacunae form osteocytes. This process usually occurs in the center of the membrane. Then, the periosteum is formed. Finally, spongy bone development occurs within the trabeculae in the center of the bone. Then, vascular tissue in the spongy bone turns into the red marrow of newly formed bone, and the flat bone is the end result.

2. Endochondral ossification: This is the process of bone growth that occurs in most bones in the human body. According to the name "endochondral ossification," the bone-forming process uses hyaline cartilage as a precursor. This process occurs in the long bones such as the femur, tibia, radius, and ulna. This process usually starts at the center of the hyaline cartilage or primary ossification center. In this stage, the periosteum of hyaline cartilage is vascularized and provides nutrition to osteoblasts. Then, osteoblasts migrate to the cartilage and start to secrete bone matrix. As the bone matrix is deposited, the cartilage begins to degrade and is replaced by bone tissue. This process continues until the entire cartilage model is replaced by bone.

9.3.2.2 Bone Remodeling

Bone remodeling is the normal process of bone metabolism in the adult skeleton. The two processes continue through human life: bone deposit and bone resorption from the function of the bone remodeling unit, or osteoblasts and osteocytes. Because of differences in shape, function, and loading approach, bone remodeling rates differ between bones; even within the same bone, the remodeling process may not be equal along the bone, such that the bone remodeling rate at the end of the radius is greater than its shaft area.

Even in conditions of bone injury, bone loss, or bone homeostasis, bone deposit occurs when bone strength is required. The site of matrix deposit is the osteoid seam, which is an unmineralized band. The calcification front is the area of bone mineralization and calcium deposit between the osteoid seam and older bone. Calcium-phosphate ion concentration (Ca^{2+}, PO_4^{3-}) and alkaline phosphatase enzyme are important trigger factors in the bone deposit process. Alkaline phosphatase is an essential chemical factor that stimulates calcium deposits in the bone site (55). When the calcium salt deposit on the bone is sufficient, it organizes as a new bone surface. Bone resorption occurs as the osteoclasts secrete lysosomal enzymes and metabolic acids like those discussed in Section 9.2.3.2. The deposited calcium salt is demineralized into a soluble form in the blood stream and carried to the target organs.

The control of the bone remodeling process can be divided into two mechanisms: hormonal control and mechanical control. Hormonal effects are derived from two associated hormones, including the parathyroid hormone, which is secreted from the parathyroid gland and takes action to stimulate osteoclasts to release calcium into the bloodstream via bone resorption. This hormone is released in the hypocalcemia state (low calcium level in the blood stream). Another opposite hormone is calcitonin, which is released from parafollicular cells (C-cells) of the thyroid gland. Calcitonin acts as an antagonist to the parathyroid hormone; that is, when the

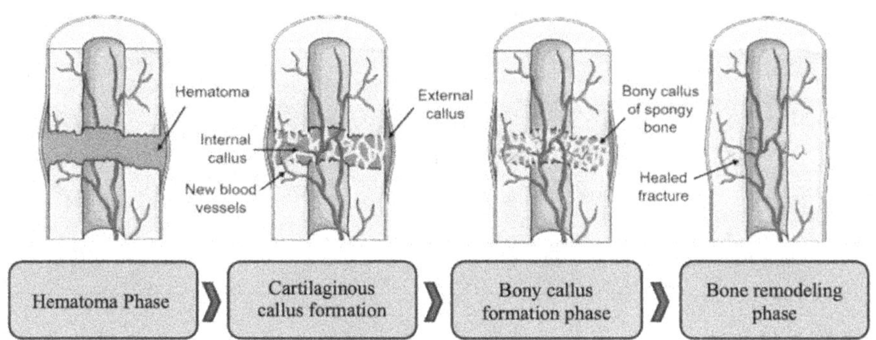

FIGURE 9.16 Bone healing in fracture (66).

body encounters hypercalcemia (high calcium in the bloodstream), this hormone is released and inhibits bone resorption and supports calcium salt deposits in the bone matrix. The mechanical effects that induce bone remodeling are mechanical force and gravity. According to Wolff's law, if mechanical loading on a particular bone increases, the bone will remodel and grow stronger over time as a response to resist that load.

A bone fracture is a traumatic injury to the bone that requires bone remodeling in the recovery process (bone healing). The bony response to fracture is hematoma formation or tearing of blood vessels around the fracture area. Massive blood clots can be found in conjunction with inflammatory symptoms (pain, swelling, redness, and skin warmth). After that, if patients get the proper medical management, like cast immobilization or surgery, the bone will turn to the fibrocartilagenius callus formation stage within three to four weeks. In this stage, phagocytes (a type of white blood cell) are secreted for cleaning up the debris at the fracture site, and the fibroblasts produce collagen fiber. The osteoblasts start to form new spongy bone and generate blood supply. In the final stage of recovery, bone remodeling promotes bone callus remodeling and bone shape reconstruction. A bone fracture is an injury condition that represents the importance of bone remodeling for humans. The process of bone healing is presented in Figure 9.16 (66). In conclusion, bone remodeling refers to the process by which bones adjust to the mechanical loads exerted on them. It is regulated by a balance between bone resorption and deposit and can occur in response to changes in mechanical, hormonal, and physiological factors.

9.3.3 Bone Mineralization

Bone mineralization is the process by which essential minerals, particularly calcium and phosphate, are deposited in bone tissue, leading to bone strength and development both in normal and pathologic bone healing. Bone composition is normally 50–70% minerals, which account for the majority of the content in human bone, especially hydroxyapatite [$Ca_5(PO_4)_3(OH)$ or $Ca_{10}(PO_4)_6(OH)_2$ in the crystal unit with two entities (43), which is the insoluble salt form of calcium and phosphorus

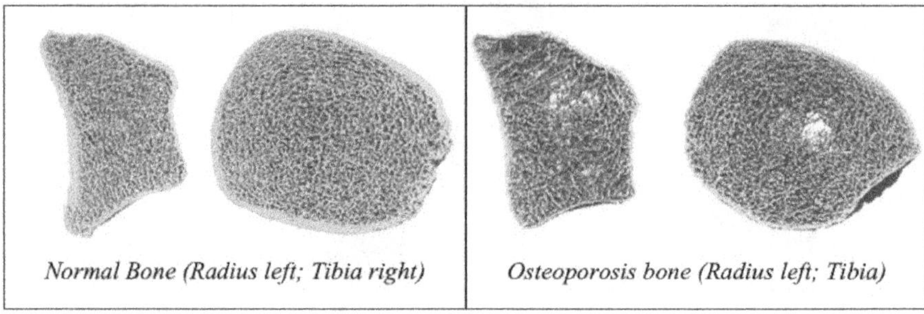

FIGURE 9.17 Illustrations comparing between normal bone and osteoporosis bone trabeculae (69).

with minor amounts of carbonate, magnesium, acid phosphate, sodium, potassium, and citrate ions. During bone mineralization, specialized cells called osteoblasts synthesize an extracellular matrix of collagen and other proteins that provide the appropriate microenvironment for bone formation. As the osteoblasts produce this matrix, they also release minerals such as calcium and phosphate, which become incorporated into the matrix and continue to increase until sufficient for crystal formation. In addition, within the extracellular matrix fluid are acidic phospholipids, inorganic phosphate, and proteins that encourage the precipitation of hydroxyapatite (67). Over time, the mineral content of the bone increases, leading to increased bone density and strength. This process is regulated by a variety of hormones, including the parathyroid hormone and vitamin D, as well as other factors such as exercise and diet. The effect of parathyroid hormone on bone mineralization is discussed in Section 9.3.2.2. In addition, vitamin D also has an indirect effect on bone formation. Vitamin D is normally synthesized beneath the skin and consequently absorbed. The vitamin can induce the liver to produce 25-hydroxyvitamin D and active 1,25-dihydroxyvitamin D (1,25-(OH)2D) from the kidney, which are the factors contributing to maintaining adequate serum calcium and phosphorus to complete the process of bone mineralization (43).

Inadequate bone mineralization can lead to various abnormal conditions, such as osteoporosis (Figure 9.17), in which bones become weak and brittle and have a higher risk of fracture. This can be caused by a variety of factors, including hormonal imbalances, nutritional deficiencies, and a lack of exercise (68).

9.3.4 Biomaterials for Bone Grafting and Joint Replacement

It has been demonstrated through substantial and continuous research that nanostructured materials stimulate new bone formation more efficiently because they promote a larger amount of specific protein interaction than conventional materials. Signs of conformity have also been found from the properties of the protein system, but subdividing the components of the glycoprotein system into nanoscale levels can result in varying intracellular variations and specificities. Some of these interactions not only induce friendly cell function, but other cells can also cause toxicity. Currently, protein tissue systems are being studied for cellular tissue engineering purposes at the nanostructure level, and important focus needs to be placed on these nanoscale materials and intercellular interaction mechanisms. Researchers argue that one of the best ways to focus on these interaction mechanism dimensions is the freeze casting method (Figure 9.18). Hydrophobic slurry circulates through channels by filling ice crystals between solutes to form the crystal structure desired by the researcher. If the solute containing ice is ceramic, setting the channel formation and discharging water forms a tissue system that forms a very dense complex (70). In the process of creating such an organizational framework, many researchers have recently developed a useful method of using hydroxyapatite. Prior to the discovery of freeze casting, a porous tissue system made of hydroxyapatite, which is too weak to withstand loads due to weak mechanical properties, was used. As an example of particulate leaching formation, gelatin spheres or salt crystals were used to form pores in the tissue system, and gelatin or salt crystals could later be leached from bubbles with water to leave a desired void (71). In order to produce a highly porous foam with high interpore connectivity by applying the same nanoparticles, pores in the empty space of the osteoblasts remain about 300 μm, but this method can be used to produce them. Furthermore, since then, freeze casting has generally enabled the manufacture of hydroxyapatite scaffolds (72), which have higher compressive strength and mechanical properties.

In this way, nanotechnology provides various specificities and benefits in the treatment of chronic diseases through goal-oriented delivery of medicines for the treatment of human diseases. However, sufficient research is needed on nanostructure toxicity, which is a disadvantage of nanotechnology, and a lot of knowledge is required in these fields, which is a major disadvantage to be overcome. These shortcomings undoubtedly require higher safety to enable safety measures associated with implementing these drug deliveries, and further research is needed to improve efficacy. Therefore, the design of nanoparticles is directly related to unexpected safety problems and should be carefully studied to solve unstable conditions, which should be studied carefully. Given these safety concerns, it is necessary to focus on different nano-based drug delivery systems, critical applications of natural compound-based nano-drug delivery, objectification of applications, controlled release of drugs, and reaction problems associated with nanomaterials in drugs. To form ideas for future revolutionary biological and medical perspectives, nanotechnology will transform through the industrial revolution and commercial implementation through many fields and stages.

Vascular grafts are essentially pipes with no biological component. They are subject to thrombosis (formation of "blood clots") on the surface with blockages. They must be flexible, designed with open porous structure and often recognized by body as foreign. Vascular grafts as main properties require achieve and maintain homeostasis, porous, permeable, good structure retention, adequate burst strength, high fatigue resistance, low thrombogenicity, and good handling properties with biostable (Figure 9.19). They have various kinds of types such as braids, weaves, and knits (Figure 9.20). Their main properties require high porosity and permeability,

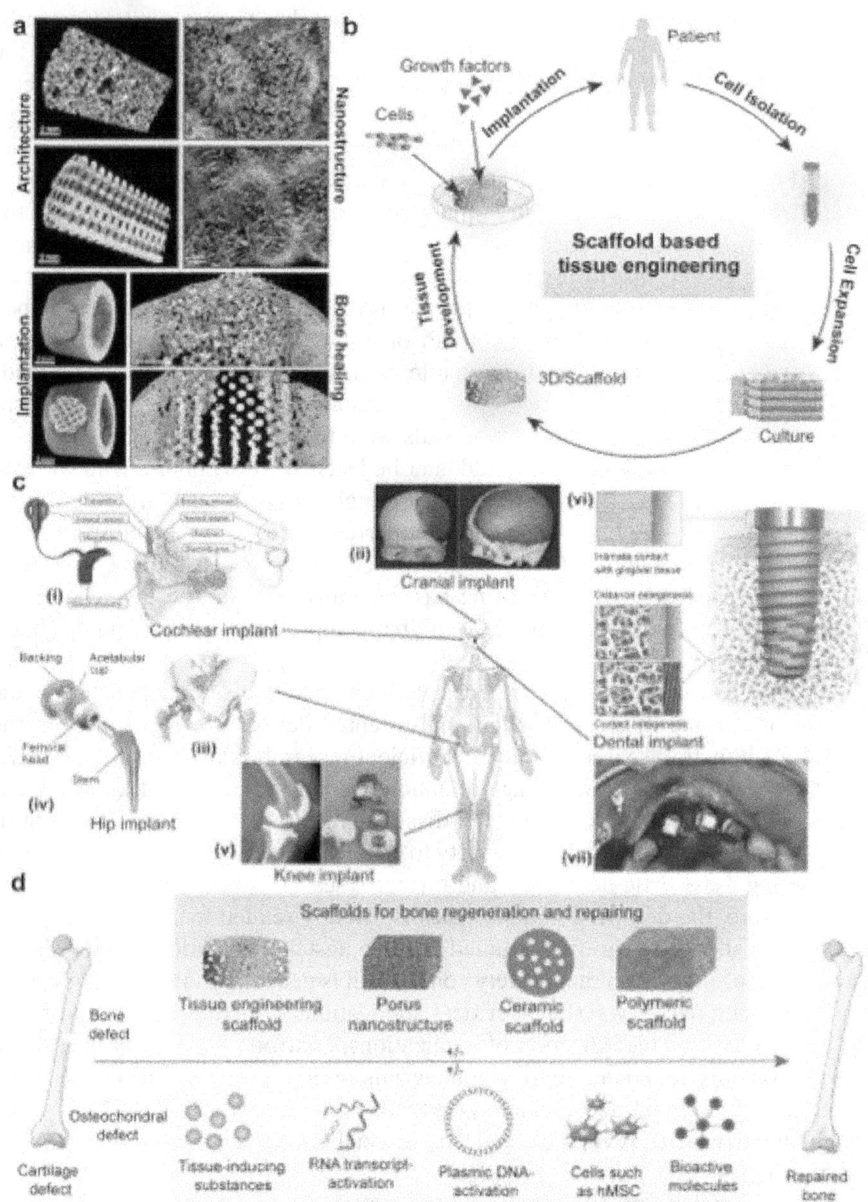

FIGURE 9.18 (a) Biomimetic calcium deficient hydroxyapatite foam group micro-3D reconstruction showing different architecture (shape and size of macropores), nanostructure scanning electron micrograph, implementation of foam scaffolds in bone, and new bone formation. Image adapted from Ref. (73). (b) Scaffold-based tissue engineering steps, cell isolation, expansion, scaffold and tissue development, and implantation. [Image modified from Ref. (74).] (c) Nanobiomaterial-based (i) cochlear implant (75). (ii) cranial implant (76). (iii & iv) hip and hip implant (77). (v) knee implant (78). (vi & vii) dental implant tissue integration (79) and insertion of implant (80). (d) Schematics of different scaffolds used in regeneration and repairing of bone, cartilage, and osteochondral defects with the integration of tissue-inducing substances, nucleic acids, stem cells, and bioactive molecules.

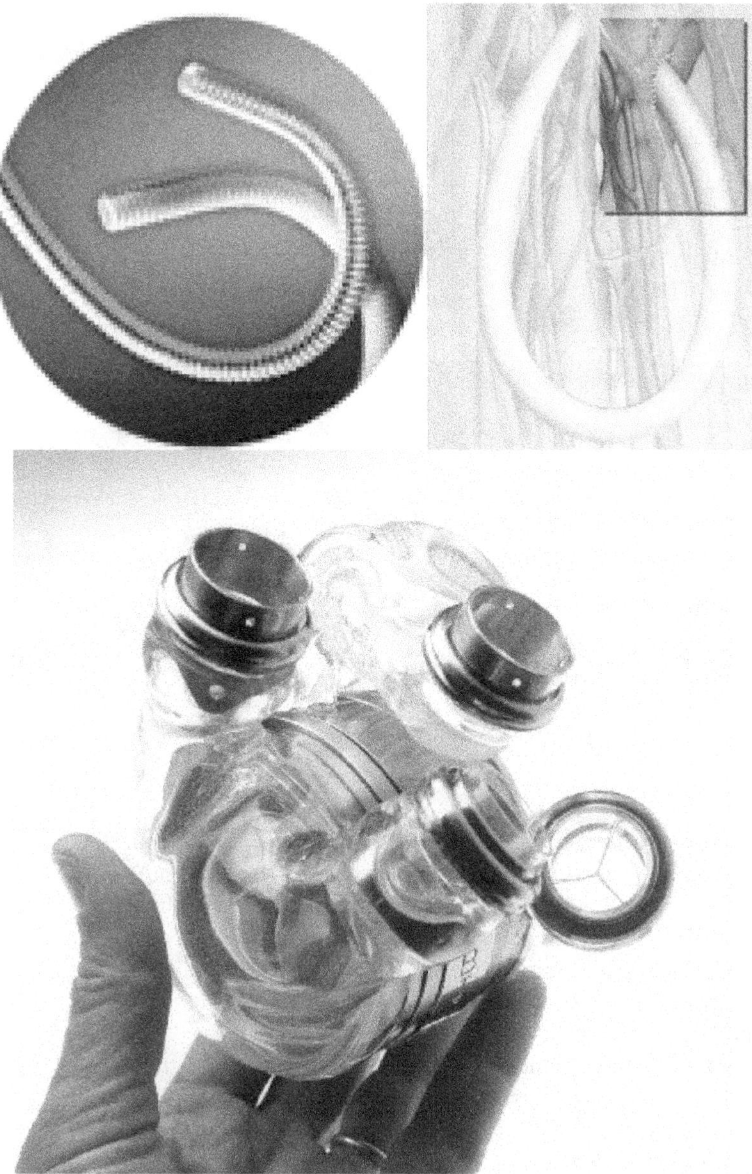

FIGURE 9.19 Typical examples of vascular grafts.

low thickness and burst strength, high kink resistance, good suture retention, good wall thickness, and good tensile properties with good ravel resistance.

Almost as soon as valve-implanted cardiac function is restored to near normal. Bileaflet tilting disk heart valve used most widely. More than 45,000 replacement valves implanted every year in the United States. Problems with heart valves are degeneration of tissue, mechanical failure, postoperative infection, and induction of blood clots.

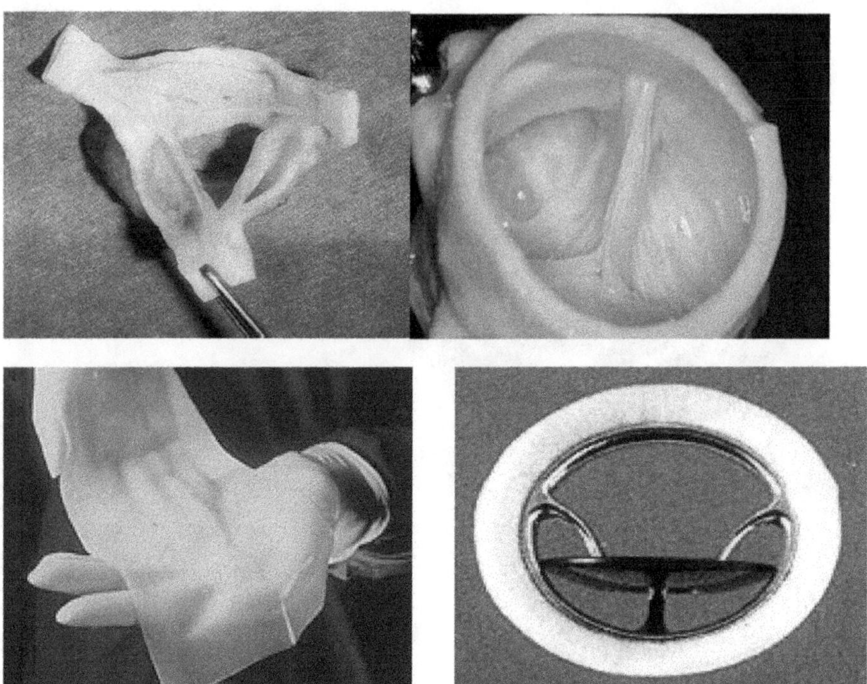

FIGURE 9.20 Vascular grafts.

Representative biomaterials for total hip replacement are the femoral ball (alumina, zirconia), acetabular cup (liner) (Co-Cr alloys, alumina), and acetabular cup (backing) (Co-Cr alloy or none). Ti porous coating allowing bone integration is used as a material for a cementless or press fit. In addition, bone cement is typically made of PMMA, a polymer compound, and bioactive glass, an inorganic material. For the determination of given device quality, one major factor is the modulus of the material (Figure 9.21).

Examples of metallic biomaterials generally used for orthopedics include stainless steel, titanium, and CoCrMo alloys. These biomaterials have high corrosion resistance due to the properties of alloys and the spontaneous formation of passive surface oxide layers at interfaces that occur in environmental aspects (Table 9.10) (81, 82). The results of chemical, physical, and mechanical interactions at the interface are determined by the surface properties of the film. Corrosion and flattening of the passive metal affect both local damage and removal of the passive film. This phenomenon is also closely related to mechanical wear due to the separation of metal particles. As the metal particles move from a de-liquid state to a surface area, a charge transfer reaction occurs at an interface to generate dissolved metal ions and solid oxides, and in this process, metal distribution and localization occur. As a result, passive film degradation appears in the stepwise tribological corrosion reaction, which induces acceleration of wear and corrosion. Repeated removal of the oxide film may cause particles and ions to be generated, which may adversely affect

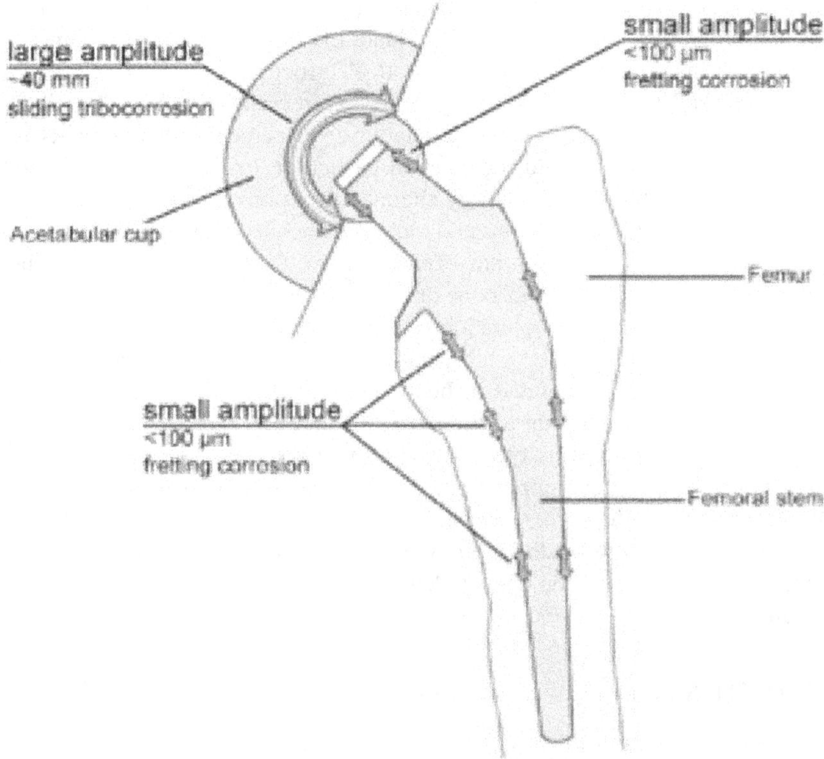

FIGURE 9.21 A schematical representation of a total hip joint replacement prosthesis. The types of motion and surface degradation mechanisms of the implant metallic components are shown (90).

TABLE 9.10
Main Biomaterials for Joint Replacement and Mechanical Properties

Material	*E* (GPa)
Silicone elastomer (rubber)	~0.002
Polyethylene (PE)	0.69
Poly(methyl methacrylate) (PMMA)	2.2–3.2
Cortical bone	17–24
Cancellous bone	0.1–4.5
Glass	73
Gold	77
Ti-6Al-4V	114
Stainless steel 316L	190
Tantalum	190
Haynes-Stellite 21 (Cast Co-Cr-Mo)	210
Aluminum oxide	380
Diamond	700–1200

the biological reaction and may cause mechanical failure of the used device (83–86). To study these complex phenomena and evaluate the biocompatibility of candidate metal materials, an approach that combines both electrochemical phenomena and tribology is required. Fleting corrosion usually results in certain forms of tribological corrosion associated with small amplitude relative displacement or vibration between surfaces (87) intended to be held together. In particular, in the case of orthopedic implants, corrosion is known to occur by chemical reactions of body fluids containing various inorganic and organic ions and molecules, while fine motion occurs at the fixed point (88). Flatting corrosion has been identified in stem/neck and neck/head contacts of modular implants, stem/bone and stem cement interfaces of cement and cement-free implants, and screw/plate joints of fixed plates (89).

The human joint is actuated by a lubrication substance on the surface of the joint cartilage, which can be regenerated to bear low friction (91, 92). When the natural joint is severely damaged due to osteoarthritis and decreased bone density due to long-term use and aging, it is often replaced by artificial implants. In the entire joint replacement, implant components are generally made up of metal-metal, metal-polymer, ceramic-ceramic, or ceramic-polymer combinations and composites, and technological advances are made by applying nanotechnology, as mentioned. As a representative example, metal on polyethylene is a very common material bond in the replacement of the entire joint (93).

9.4 CONCLUSION

In clinical settings, a variety of nanomaterials have been applied for various purposes. The general concerns of the long-term requirements for biomaterials for implants in the human body include hemocompatibility (blood compatibility), nontoxicity, durability, irritation of biological tissue, resistance to platelet and thrombosis deposition, non-degradability in physiological settings, non-absorption of blood elements, and the ability to not release foreign substances. Nanostructured materials are proposed to be more effective at stimulating the production of new bones than conventional materials because they encourage a greater quantity of specialized protein interaction. This can be applied in the treatment of several degenerative joint and bone diseases, which can support the use of nanomaterials for musculoskeletal implants in the human body and the need for further investigation of advanced materials for bone and supportive structure modification.

REFERENCES

[1] World Health Organization. *Musculoskeletal health 2022*. Available from: www.who.int/news-room/fact-sheets/detail/musculoskeletal-conditions#:~:text=Approximately%201.71%20billion%20people%20have,of%20disability%20in%20160%20countries.

[2] Welsh TP, Yang AE, Makris UE. Musculoskeletal pain in older adults: a clinical review. *Medical Clinics*. 2020;104(5):855–872.

[3] Thomas E, Peat G, Harris L, Wilkie R, Croft PR. The prevalence of pain and pain interference in a general population of older adults: cross-sectional findings from the North Staffordshire Osteoarthritis Project (NorStOP). *Pain*. 2004;110(1–2):361–368.

[4] Roux CH, Guillemin F, Boini S, Longuetaud F, Arnault N, Hercberg S, et al. Impact of musculoskeletal disorders on quality of life: an inception cohort study. *Annals of the Rheumatic Diseases*. 2005;64(4):606.
[5] Girón J, Kerstner E, Medeiros T, Oliveira L, Machado GM, Malfatti CF, et al. Biomaterials for bone regeneration: an orthopedic and dentistry overview. *Brazilian Journal of Medical and Biological Research*. 2021;54(9):e11055.
[6] Stevens MM. Biomaterials for bone tissue engineering. *Materials Today*. 2008; 11(5):18–25.
[7] Travlos GS. Normal structure, function, and histology of the bone marrow. *Toxicol Pathology*. 2006;34(5):548–565.
[8] Janssen I, Heymsfield SB, Wang ZM, Ross R. Skeletal muscle mass and distribution in 468 men and women aged 18–88 yr. *Journal of Applied Physiology (1985)*. 2000;89(1):81–88.
[9] Velleman SG, McFarland DC. Chapter 16—Skeletal muscle. In: Scanes CG, editor. *Sturkie's Avian Physiology* (Sixth Edition). San Diego: Academic Press; 2015. p. 379–402.
[10] Gollapudi SK, Michael JJ, Chandra M. Striated muscle dynamics. In: *Reference Module in Biomedical Sciences*. Amsterdam: Elsevier; 2014.
[11] Marieb EN. *Human Anatomy and Physiolofgy* (3rd ed.). Redwood City: California the Benjamin/Cummings Publishing Company; 1991.
[12] Fernández Costa JM, Fernández-Garibay X, Velasco F, Ramón-Azcón J. Bioengineered in vitro skeletal muscles as new tools for muscular dystrophies preclinical studies. *Journal of Tissue Engineering*. 2021;12:1–19.
[13] Tarafder S, Lee CH. Chapter 14—Synovial joint: in situ regeneration of osteochondral and fibrocartilaginous tissues by homing of endogenous cells. In: Lee SJ, Yoo JJ, Atala A, editors. *In Situ Tissue Regeneration*. Boston: Academic Press; 2016. p. 253–273.
[14] Lawry GV, Bewyer D. Anatomy of joints, general considerations, and principles of joint examination. In: Lawry GV, Kreder HJ, Hawker GA, Jerome D, editors. *Fam's Musculoskeletal Examination and Joint Injection Techniques* (2nd ed.). Philadelphia: Mosby; 2008. p. 1–5.
[15] Tamer TM. Hyaluronan and synovial joint: function, distribution and healing. *Interdiscip Toxicology*. 2013;6(3):111–125.
[16] Navarro M, Ruberte J, Carretero A, López-Luppo M. 3—Arthrology. In: Ruberte J, Carretero A, Navarro M, editors. *Morphological Mouse Phenotyping*. Cambridge: Academic Press; 2017. p. 55–62.
[17] Kalvaityte U, Matta C, Bernotiene E, Pushparaj P, Kiapour A, Mobasheri A. Exploring the translational potential of clusterin as a biomarker of early osteoarthritis. *Journal of Orthopaedic Translation*. 2022;32:77–84.
[18] Sophia Fox AJ, Bedi A, Rodeo SA. The basic science of articular cartilage: structure, composition, and function. *Sports Health*. 2009;1(6):461–468.
[19] Responte DJ, Natoli RM, Athanasiou KA. Collagens of articular cartilage: structure, function, and importance in tissue engineering. *Critical Reviews in Biomedical Engineering*. 2007;35(5):363–411.
[20] Alford JW, Cole BJ. Cartilage restoration, part 1: basic science, historical perspective, patient evaluation, and treatment options. *American Journal of Sports Medicine*. 2005;33(2):295–306.
[21] Wu Y, Li J, Zeng Y, Pu W, Mu X, Sun K, et al. Exosomes rewire the cartilage microenvironment in osteoarthritis: from intercellular communication to therapeutic strategies. *International Journal of Oral Science*. 2022;14(1):40.
[22] Braun HJ, Gold GE. Advanced MRI of articular cartilage. *Imaging Medicine*. 2011;3(5):541–555.

[23] Roughley PJ. Articular cartilage and changes in arthritis: noncollagenous proteins and proteoglycans in the extracellular matrix of cartilage. *Arthritis Research*. 2001;3(6):342–347.

[24] Armiento AR, Alini M, Stoddart MJ. Articular fibrocartilage—why does hyaline cartilage fail to repair? *Advanced Drug Delivery Reviews*. 2019;146:289–305.

[25] Chung EJ, Shah N, Shah RN. 11—Nanomaterials for cartilage tissue engineering. In: Gaharwar AK, Sant S, Hancock MJ, Hacking SA, editors. *Nanomaterials in Tissue Engineering*. Cambridge: Woodhead Publishing; 2013. p. 301–334.

[26] Eschweiler J, Horn N, Rath B, Betsch M, Baroncini A, Tingart M, et al. The biomechanics of cartilage—an overview. *Life*. 2021;11(4):302.

[27] Mow VC, Ratcliffe A, Poole AR. Cartilage and diarthrodial joints as paradigms for hierarchical materials and structures. *Biomaterials*. 1992;13(2):67–97.

[28] Buckwalter JA, Mankin HJ. Articular cartilage: part I. *Journal of Bone and Joint Surgery*. 1997;79(4):600.

[29] Luo Y, Sinkeviciute D, He Y, Karsdal M, Henrotin Y, Mobasheri A, et al. The minor collagens in articular cartilage. *Protein Cell*. 2017;8(8):560–572.

[30] Melching LI, Roughley PJ. Modulation of keratan sulfate synthesis on lumican by the action of cytokines on human articular chondrocytes. *Matrix Biology*. 1999;18(4):381–390.

[31] Zhang Y, Jordan JM. Epidemiology of osteoarthritis. *Clin Geriatric Medicine*. 2010; 26(3):355–369.

[32] Buckwalter JA, Mankin HJ, Grodzinsky AJ. Articular cartilage and osteoarthritis. *Instructional Course Lectures*. 2005;54:465–480.

[33] Abhishek A, Doherty M. Diagnosis and clinical presentation of osteoarthritis. *Rheumatic Disease Clinics of North America*. 2013;39(1):45–66.

[34] Li X, Majumdar S. Quantitative MRI of articular cartilage and its clinical applications. *Journal of Magnetic Resonance Imaging*. 2013;38(5):991–1008.

[35] Gornale SS, Patravali PU, Hiremath PS. Automatic detection and classification of Knee Osteoarthritis using Hu's invariant moments. *Front Robot AI*. 2020;7:591827.

[36] Fournier DE, Kiser PK, Shoemaker JK, Battié MC, Séguin CA. Vascularization of the human intervertebral disc: a scoping review. *JOR Spine*. 2020;3(4):e1123.

[37] Bibby SRS, Jones DA, Lee RB, Yu J, Urban JPG. The pathophysiology of the intervertebral disc. *Joint Bone Spine*. 2001;68(6):537–542.

[38] Bowles RD, Setton LA. Biomaterials for intervertebral disc regeneration and repair. *Biomaterials*. 2017;129:54–67.

[39] Guardado AA, Baker A, Weightman A, Hoyland JA, Cooper G. Lumbar intervertebral disc herniation: annular closure devices and key design requirements. *Bioengineering*. 2022;9(2):47.

[40] Mader JT, Stevens CM, Stevens JH, Ruble R, Lathrop JT, Calhoun JH. Treatment of experimental osteomyelitis with a fibrin sealant antibiotic implant. *Clinical Orthopaedics and Related Research*. 2002;403:58–72.

[41] Atala A, Bauer SB, Soker S, Yoo JJ, Retik AB. Tissue-engineered autologous bladders for patients needing cystoplasty. *Lancet*. 2006;367(9518):1241–1246.

[42] Cao Y, Vacanti JP, Paige KT, Upton J, Vacanti CA. Transplantation of chondrocytes utilizing a polymer-cell construct to produce tissue-engineered cartilage in the shape of a human ear. *Plast Reconstructive Surgery*. 1997;100(2):297–302. discussion 3–4.

[43] Clarke B. Normal bone anatomy and physiology. *Clinical Journal of the American Society of Nephrology*. 2008;3(Suppl 3):S131–S139.

[44] Dallas SL, Prideaux M, Bonewald LF. The osteocyte: an endocrine cell . . . and more. *Endocrine Reviews*. 2013;34(5):658–690.

[45] Iacobini C, Fantauzzi CB, Pugliese G, Menini S. Role of galectin-3 in bone cell differentiation, bone pathophysiology and vascular osteogenesis. *International Journal of Molecular Sciences*. 2017;18(11):2481.

[46] Jiang JX, Siller-Jackson AJ, Burra S. Roles of gap junctions and hemichannels in bone cell functions and in signal transmission of mechanical stress. *Frontiers in Bioscience.* 2007;12:1450–1462.
[47] Ru J-Y, Wang Y-F. Osteocyte apoptosis: the roles and key molecular mechanisms in resorption-related bone diseases. *Cell Death Disease.* 2020;11(10):846.
[48] Knothe Tate ML, Adamson JR, Tami AE, Bauer TW. The osteocyte. *The International Journal of Biochemistry & Cell Biology.* 2004;36(1):1–8.
[49] Wei W, Wan Y. Chapter 10—Regulation of bone resorption by PPARγ. In: Karsenty G, editor. *Translational Endocrinology of Bone.* San Diego: Academic Press; 2013. p. 103–122.
[50] Wilson SR, Peters C, Saftig P, Brömme D. Cathepsin K activity-dependent regulation of osteoclast actin ring formation and bone resorption. *Journal of Biological Chemistry.* 2009;284(4):2584–2592.
[51] Ettinger S. Chapter 9—Osteoporosis and fracture risk. In: Ettinger S, editor. *Nutritional Pathophysiology of Obesity and its Comorbidities.* Cambridge: Academic Press; 2017. p. 209–234.
[52] Barrère F, van Blitterswijk CA, de Groot K. Bone regeneration: molecular and cellular interactions with calcium phosphate ceramics. *International Journal Nanomedicine.* 2006;1(3):317–332.
[53] Weitzmann MN. The role of inflammatory cytokines, the RANKL/OPG Axis, and the immunoskeletal interface in physiological bone turnover and osteoporosis. *Scientifica (Cairo).* 2013;2013:125705.
[54] Feng X. Chemical and biochemical basis of cell-bone matrix interaction in health and disease. *Current Chemical Biology.* 2009;3(2):189–196.
[55] Carvalho MS, Cabral JMS, da Silva CL, Vashishth D. Bone matrix non-collagenous proteins in tissue engineering: creating new bone by mimicking the extracellular matrix. *Polymers (Basel).* 2021;13(7):1095.
[56] Arden NK, Perry TA, Bannuru RR, Bruyère O, Cooper C, Haugen IK, et al. Non-surgical management of knee osteoarthritis: comparison of ESCEO and OARSI 2019 guidelines. *Nature Reviews Rheumatology.* 2021;17(1):59–66.
[57] Dornacher D, Leitz F, Kappe T, Reichel H, Faschingbauer M. The degree of correction in open-wedge high tibial osteotomy compromises bone healing: a consecutive review of 101 cases. *Knee.* 2021;29:478–485.
[58] Yang Z-W, Wei X-B, Fu B-Q, Chen J-Y, Yu D-Q. Prevalence and prognostic significance of malnutrition in hypertensive patients in a community setting. *Frontiers in Nutrition.* 2022;9.
[59] Kim HJ, Seo I, Shin JY, Lee KS, Park KH, Kyung HS. Comparison of bone healing in open-wedge high tibial osteotomy between the use of allograft bone chips with autologous bone marrow and the use of allograft bone chips alone for gap filling. *Journal of Knee Surgery.* 2020;33(6):576–581.
[60] Jung WH, Takeuchi R, Kim DH, Nag R. Faster union rate and better clinical outcomes using autologous bone graft after medial opening wedge high tibial osteotomy. *Knee Surg Sports Traumatol Arthrosc.* 2020;28(5):1380–1387.
[61] Drogo P, Andreozzi V, Rossini M, Caperna L, Iorio R, Mazza D, et al. Mid-term CT assessment of bone healing after nanohydroxyapatite augmentation in open-wedge high tibial osteotomy. *Knee.* 2020;27(4):1167–1175.
[62] Bei T, Yang L, Huang Q, Wu J, Liu J. Effectiveness of bone substitute materials in opening wedge high tibial osteotomy: a systematic review and meta-analysis. *Annals of Medicine.* 2022;54(1):565–577.
[63] Putnis S, Neri T, Klasan A, Coolican M. The outcome of biphasic calcium phosphate bone substitute in a medial opening wedge high tibial osteotomy. *Journal of Materials Science: Materials in Medicine.* 2020;31(6):53.

[64] Pan CS, Wang X, Ding LZ, Zhu XP, Xu WF, Huang LX. The best position of bone grafts in the medial open-wedge high tibial osteotomy: a finite element analysis. *Computer Methods and Programs in Biomedicine*. 2023;228:107253.

[65] Zhou L, Pomerantseva I, Bassett EK, Bowley CM, Zhao X, Bichara DA, et al. Engineering ear constructs with a composite scaffold to maintain dimensions. *Tissue Engineering Part A*. 2011;17(11–12):1573–1581.

[66] Pfeiffenberger M, Damerau A, Lang A, Buttgereit F, Hoff P, Gaber T. Fracture healing research—shift towards in vitro modeling? *Biomedicines*. 2021;9(7):748.

[67] Blair HC, Larrouture QC, Li Y, Lin H, Beer-Stoltz D, Liu L, et al. Osteoblast differentiation and bone matrix formation in vivo and in vitro. *Tissue Engineering Part B Reviews*. 2017;23(3):268–280.

[68] Pinheiro MB, Oliveira J, Bauman A, Fairhall N, Kwok W, Sherrington C. Evidence on physical activity and osteoporosis prevention for people aged 65+ years: a systematic review to inform the WHO guidelines on physical activity and sedentary behaviour. *International Journal of Behavioral Nutrition and Physical Activity*. 2020;17(1):150.

[69] Eastell R, O'Neill TW, Hofbauer LC, Langdahl B, Reid IR, Gold DT, et al. Postmenopausal osteoporosis. *Nature Reviews Disease Primers*. 2016;2(1):16069.

[70] Sumitomo T, Kakisawa H, Owaki Y, Kagawa Y. Transmission electron microscopy observation of nanoscale deformation structures in nacre. *Journal of Materials Research*. 2008;23(12):3213–3221.

[71] Yao N, Epstein A, Akey A. Crystal growth via spiral motion in abalone shell nacre. *Journal of Materials Research*. 2006;21(8):1939–1946.

[72] Deville S, Saiz E, Tomsia AP. Freeze casting of hydroxyapatite scaffolds for bone tissue engineering. *Biomaterials*. 2006;27(32):5480–5489.

[73] Barba A, Maazouz Y, Diez-Escudero A, Rappe K, Espanol M, Montufar EB, et al. Osteogenesis by foamed and 3D-printed nanostructured calcium phosphate scaffolds: effect of pore architecture. *Acta Biomater*. 2018;79:135–147.

[74] Asadian M, Chan KV, Norouzi M, Grande S, Cools P, Morent R, et al. Fabrication and plasma modification of nanofibrous tissue engineering scaffolds. *Nanomaterials*. 2020;10(1):119.

[75] Brand Y, Senn P, Kompis M, Dillier N, Allum JH. Cochlear implantation in children and adults in Switzerland. *Swiss Medical Weekly*. 2014;144:w13909.

[76] Jardini AL, Larosa MA, Macedo MF, Bernardes LF, Lambert CS, Zavaglia CAC, et al. Improvement in cranioplasty: advanced prosthesis biomanufacturing. *Procedia CIRP*. 2016;49:203–208.

[77] Di Puccio F, Mattei L. Biotribology of artificial hip joints. *World Journal of Orthopedics*. 2015;6(1):77–94.

[78] Chithartha K, Thilak J, Sukesh AN, Theruvil B. Fatigue fracture of the femoral component in total knee replacement. *Knee*. 2020;27(5):1439–1445.

[79] Lavenus S, Louarn G, Layrolle P. Nanotechnology and dental implants. *International Journal of Biomaterials*. 2010;2010:915327.

[80] Elias CN, Meyers MA, Valiev RZ, Monteiro SN. Ultrafine grained titanium for biomedical applications: an overview of performance. *Journal of Materials Research and Technology*. 2013;2(4):340–350.

[81] Virtanen S, Milosev I, Gomez-Barrena E, Trebse R, Salo J, Konttinen YT. Special modes of corrosion under physiological and simulated physiological conditions. *Acta Biomaterialia*. 2008;4(3):468–476.

[82] Milošev I. Metallic materials for biomedical applications: laboratory and clinical studies. *Pure and Applied Chemistry*. 2010;83(2):309–324.

[83] Jones MR, Ehrhardt KP, Ripoll JG, Sharma B, Padnos IW, Kaye RJ, et al. Pain in the Elderly. *Current Pain and Headache Reports*. 2016;20(4):23.

[84] Marino CEB, Mascaro LH. EIS characterization of a Ti-dental implant in artificial saliva media: dissolution process of the oxide barrier. *Journal of Electroanalytical Chemistry*. 2004;568:115–120.
[85] Hodgson AWE, Mueller Y, Forster D, Virtanen S. Electrochemical characterisation of passive films on Ti alloys under simulated biological conditions. *Electrochimica Acta*. 2002;47(12):1913–1923.
[86] Merritt K, Brown SA. Biological effects of corrosion products from metals. In: Fraker AC, Griffin CD, editors. *Corrosion and Degradation of Implant Materials, STP 859*. West Conshohocken, PA: ASTM International; 1983. p. 195–207.
[87] Barril S, Mischler S, Landolt D. Electrochemical effects on the fretting corrosion behaviour of Ti6Al4V in 0.9% sodium chloride solution. *Wear*. 2005;259(1):282–291.
[88] Windler M., Klabunde R. Titanium for hip and knee prostheses. In: Brunette DM, Tengvall P, Textor M, Thomsen P, editors. *Material Science, Surface Science, Engineering, Biological Responses and Medical Applications*. Berlin: Springer-Verlag; 2001.
[89] Matsumoto K, Ogawa H, Yoshioka H, Akiyama H. Differences in patient-reported outcomes between medial opening-wedge high tibial osteotomy and total knee arthroplasty. *Journal of Orthopaedic Surgery (Hong Kong)*. 2020;28(1):2309499019895636.
[90] Diomidis N, Mischler S, More NS, Roy M. Tribo-electrochemical characterization of metallic biomaterials for total joint replacement. *Acta Biomaterialia*. 2012;8(2):852–859.
[91] Podsiadlo P, Kuster M, Stachowiak GW. Numerical analysis of wear particles from non-arthritic and osteoarthritic human knee joints. *Wear*. 1997;210(1):318–325.
[92] Podsiadlo P, Stachowiak GW. 3-D imaging of surface topography of wear particles found in synovial joints. *Wear*. 1999;230(2):184–193.
[93] Wilches LV, Uribe JA, Toro A. Wear of materials used for artificial joints in total hip replacements. *Wear*. 2008;265:143–149.

10 Bionanomaterials for Sensors, Actuators, Drug Delivery, and Their Medical Applications

Ayesha Malik, Saba Iqbal, Hassan Akbar, Ashfaq Ahmad, Asghar Ali, and Won-Chun Oh

10.1 INTRODUCTION

Compatibility and new effects brought by nanoscale materials are just two of the factors that have made nanobiomaterials an effective tool for medicinal applications. The possibilities to add biomaterials to the development of nanostructures have opened the door for innovative applications in several fields [1, 2]. Bionanomaterials are the nanoscale substances that are produced biologically. They can display distinctive structural, chemical, physical, optical, biological, mechanical, and electrical features that set them apart from bulk matter thanks to their incredibly small size. With the aid of these special qualities, they are able to fulfill a variety of functions in the biomedical field, including tissue engineering, medication and gene delivery, cancer treatment, the treatment of neurological illnesses, inflammation, etc. In addition to therapeutic applications, bionanomaterials are employed in the imaging and diagnostics of numerous biological molecules that aid in the detection of a variety of disorders [3, 4].

The term "bionanomaterials" refers to a group of materials that are either produced by living things or have a high degree of compatibility. Numerous studies have been conducted in this area to examine and utilize the materials in modern medicine [5, 6]. The ability of biocompatibility encouraged researchers to investigate the properties of nanotechnology [7]. Nanomaterials and biomaterials are combined under the umbrella name "nanobiomaterials." Functional materials made up of elements with at least one dimension *n* less than 100 nm are known as nanomaterials [8]. The term "biomaterials" refers to biological materials, whether they are synthetic or natural, that can be used to repair, change damaged organs, sustain, tissues, or human functions. The utilization of materials as well as arrangements at the nanoscopic size 1–100 nm for developing useful tenders in biology, physics, chemistry, as well as engineering is known as nanotechnology [5, 9]. Nanobiomaterials cure disorders within the human body using nanotechnology and nanomaterials. Numerous biomedical uses in tissue regeneration and repair, gene delivery and gene delivery,

DOI: 10.1201/9781003425427-13

medical imaging, cancer therapy, as well as theragnostics have driven interest in nanostructured biomaterials like nanofibers, nanoparticles, nanosurfaces, nanocomposites, and nanowires [10].

All nanobiomaterials are constructed from nanoparticles. They can be made of polymers (synthetic or natural) and inorganic materials, and they can be utilized as molecular association (metals or ceramics). Nanobiomaterials can be found in a variety of shapes, with spheres being the most prevalent, including rods, needles, platelets, and polygons. Nanobiomaterials are created via preparation processes in a variety of ways, such as dispersions, deposited layers, colloids, agglomerates, or suspensions [11]. For better performance and modulation, nanobiomaterials' shape and size affect their electrical, physical, mechanical, chemical, and optical characteristics. This contrasts with typical bulk materials. The nanobiomaterials' intended use and application, as well as desired interface with the cell membrane, all influence their size and shape. To achieve synthesis of nanobiomaterial design and precision control in the design-restricted size distribution and form, parameters are therefore crucial [12].

The usage of bio-active materials has replaced that of bio-inert ones in the treatment of disease and tissue abnormalities. In implantation uses such mechanical heart valves, dental implants, hip joints, intraocular lenses, shoulder joints, and knee joints, nanobiomaterials are employed.

Additionally, they are employed for cell cultures, blood protein analysis in medical laboratories, biomolecule dispensation for technological uses, implants for diagnostic gene arrays, and controlling cattle fertility [13]. For the particles to be able to adhere to different surfaces, like polymeric carriers, drug molecules, fluorophores, cell membranes, antibodies, and other biological machineries, as compulsory for the specific use, it is important to surface-modify and functionalize nanobiomaterials with monolayers of linker molecules and biocompatible material. Any foreign substance, including nanobiomaterials, that is ingested into the body triggers a congregation immune reaction that impairs the function of the substance [14]. The idea of nanotoxicity and biopersistence is a growing difficulty involving security concerns of body tissues that conjugate and internalize to nanoparticles due to the nature of nanobiomaterials [15].

10.1.1 Classification of Bionanomaterials

Depending on their use, nanobiomaterials can be categorized based on a variety of physical or chemical characteristics, including electrical and composition, magnetic, or optical capabilities. Nanobiomaterials are frequently categorized according to their content, dimension, and form. These characteristics control release kinetics and drug loading, biocompatibility, cellular interactions, and cytotoxicity in biomedical and pharmaceutical applications. Similar to this, further research and attention are needed for morphological and the dimensionality impacts of shape, size, tissue penetration, cytotoxicity, and cellular uptake [16].

10.1.2 Composition

Nanobiomaterials can be made of several morphologies of materials in terms of size and shape, including [8] inorganic, [17] organic, [9] composite based, or [5]

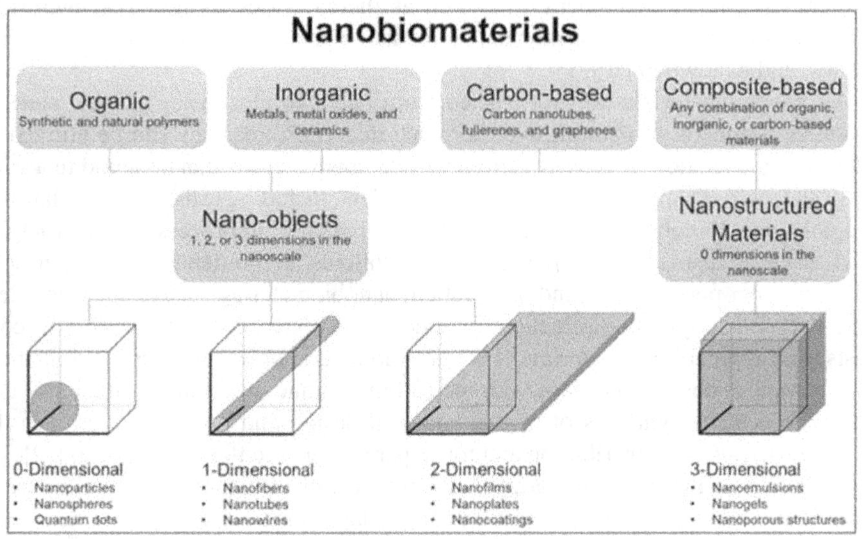

FIGURE 10.1 Division of nanobiomaterials into different categories based on their chemical makeup and dimensionality; the cube lines are 100 nm in all dimensions.

carbon-based materials (Figure 10.1). Nanobiomaterials' composition includes the material's interior, exterior, and any surface alterations. The chemical composition of nanobiomaterials is typically the first aspect to be changed for biocompatibility and decreased cytotoxicity.

Organic nanobiomaterials consist of organic components with carbon sourced from natural or synthetic polymers.

Polymeric nanobiomaterials are extensively researched and employed in medical uses for targeted drug release and sustained or controlled drug delivery. Nanoparticles, cyclodextrins, dendrimers, liposomes, solid lipids, and micelle nanoparticles are among the nanobiomaterials in this category [11]. Because of their low cytotoxicity, high biocompatibility, customizable biodegradability, and mechanical qualities, polymers including chitosan, silk, polylactide-co-glycoside acid, and polycaprolactone are popular for modifying drug relief kinetics and cellular interaction cytotoxicity.

Inorganic nanobiomaterials contain metals (bimetallic, metallics, or metal oxides) as well as nonmetals but do not contain the element carbon (ceramics and silicates). Gold, silver, iron, and copper nanoparticles have all been investigated as metallic nanobiomaterials for use in drug delivery, diagnostics, and restorative dentistry.

Carbon nanotubes (CNTs), graphene and its derivatives, and carbon-based quantum dots are examples of carbon-based nanobiomaterials. They have sparked a lot of attention in biomedical uses such tissue engineering, drug administration, biosensing, and imaging due to their good mechanical, thermal, electrical, and optical properties [18]. Theragnostic applications are a good fit for carbon-based nanobiomaterials because of their superior optical qualities, high surface areas and high electrical conductivity [11, 18]. Despite these special qualities, biodegradation, biocompatibility, and cytotoxicity need to be thoroughly investigated before any carbon-based biomedical modalities are used in medical settings [19, 20]. Depending on the kind of biomedical application required, the mixing of multiple distinct materials can provide hybrid or composite materials with adjustable biocompatibility, mechanical strength, or biodegradation rate [21]. For instance, natural and synthetic polymers are frequently blended to strengthen the natural material's weaker mechanical properties, whereas natural materials are also employed to increase the biocompatibility and cellular interfaces of synthetic materials. Similarly, metal nanoparticles and natural polymers can be used to enhance cellular interactions, biocompatibility, and drug loading capacity. As an illustration, hydroxyapatite and zinc oxide nanoparticles implanted in an alginate matrix to generate a biphasic mixture nanobiomaterial for bone-tissue engineering have improved biocompatibility and maintained their antibacterial properties [21].

10.1.3 Dimensionality

The number of exterior dimensions (height, depth, and width) that fall under the nanoscale division is a dimensional quality of nanobiomaterials (Figure 10.1). Zero-dimensional (0D) nanobiomaterials, which include quantum dots and fundamental solid or hollow nanoparticle forms, have all three exterior dimensions at the nanoscale. Three types of nanobiomaterials exist: one dimensional (1D), two dimensional (2D), and three dimensional (3D). One-dimensional nanobiomaterials have dual external dimensions at the nanoscale and consist of extended shapes such as nanofibers, nanotubes, or nanowires; two-dimensional nanobiomaterials have a single external dimension at the nanoscale and contain nanofilms, nanocoatings, and nanolayers; and three-dimensional nanomaterials [11, 22]. Time has been considered a fourth dimension in recent advances in nanomedicine for cancer identification and treatment. This shows how nanobiomaterials can alter over time, notably how they might deteriorate in the body rather than building up [23]. The 1D and 2D nanomaterials are advancing quickly as a potential class of nanomaterials in tissue engineering, drug delivery, diagnostics, biosensing, and antimicrobial materials. Biomolecular and cellular interactions at the material interface are facilitated by high surface-to-volume ratio and the flat topography of the 2D and 1D nanomaterials [24, 25].

10.2 APPLICATIONS OF BIONANOMATERIALS

10.2.1 Bionanomaterials for Drug Delivery

Humans have long employed natural remedies derived from plants to treat a variety of illnesses. Most modern medications are made from herbs using conventional wisdom and methods. Almost 25% of the most important pharmaceutical chemicals and their derivatives that are now on the market come from natural resources [26, 27]. To create nanotechnology, nanomedicine uses beneficial molecules at the nanoscale level. Nanoparticles have been the dynamic force behind the development of drug delivery, nanobiotechnology, tissue engineering, and biosensors in the biomedical field [28].

A developing subject called nanomedicine applies nanoscience knowledge and approaches to medical biology, treatment, and infection avoidance. It suggests the use of nanodimensional assets like nanosensors for delivery, nanotubes, diagnosis, and sensory functions, as well as actuating resources in existing cells. For example, a nanoparticle-based approach that combined cancer detection and treatment modalities has been developed [29]. Currently FDA-approved lipid schemes like micelles and liposomes were a part of the main generation of nanoparticle-based therapies [30]. Inorganic nanoparticles like magnetic nanoparticles or gold may be present in these micelles and liposomes [31].

These features have led to rise in the usage of nonliving nanoparticles, imaging, and drug delivery determinations. Additionally, according to reports, nanostructures improve the transportation of sparsely water-soluble medicines to their intended place and prevent drugs from becoming tainted in the gastrointestinal area.

While using nanomaterials for drug delivery, the physicochemical characteristics of the drug are taken into consideration when choosing the nanoparticle. When it derives from the supply of usual remedies for the dealing of cancer and many other infections, it offers several benefits. Because of their numerous distinctive properties, including their ability to induce tumor-suppressing and function as antimicrobial mediators, natural chemicals have been thoroughly researched in the handling of diseases. Caffeine and curcumin have been linked to autophagy [32], whereas caracole, cinnamaldehyde, and curcumin have been linked to antimicrobial characteristics. In addition to these elements, their features, such as targeting, bioavailability and well-ordered release, were enhanced.

When creating target definite drug delivery organizations, metallic, inorganic, biological, and polymeric nanostructures, such as micelles, dendrimers, and liposomes, are normally taken into account. These nanoparticles are specifically added to medicines that have poor solubility and absorption [28, 33]. The efficiency of these nanostructures as drug delivery schemes differs based on their shape, size, and additional inherent biophysical properties. For instance, polymeric nanoparticle diameters between 10 and 1,000 nm have characteristics that make them excellent delivery systems [34]. Many synthetic polymers, including poly-L-lactic acid, polyvinyl alcohol, polyethylene glycol, and normal polymers, like chitosan and alginate, are broadly used in the nanofabrication of nanoparticles because of their high biodegradability and biocompatibility characteristics [35–38].

Nanospheres and nanocapsules are dual categories of polymeric nanoparticles that provide excellent drug transfer devices. Like micelles, liposomes, compact lipid nanostructures, and phospholipids are mainly cooperative in the transfer of targeted drugs. Based primarily on the biochemical and biophysical features of the targeted medications selected for therapy, the usage of the optimum nanodrug delivery system is chosen [38]. In nanomedicine, issues like the poisonousness demonstrated by nanoparticles must be disregarded. To reduce toxicity issues, nanoparticles have primarily been used recently in conjunction with normal products.

Drug targeting, which can be active or passive, is another significant element that uses nanoformulations or nanomaterials as drug transfer systems. In active pointing, drug delivery schemes are joint by moieties, such as peptides and antibodies, to fix them to the receptor constructions conveyed at the target area. Currently, most drugs transported via nanotechnology are envisioned to treat cancer, and there are many biopolymeric resources that are used in drug transfer schemes. These resources and their characteristics are elaborated upon in the following.

10.2.2 Chitosan

Chitosan can be used to function at the constrictive epithelial junctions since it has muco-adhesive qualities. As a outcome, chitosan-based nanoparticles are frequently active to create sustained drug delivery schemes for a variety of epithelia, with the buccal [39], intestinal [40], nasal [41], pulmonary, and ocular [42] epithelia. In order to deliver the antibiotic ceftazidime to the eye, Silva et al. produced and assessed the effectiveness of a 0.75% w/w isotonic solution of hydroxypropyl methylcellulose (HPMC) containing chitosan/sodium tripolyphosphate/hyaluronic acid nanoparticles.

10.2.3 Xanthan Gum

Xanthomonas campestris produces xanthan gum, a strong molecular mass heteropolysaccharide. It has excellent bioadhesive capabilities and is a polyanionic polysaccharide. Xanthan gum is regularly used as a therapeutic excipient since it is thought to be nontoxic and non-irritating [43].

10.2.4 Cellulose

In drug transfer approaches, cellulose and its results are extensively used, typically to change the gelation and solubility of medicines, which controls their statement profile [44]. Repaglinide, an anti-hyperglycemic (RPG), was studied by Elseoud et al. [45] using chitosan nanoparticles and cellulose nanocrystals for oral drug delivery.

10.2.5 Liposomes

Liposomes are unique among extensively explored drug transfer carriers, and they are used in the cosmetic and pharmaceutical sectors to transport a wide range of chemicals. Liposomes are a well-known formulation technology for improving medicine distribution. These are assumed to be improved drug delivery system since they are simpler to integrate medications into and have a membrane structure, like cell membranes [46].

10.2.6 Inorganic Nanoparticles

Iron oxide, silver, gold, and nanocomposites are examples of nonliving nanoparticles. Only a very few nanomaterials have been accepted for therapeutic usage, with the majority remaining in the medical testing stage. Silver or gold nanomaterials exhibit features like surface plasmon significance that dendrimers and microspheres do not. They demonstrate various advantages, including strong biocompatibility and adaptability in surface modification.

10.2.7 Metallic Nanoparticles

Metallic nanoparticles have attracted increasing interest in recent years for a variety of medical uses, with biosensors, bioimaging, sustained/target drug transfer, photoablation, and hyperthermia treatment [47, 48]. Additionally, these nanoparticles have been modified and functionalized with functional groups that enable them to fix to medicines, ligands, and other antibodies, making these structures usable for procedures in biomedical uses [49].

10.3 BIONANOMATERIALS FOR SENSORS

A biosensor is described as "a device that uses specific biochemical reactions mediated by isolated enzymes, immune systems, tissues, organelles or whole cells to detect chemical compounds, typically by electrical, thermal or optical signals" [50] by the International Union of Pure and Applied Chemistry (IUPAC). This makes it easier to understand what a biosensor does. Three essential components are present in biosensors [50]:

- A specific receptor that binds analytes.
- A transducer that produces a sign after the tie occasion.
- A signal-detecting scheme that can calculate the signal and turn it into useful data.

Medical diagnostics [50, 51], fermentation, biodefense, and plant biology are only a few of the many disciplines where biosensing is crucial. With the use of contemporary biosensors, the time needed to identify infections like anthrax has decreased from two to three days to five minutes [52]. Bionanomaterials have demonstrated not only their potential for use in treatments but also their enormous promise for use in diagnostics.

10.3.1 Carbon Nanotubes in Biosensors

A biosensor is a device that uses physiochemical mechanisms to provide a quantifiable signal and biological reactions to detect an analyte. Immobilization of biological components affects biosensors' sensitivity and selectivity. Carbon nanotubes are one of the best components for the transduction of signals connected with the acknowledgement of analytes, disease biomarkers, or metabolites [53] because of their special properties. The carbon nanotubes restrained on a sensor surface to improve the electrochemical recognition of cancer cells, as shown in Figure 10.2 [54]. For instance, electrochemical biosensors based on oxidation and reduction processes among analytes and biomolecules have been created using CNTs. It has been discovered that adding CNTs to polymers improves the hybrid material's mechanical and electrical conductivity, increasing the sensitivity of the biosensor [55]. Biosensors for the recognition of uric acid, glucose, ascorbic acid, hydrogen peroxide, folic acid, deoxyribonucleic acid (DNA), and cancer cells have been reported to use such combinations [56].

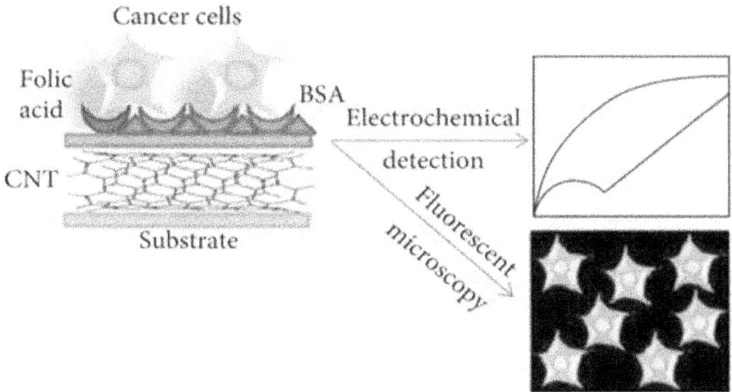

FIGURE 10.2 Schematic diagram of carbon nanotubes restrained on a sensor surface for improved electrochemical recognition of cancer cells.

It has also been claimed that optical carbon nanotube-based biosensors can identify cancer cells by observing changes in how light (ultraviolet, visible, or infrared) is emitted [57].

10.3.2 Inorganic Nanomaterials in Biosensors

Inorganic nanomaterials may possess changed anisotropies, like spherical, triangular, and nanohole [58]. They are seen in diverse forms, like core-shell structures, bimetallic alloys, metal organic frameworks (MOFs), nanowire arrays, and nanotubes. In addition to their huge surface area, some nonliving nanomaterials, like Fe_3O_4, also exhibit a magnetic character and may be simply operated by an outside magnetic field [59, 60]. They can be utilized particularly in point of care (POC) testing, homogenizing, trapping, enriching, conveying, and labeling analytes. They can be used for the microfluidic mixing required for lab-on-a-chip biosensing. This core typically has an inorganic [61] or polymeric coating that serves as a biofunctionalization site. Superparamagnetic behavior may be produced when numerous MNPs are embedded in a non-magnetic matrix [62]. Inorganic nanoparticles of several different types have been applied to medical diagnosis.

10.3.3 Quantum Dots

Nonliving quantum dots (QDs) typically have a bimetallic alloy fundamental and a shell coating made of a material like metal chalcogenide. When the diameter of QDs is smaller than the electron hole Bohr radius, quantum confinement effects predominate, giving rise to distinct visual features. Due to the QDs' emission in the near-infrared region (650 nm), which is different from conventional organic dyes and is highly desirable in the field of biomedical images due to the low tissue absorption and decreased light scattering, the QDs have attracted a lot of attention in the field of nanomedicine. Numerous compounds, including proteins [63], pathogens [64], lung cancer biomarkers [65], and nucleic acids, have been detected using QDs. For fluorescence imaging, streptavidin-coated quantum dots (Str-QDs) with immobilized biotinylated DNA served as labels, while superparamagnetic beads (SMBs) with DNA immobilized served as capture probes. Photostable QDs improve the exposure efficacy of the biosensor.

10.3.4 Magnetic Nanoparticles

Several fields, including the food industry, medical diagnostics, and ecological research, can benefit from MNP biosensors [66]. To use MNPs in healthcare biosensing, three conditions must be considered: (1) MNPs must maintain a strong saturation magnetization level to be controlled in blood without the use of extremely powerful magnetic fields. (2) The size of MNPs should fall between 10 and 50 nm to prevent aggregation or precipitation due to gravitational forces and to ensure colloidal stability, especially in water at pH 7.0, yielding a large surface area for a specific volume of the material [59, 67]. (3) MNPs must also be biocompatible and non-toxic. MNPs have shown enormous promise for use in early stage cancer sensing.

Polyclonal antibodies were utilized in the development of a sandwich assay. The sensor's 94% sensitivity and 98% specificity allowed it to successfully distinguish between ovarian cancer patients and healthy people.

10.3.5 Gold Nanoparticles

Conductive gold nanoparticles (AuNPs) have a huge surface area, distinct optical characteristics, and are conductive materials. A surface plasmon is constrained in AuNPs, resulting in localized surface plasmon resonance (LSPR). As a result, as they grow larger than 100 nm, their color switches from red to yellow [50]. Multifunctional AuNPs are currently frequently employed to identify a wide range of infections [68], neurological disorders [69], diabetes mellitus [70], amino acids, nucleic acids [71], and cancer biomarkers [72, 73].

10.3.6 Organic Nanomaterials in Biosensors

Except for a few of the most advanced molecular machines, most biological nanomaterials are polymeric by environment. For biomedical uses, including medication delivery and medical diagnostics, interest in polymeric nanoparticles has increased. This is explained by their biocompatibility, innate inertness, and design flexibility. These nanoparticles can be made relatively cheaply and are thermally stable [74].

10.3.7 Nanostructured Films

All biosensor components must be biocompatible to create sensors for detecting different biological parameters. The materials must to be resistant to bodily fluids and tissues and harmless. Nanostructured films can serve as the perfect transducing component, transmitting the analyte's stimuli and allowing covalent functionalization. ZnO-poly (vinyl alcohol) mixture films were used to place over an FTO electrode to take advantage of this property of nanofilms to create an enzymatic urea biosensor. The hybrid films were used by the impedimetric biosensor as transducers, and the hydroxyl groups from polyvinyl alcohol (PVA) were used to covalently immobilize the urease enzyme. Proteins in general and enzymes are significant structural components that have found use in biosensing [75]. Based on carbohydrates or glycol polymers, a different class of potential sensing components [76] is available. The sensor showed a high affinity for cholesterol.

10.3.8 Dendrimers

Due to their distinct bionanomaterials, high branching, and functionalized surface groups, dendrimers have drawn a lot of interest in the field of sensors. Dendrimers are adaptable for sensor applications thanks to these properties, enabling the inclusion of different sensing components and signal transduction methods. The three architectural components of dendrimers, which are molecular nanoparticles, are as follows: (1) a central core of initiator and metallic ions; (2) generations, which represent interior layers of repeating units bound to the core; and (3) terminal functionality, which represents the particle's exterior and is involved in the outermost generation of the interior layers. Despite the ease with which analytes like proteins and dendrimers bind to inorganic biosensor surfaces, denaturation could cause them to lose their function. Dendrimers being immobilized on the biosensor surface is an intriguing method to avoid this occurrence [77].

10.3.9 Covalent Organic Frameworks

The potential for COF-based sensors in the recognition of different antibiotics is quite positive. With the aid of rigidifying-induced fluorescence improvement and adsorption-triggered pre-concentration, COFs' fluorescence quenching efficacy can be increased because of their high porosity and plenty of N sites. COFs are helpful for the detection of several illness biomarkers, particularly those reveal cancer, in addition to antibiotics. As a platform, they exhibit considerable adaptability to different detection techniques. A human epidermal growth factor receptor (HEGFR) and Michigan cancer foundation cell line (MCF-7) electrochemical sensor was created by Yan et al. [78].

10.3.10 Molecular Machines

Nanomotors, often referred to as molecular machines (MoMas), are tiny machines that can propel themselves or be propelled by an external power source through a liquid phase. According to the definition given by an author, they are "an assembly of a discrete number of molecular components designed to produce mechanical-like movements (output) because of suitable external stimuli (input)" [79]. In this biosensor, biotinylated antibodies were initially applied asymmetrically to avidin (AVN) functionalized Fe3O4 MNPs, and then biotinylated BSA was applied to the remaining free sites.

10.3.11 Polymer Nanocomposites

Polymer nanocomposites (PNCs) have a diversity of morphologies, clever functions, and simple fabrication processes. The biocompatibility, environmental stability, improved electrical behavior, and cost-effectiveness of PNCs are all due to these unusual properties. To facilitate biorecognition, PNCs are compatible with a wide range of biomolecules, including ODN, aptamers, proteins, and enzymes. They are the perfect candidates for signal transduction in biosensing systems due to their huge surface area, quick electron transfer rate, and combination of the features [80]. A colorimetric bacterial biosensor was created by Robby and Park [81] using a nanocomposite of polymer dots (PDs) intercalated into montmorillonite (MMT).

10.4 BIONANOMATERIALS FOR ACTUATORS

Actuators, which are machines that transform different types of energy into mechanical movement, have shown significant possibilities for bionanomaterials. Actuators made of bionanomaterials have benefits such as being able to mimic natural systems and being biocompatible and responsive to biological stimuli. Actuators essentially respond to an electrical signal or stimuli generated from a processing unit or a signal directly fetched from a sensor. Nanomaterial sensors sense an external stimulus and convert it into measurable signal, which can then be transferred to a processing unit or a monitoring device. Such stimuli may be created mechanically, thermally, chemically, or magnetically, and the reactions may take the shape of structural deformations, heating, noise, or even the release of substances [82]. Such a response strategy is important in the field of wearable technology because it can turn "passive" wearables into "active" or "smart" wearables. Wearable actuators have expanded their applications in the textile industry thanks to the development of e-textiles and smart fabrics [83], while heating elements built into wearables, nanocomposite-based therapeutic devices [84, 85], artificial muscles and muscular actuators, rehabilitation devices [86], and wearable drug delivery systems [87] have also contributed to their development. Depending on their intended uses, they ought to be strong enough, solid enough structurally, flexible enough, and able to exert enough actuating force. Additionally, they should be comfortable and biocompatible, particularly for therapeutic and medication delivery systems.

10.4.1 Types of Actuators

Actuators often fall under the category of transducers, which transform any kind of electrical signal into a measurable mechanical output. Actuators are a category of devices that are used in clothing and other wearable applications that, upon receiving an electronic signal, cause mechanical movement, noise, or drug release. Wearable warmers, therapeutic devices, artificial muscles [88, 89], and wearable drug delivery systems are just a few of the intriguing uses for wearable actuators that have been made possible by the invention of smart fabrics [87, 90].

10.4.2 Wearable Heating Elements

One of the main topics covered by wearable electronics is heating and thermal applications in wearable technology. Personal thermal comfort and thermotherapeutic applications are generally the two main fields in which wearable heating components are used. Clothing and jackets with heating and cooling capabilities have emerged as key players in applications for human thermal management and even as youth fashion trends. Thermotherapeutic wearables have a significant role in the medical industry [91].

10.4.2.1 Shape Memory Polymers

A group of bionanomaterials known as shape memory polymers are capable of reversible form changes in response to environmental stimuli like heat, light, or pH. They have been used in actuators for soft robotics, controlled drug delivery, and scaffolds for tissue engineering [92].

10.4.3 Biomolecular Motors

Biomolecular motors use biological elements to produce mechanical motion, such as those built on naturally occurring molecular motors (such myosin or kinesin). These motors can be incorporated into bionanomaterial-based actuators for use in medication delivery systems and nanoscale devices [93].

10.4.4 Actuators and Artificial Muscles

Different material classes, such as dielectric elastomers [94], fabric-based materials [95], and electroactive polymers, have been developed by researchers to be employed as wearable and flexible actuator. In order to create a "texactuator" (textile plus actuator), Maziz et al. effectively formed conducting polymers into fabric-like yarns. These yarns have demonstrated strain enhancement when knitted together 53 times better than that of woven yarn-based materials. This fabric can be used to create a suit that can be worn like an exoskeleton and utilized as assistive clothing or artificial muscle. Additionally, this kind of fabric has application in compression therapy. Kim et al. have proposed a muscle-like bioelectronic soft actuator with a tensile strength increase of up to 75%. The changes in materials can be incorporated with the high-frequency response [96]. The mat has been modified to function as a flexible actuator that can react to a variety of factors, including humidity, light, and electrical current.

10.5 CONCLUSION AND FUTURE DIRECTIONS

Over the previous ten years, the science of biotechnology has become a revolutionary subject, and the development of nanotechnology has made it more well known. Biotechnology and nanotechnology are combined to create nanobiotechnology, which is used to benefit human welfare. Bionanomaterials are nanoscale substances composed of biological elements such as peptides, plants, nucleic acids, and others.

These bionanomaterials have excellent biocompatibility, which enables them to realize their therapeutic, diagnostic, and imaging potential in the biomedical field. Nanomedicines, medication delivery, and bone and tissue engineering are all therapeutic uses. The characteristics of bionanomaterials are powerfully influenced by their size and are examined using a variability of advanced analytical techniques to help them find their ideal applications. Nanobiomaterials have grown in importance over the past few decades and have been proven to be a valuable resource in the biomedical industry. This is since they are simple to adjust and have kept pace with the biomedical industry. Due to their small size and ability to attain a degree of accuracy, nanobiomaterials are more effective than traditional pharmaceuticals at treating disease and enhancing patient quality of life. Due to their ability to target certain medications, such as cytotoxic chemotherapeutics, these treatments' negative side effects are considerably reduced. These properties of nanobiomaterials, which have been used in modern medicine in two key areas, tissue engineering and the therapy of new viruses, can aid in many problems relating to human health. Nanobiomaterials are extremely significant in the biomedical industry because they may be engineered to be reliable, trustworthy, and compatible with human tissue. To demonstrate concerns about bioelimination, nanotoxicity, and translations to therapeutic uses, further clinical trials and research are compulsory. To ensure the trustworthy and secure use of these goods, there is a need for worldwide action to develop principles, certifications, and rules for the projected wave of nanomedicines.

REFERENCES

[1] Karagkiozaki, V., et al., *Novel nanostructured biomaterials: implications for coronary stent thrombosis.* International Journal of Nanomedicine, 2012: p. 6063–6076.

[2] Elmowafy, E., et al., *Polyhydroxyalkanoate (PHA): applications in drug delivery and tissue engineering.* Expert Review of Medical Devices, 2019. **16**(6): p. 467–482.

[3] Torres-Sangiao, E., A.M. Holban, and M.C. Gestal, *Advanced nanobiomaterials: vaccines, diagnosis and treatment of infectious diseases.* Molecules, 2016. **21**(7): p. 867.

[4] Kapat, K., et al., *Piezoelectric nano-biomaterials for biomedicine and tissue regeneration.* Advanced Functional Materials, 2020. **30**(44): p. 1909045.

[5] Bayda, S., et al., *The history of nanoscience and nanotechnology: from chemical—physical applications to nanomedicine.* Molecules, 2019. **25**(1): p. 112.

[6] Álvarez-Suárez, A.S., et al., *Electrospun fibers and sorbents as a possible basis for effective composite wound dressings.* Micromachines, 2020. **11**(4): p. 441.

[7] Song, S., et al., *Biomedical application of graphene: from drug delivery, tumor therapy, to theranostics.* Colloids and Surfaces B: Biointerfaces, 2020. **185**: p. 110596.

[8] Hochella Jr, M.F., *There's plenty of room at the bottom: nanoscience in geochemistry.* Geochimica et Cosmochimica Acta, 2002. **66**(5): p. 735–743.

[9] Thakur, R.S., and R. Agrawal, *Application of nanotechnology in pharmaceutical formulation design and development.* Current Drug Therapy, 2015. **10**(1): p. 20–34.

[10] Ramburrun, P., R.A. Khan, and Y.E. Choonara, *Design, preparation, and functionalization of nanobiomaterials for enhanced efficacy in current and future biomedical applications.* Nanotechnology Reviews, 2022. **11**(1): p. 1802–1826.

[11] Saleh, T.A., *Nanomaterials: classification, properties, and environmental toxicities.* Environmental Technology & Innovation, 2020. **20**: p. 101067.

[12] Salata, O.V., *Applications of nanoparticles in biology and medicine.* Journal of Nanobiotechnology, 2004. **2**(1): p. 1–6.

[13] Ratner, B.D., et al., *Biomaterials science: an introduction to materials in medicine*. MRS Bulletin, 2006. **31**: p. 59.
[14] Balasundaram, G., and T.J. Webster, *Nanotechnology and biomaterials for orthopedic medical applications*. Nanomedicine (Lond), 2006. **1**(2): p. 169–176.
[15] Saji, V.S., H.C. Choe, and K.W. Yeung, *Nanotechnology in biomedical applications: a review*. International Journal of Nano and Biomaterials, 2010. **3**(2): p. 119–139.
[16] Schaeublin, N.M., et al., *Does shape matter? Bioeffects of gold nanomaterials in a human skin cell model*. Langmuir, 2012. **28**(6): p. 3248–3258.
[17] De Aza, P., A. De Aza, and S. De Aza, *Crystalline bioceramic materials*. Boletin de la Sociedad Espanola de Ceramica y Vidrio, 2005. **44**(3): p. 135–145.
[18] Patel, K.D., R.K. Singh, and H.-W. Kim, *Carbon-based nanomaterials as an emerging platform for theranostics*. Materials Horizons, 2019. **6**(3): p. 434–469.
[19] Maiti, D., et al., *Carbon-based nanomaterials for biomedical applications: a recent study*. Frontiers in Pharmacology, 2019. **9**: p. 1401.
[20] Jacob, P.J.S., Cotton based cellulose nanocomposites: synthesis and application, in *Cotton*. 2022, IntechOpen.
[21] Turlybekuly, A., et al., *Synthesis, characterization, in vitro biocompatibility and anti-bacterial properties study of nanocomposite materials based on hydroxyapatite-biphasic ZnO micro-and nanoparticles embedded in Alginate matrix*. Materials Science and Engineering: C, 2019. **104**: p. 109965.
[22] Liu, Y., et al., *Nanobiomaterials: from 0D to 3D for tumor therapy and tissue regeneration*. Nanoscale, 2019. **11**(29): p. 13678–13708.
[23] Torresan, V., et al., *4D multimodal nanomedicines made of Nonequilibrium Au—Fe alloy nanoparticles*. ACS Nano, 2020. **14**(10): p. 12840–12853.
[24] Murali, A., et al., *Emerging 2D nanomaterials for biomedical applications*. Materials Today, 2021. **50**: p. 276–302.
[25] Liu, M., et al., *Tubule nanoclay-organic heterostructures for biomedical applications*. Macromolecular Bioscience, 2019. **19**(4): p. 1800419.
[26] Swamy, M.K., and U.R. Sinniah, *Patchouli (Pogostemon cablin Benth.): botany, agrotechnology and biotechnological aspects*. Industrial Crops and Products, 2016. **87**: p. 161–176.
[27] Mohanty, S.K., et al., *Leptadenia reticulata (Retz.) Wight & Arn.(Jivanti): botanical, agronomical, phytochemical, pharmacological, and biotechnological aspects*. Molecules, 2017. **22**(6): p. 1019.
[28] Mirza, A.Z., and F.A. Siddiqui, *Nanomedicine and drug delivery: a mini review*. International Nano Letters, 2014. **4**(1): p. 1–7.
[29] Haba, Y., et al., *Preparation of poly (ethylene glycol)-modified poly (amido amine) dendrimers encapsulating gold nanoparticles and their heat-generating ability*. Langmuir, 2007. **23**(10): p. 5243–5246.
[30] Shi, X., K. Sun, and J.R. Baker Jr, *Spontaneous formation of functionalized dendrimer-stabilized gold nanoparticles*. The Journal of Physical Chemistry C, 2008. **112**(22): p. 8251–8258.
[31] Park, S.-H., et al., *Loading of gold nanoparticles inside the DPPC bilayers of liposome and their effects on membrane fluidities*. Colloids and Surfaces B: Biointerfaces, 2006. **48**(2): p. 112–118.
[32] Wang, N., and Y. Feng, *Elaborating the role of natural products-induced autophagy in cancer treatment: achievements and artifacts in the state of the art*. BioMed Research International, 2015. **2015**.
[33] Krauel, K., et al., *Entrapment of bioactive molecules in poly (alkylcyanoacrylate) nanoparticles*. American Journal of Drug Delivery, 2004. **2**(4): p. 251–259.
[34] Bonifacio, B.V., et al., *Nanotechnology-based drug delivery systems and herbal medicines: a review*. International Journal of Nanomedicine, 2014. **9**: p. 1.

[35] Watkins, R., et al., *Natural product-based nanomedicine: recent advances and issues.* International Journal of Nanomedicine, 2015. **10**: p. 6055.
[36] Tan, Q., et al., *Preparation and evaluation of quercetin-loaded lecithin-chitosan nanoparticles for topical delivery.* International Journal of Nanomedicine, 2011. **6**: p. 1621.
[37] Sanna, V., et al., *Development of novel cationic chitosan-and anionic alginate—coated poly (d, l-lactide-co-glycolide) nanoparticles for controlled release and light protection of resveratrol.* International Journal of Nanomedicine, 2012. **7**: p. 5501.
[38] Casettari, L., and L. Illum, *Chitosan in nasal delivery systems for therapeutic drugs.* Journal of Controlled Release, 2014. **190**: p. 189–200.
[39] Portero, A., et al., *Reacetylated chitosan microspheres for controlled delivery of antimicrobial agents to the gastric mucosa.* Journal of Microencapsulation, 2002. **19**(6): p. 797–809.
[40] Artursson, P., et al., *Effect of chitosan on the permeability of monolayers of intestinal epithelial cells (Caco-2).* Pharmaceutical Research, 1994. **11**(9): p. 1358–1361.
[41] Fernández-Urrusuno, R., et al., *Enhancement of nasal absorption of insulin using chitosan nanoparticles.* Pharmaceutical Research, 1999. **16**(10): p. 1576–1581.
[42] Al-Qadi, S., et al., *Microencapsulated chitosan nanoparticles for pulmonary protein delivery: in vivo evaluation of insulin-loaded formulations.* Journal of Controlled Release, 2012. **157**(3): p. 383–390.
[43] Goswami, S., and S. Naik, *Natural gums and its pharmaceutical application.* Journal of Scientific and Innovative Research, 2014. **3**(1): p. 112–121.
[44] Sun, B., et al., *Applications of cellulose-based materials in sustained drug delivery systems.* Current Medicinal Chemistry, 2019. **26**(14): p. 2485–2501.
[45] Abo-Elseoud, W.S., et al., *Chitosan nanoparticles/cellulose nanocrystals nanocomposites as a carrier system for the controlled release of repaglinide.* International Journal of Biological Macromolecules, 2018. **111**: p. 604–613.
[46] Bozzuto, G., and A. Molinari, *Liposomes as nanomedical devices.* International Journal of Nanomedicine, 2015. **10**: p. 975.
[47] McNamara, K., and S.A. Tofail, *Nanoparticles in biomedical applications.* Advances in Physics: X, 2017. **2**(1): p. 54–88.
[48] McNamara, K., and S.A. Tofail, *Nanosystems: the use of nanoalloys, metallic, bimetallic, and magnetic nanoparticles in biomedical applications.* Physical Chemistry Chemical Physics, 2015. **17**(42): p. 27981–27995.
[49] Kudr, J., et al., *Magnetic nanoparticles: from design and synthesis to real world applications.* Nanomaterials, 2017. **7**(9): p. 243.
[50] Altintas, Z., *Biosensors and nanotechnology: applications in health care diagnostics.* 2017, John Wiley & Sons.
[51] Mehrotra, P., *Biosensors and their applications—A review.* Journal of Oral Biology and Craniofacial Research, 2016. **6**(2): p. 153–159.
[52] Saito, M., et al., *Field-deployable rapid multiple biosensing system for detection of chemical and biological warfare agents.* Microsystems & Nanoengineering, 2018. **4**(1): p. 1–11.
[53] Camilli, L., and M. Passacantando, *Advances on sensors based on carbon nanotubes.* Chemosensors, 2018. **6**(4): p. 62.
[54] Tîlmaciu, C.-M., and M.C. Morris, *Carbon nanotube biosensors.* Frontiers in Chemistry, 2015. **3**: p. 59.
[55] Barsan, M.M., M.E. Ghica, and C.M. Brett, *Electrochemical sensors and biosensors based on redox polymer/carbon nanotube modified electrodes: a review.* Analytica Chimica Acta, 2015. **881**: p. 1–23.
[56] Ghica, M.E., and C.M. Brett, *Poly (brilliant green) and poly (thionine) modified carbon nanotube coated carbon film electrodes for glucose and uric acid biosensors.* Talanta, 2014. **130**: p. 198–206.

[57] Wang, X., et al., *Ultrasensitive and selective detection of a prognostic indicator in early-stage cancer using graphene oxide and carbon nanotubes*. Advanced Functional Materials, 2010. **20**(22): p. 3967–3971.

[58] Malekzad, H., et al., *Noble metal nanoparticles in biosensors: recent studies and applications*. Nanotechnology Reviews, 2017. **6**(3): p. 301–329.

[59] Altintas, Z., *Applications of magnetic nanomaterials in biosensors and diagnostics*. Biosensors and Nanotechnology: Applications in Health Care Diagnostics, 2018: p. 277–296.

[60] van Reenen, A., A.M. de Jong, and M.W. Prins, *How actuated particles effectively capture biomolecular targets*. Analytical Chemistry, 2017. **89**(6): p. 3402–3410.

[61] Egan, J.G., et al., *A novel material for the detection and removal of mercury (ii) based on a 2, 6-bis (2-thienyl) pyridine receptor*. Journal of Materials Chemistry C, 2019. **7**(33): p. 10187–10195.

[62] Tian, B., *Magnetic nanoparticle based biosensors for pathogen detection and cancer diagnostics*. 2018, Acta Universitatis Upsaliensis.

[63] Wang, K., et al., *Differentiation of proteins and cancer cells using metal oxide and metal nanoparticles-quantum dots sensor array*. Sensors and Actuators B: Chemical, 2017. **250**: p. 69–75.

[64] Dogan, Ü., et al., *Rapid detection of bacteria based on homogenous immunoassay using chitosan modified quantum dots*. Sensors and Actuators B: Chemical, 2016. **233**: p. 369–378.

[65] Wu, S., et al., *Multiplexed detection of lung cancer biomarkers based on quantum dots and microbeads*. Talanta, 2016. **156**: p. 48–54.

[66] Altintas, Z., et al., *A novel magnetic particle-modified electrochemical sensor for immunosensor applications*. Sensors and Actuators B: Chemical, 2012. **174**: p. 187–194.

[67] Farka, Z., et al., *Nanoparticle-based immunochemical biosensors and assays: recent advances and challenges*. Chemical Reviews, 2017. **117**(15): p. 9973–10042.

[68] Savas, S., et al., *Nanoparticle enhanced antibody and DNA biosensors for sensitive detection of Salmonella*. Materials, 2018. 11(9): p. 1541.

[69] Ji, D., et al., *Smartphone-based differential pulse amperometry system for real-time monitoring of levodopa with carbon nanotubes and gold nanoparticles modified screen-printing electrodes*. Biosensors and Bioelectronics, 2019. **129**: p. 216–223.

[70] Guo, X., et al., *Glucose biosensor based on a platinum electrode modified with rhodium nanoparticles and with glucose oxidase immobilized on gold nanoparticles*. Microchimica Acta, 2014. **181**: p. 519–525.

[71] Thirumalraj, B., et al., *Highly sensitive fluorogenic sensing of L-Cysteine in live cells using gelatin-stabilized gold nanoparticles decorated graphene nanosheets*. Sensors and Actuators B: Chemical, 2018. **259**: p. 339–346.

[72] Yan, Z., et al., *A label-free immunosensor for detecting common acute lymphoblastic leukemia antigen (CD10) based on gold nanoparticles by quartz crystal microbalance*. Sensors and Actuators B: Chemical, 2015. **210**: p. 248–253.

[73] Altintas, Z., S.S. Kallempudi, and Y. Gurbuz, *Gold nanoparticle modified capacitive sensor platform for multiple marker detection*. Talanta, 2014. **118**: p. 270–276.

[74] Pedro, G.C., et al., *A novel nucleic acid fluorescent sensing platform based on nanostructured films of intrinsically conducting polymers*. Analytica Chimica Acta, 2019. **1047**: p. 214–224.

[75] Faccio, G., Proteins as nanosized components of biosensors, in *Nanomaterials design for sensing applications*. 2019, Elsevier. p. 229–255.

[76] Crucho, C.I., and M.T. Barros, Stimuli-responsive glyconanomaterials for sensing applications, in *Nanomaterials Design for Sensing Applications*. 2019, Elsevier. p. 257–279.

[77] Altintas, Z., et al., *Development of surface chemistry for surface plasmon resonance based sensors for the detection of proteins and DNA molecules*. Analytica Chimica Acta, 2012. **712**: p. 138–144.

[78] Yan, X., et al., *Two-dimensional porphyrin-based covalent organic framework: a novel platform for sensitive epidermal growth factor receptor and living cancer cell detection.* Biosensors and Bioelectronics, 2019. **126**: p. 734–742.

[79] Balzani, V., et al., *Artificial molecular machines.* Angewandte Chemie International Edition, 2000. **39**(19): p. 3348–3391.

[80] Shrivastava, S., N. Jadon, and R. Jain, *Next-generation polymer nanocomposite-based electrochemical sensors and biosensors: a review.* TrAC Trends in Analytical Chemistry, 2016. **82**: p. 55–67.

[81] Robby, A.I., and S.Y. Park, *Recyclable metal nanoparticle-immobilized polymer dot on montmorillonite for alkaline phosphatase-based colorimetric sensor with photothermal ablation of Bacteria.* Analytica Chimica Acta, 2019. **1082**: p. 152–164.

[82] Tao, X., *Handbook of smart textiles.* 2015, Springer.

[83] Zakharov, A., and L. Pismen, *Active textiles with Janus fibres.* Soft Matter, 2018. **14**(5): p. 676–680.

[84] An, B.W., et al., *Stretchable, transparent electrodes as wearable heaters using nanotrough networks of metallic glasses with superior mechanical properties and thermal stability.* Nano Letters, 2016. **16**(1): p. 471–478.

[85] Jang, N.-S., et al., *Simple approach to high-performance stretchable heaters based on kirigami patterning of conductive paper for wearable thermotherapy applications.* ACS Applied Materials & Interfaces, 2017. **9**(23): p. 19612–19621.

[86] Kim, S.S., et al., *High-fidelity bioelectronic muscular actuator based on graphene-mediated and TEMPO-oxidized bacterial cellulose.* Advanced Functional Materials, 2015. **25**(23): p. 3560–3570.

[87] Amjadi, M., et al., *Recent advances in wearable transdermal delivery systems.* Advanced Materials, 2018. **30**(7): p. 1704530.

[88] Son, D., et al., *Multifunctional wearable devices for diagnosis and therapy of movement disorders.* Nature Nanotechnology, 2014. **9**(5): p. 397–404.

[89] Chen, W.-H., et al., *Biomedical polymers: synthesis, properties, and applications.* Science China Chemistry, 2022. **65**(6): p. 1010–1075.

[90] Di, J., et al., *Stretch-triggered drug delivery from wearable elastomer films containing therapeutic depots.* ACS Nano, 2015. **9**(9): p. 9407–9415.

[91] Choi, S., et al., *Stretchable heater using ligand-exchanged silver nanowire nanocomposite for wearable articular thermotherapy.* ACS Nano, 2015. **9**(6): p. 6626–6633.

[92] Andreas Lendlein, S., *Shape-memory effect.* Angewandte Chemie International Edition, 2002. **41**: p. 2034–2057.

[93] Bachand, G.D., et al., *Biomolecular motors in nanoscale materials, devices, and systems.* Wiley Interdisciplinary Reviews: Nanomedicine and Nanobiotechnology, 2014. **6**(2): p. 163–77.

[94] Brochu, P., and Q. Pei, *Dielectric elastomers for actuators and artificial muscles.* Electroactivity in Polymeric Materials, 2012: p. 1–56.

[95] Wang, F., et al., *An eco-friendly ultra-high performance ionic artificial muscle based on poly (2-acrylamido-2-methyl-1-propanesulfonic acid) and carboxylated bacterial cellulose.* Journal of Materials Chemistry B, 2016. **4**(29): p. 5015–5024.

[96] Okuzaki, H., et al., *Ionic liquid/polyurethane/PEDOT: PSS composites for electro-active polymer actuators.* Sensors and Actuators B: Chemical, 2014. **194**: p. 59–63.

Index

For Product Safety Concerns and Information please contact our EU representative GPSR@taylorandfrancis.com
Taylor & Francis Verlag GmbH, Kaufingerstraße 24, 80331 München, Germany

www.ingramcontent.com/pod-product-compliance
Lightning Source LLC
Chambersburg PA
CBHW060345220326
41598CB00023B/2812

* 9 7 8 1 0 3 2 5 4 5 5 7 8 *